Differential & Difference Equations for Engineers and Scientists

Differential & Difference Equations for Engineers and Scientists

Edited by **Charlin Chester**

Willford Press
New York

Published by Willford Press,
118-35 Queens Blvd., Suite 400,
Forest Hills, NY 11375, USA
www.willfordpress.com

Differential & Difference Equations for Engineers and Scientists
Edited by Charlin Chester

International Standard Book Number: 978-1-68285-202-6 (Hardback)

Printed in China.

Contents

Preface

Differential and difference equations are one of the most apt methods of measuring change. Be it for measuring the changes in investment returns or measurement of growth and decay, one might find the relevance of these equations everywhere. This book unfolds the innovative aspects of differential and difference equations which will be crucial for the progress of mathematics and physics in the future. This book strives to provide a fair idea about the types of equations like ordinary differential equations, partial differential equations, abstract differential equations, etc. This book is ideal for engineering students and scientists as well.

This book unites the global concepts and researches in an organized manner for a comprehensive understanding of the subject. It is a ripe text for all researchers, students, scientists or anyone else who is interested in acquiring a better knowledge of this dynamic field.

I extend my sincere thanks to the contributors for such eloquent research chapters. Finally, I thank my family for being a source of support and help.

Editor

1

Some properties of Fibonacci and Chebyshev polynomials

Yang Li*

*Correspondence:
ly13119143216@126.com
Department of Mathematics,
Northwest University, Xi'an, Shaanxi,
P.R. China

Abstract

In this paper, we study the properties of Chebyshev polynomials of the first and second kind and those of Fibonacci polynomials and use an elementary method to give Chebyshev polynomials of the first and second kind in terms of Fibonacci polynomials and *vice versa*. Finally, we get some identities involving the Fibonacci numbers and the Lucas numbers.

Keywords: elementary method; Chebyshev polynomials; Fibonacci polynomials; Fibonacci numbers; Lucas numbers

1 Introduction

As we know, the Chebyshev polynomials and Fibonacci polynomials are usually defined as follows: Chebyshev polynomials of the first kind are $T_{n+2}(x) = 2xT_{n+1}(x) - T_n(x)$, $n \geq 0$, with the initial values $T_0(x) = 1$, $T_1(x) = x$; Chebyshev polynomials of the second kind are $U_{n+2}(x) = 2xU_{n+1}(x) - U_n(x)$, $n \geq 0$, with the initial values $U_0(x) = 1$, $U_1(x) = 2x$; Fibonacci polynomials are $F_{n+2}(x) = xF_{n+1}(x) + F_n(x)$, $n \geq 0$, with the initial values $F_0(x) = 0$, $F_1(x) = 1$. From the second-order linear recurrence sequences, we have

$$T_n(x) = \frac{1}{2}\left[\left(x + \sqrt{x^2 - 1}\right)^n + \left(x - \sqrt{x^2 - 1}\right)^n\right],$$

$$U_n(x) = \frac{1}{2\sqrt{x^2 - 1}}\left[\left(x + \sqrt{x^2 - 1}\right)^{n+1} - \left(x - \sqrt{x^2 - 1}\right)^{n+1}\right],$$

$$F_n(x) = \frac{1}{2^n\sqrt{x^2 + 4}}\left[\left(x + \sqrt{x^2 + 4}\right)^n - \left(x - \sqrt{x^2 + 4}\right)^n\right].$$

These polynomials play a very important role in the study of the theory and application of mathematics, and they are closely related to the famous Fibonacci numbers $\{F_n\}$ and Lucas numbers $\{L_n\}$ which are defined by the second-order linear recurrence sequences

$$F_{n+2} = F_{n+1} + F_n,$$

$$L_{n+2} = L_{n+1} + L_n,$$

where $n \geq 0$, $F_0 = 0$, $F_1 = 1$, $L_0 = 2$ and $L_1 = 1$. Therefore, many authors have investigated these polynomials and got many properties and corollaries. For example, Zhang [1] uses the Chebyshev polynomials and has obtained the general formulas involving F_n and L_n,

$$\sum_{a_1+a_2+\cdots+a_{k+1}=n} F_{m(a_1+1)} \cdot F_{m(a_2+1)} \cdots F_{m(a_{k+1}+1)}$$

$$= (-i)^{mn} \frac{F_m^{k+1}}{2^k \cdot k!} U_{n+k}^{(k)}\left(\frac{i^m L_m}{2}\right),$$

$$\sum_{a_1+a_2+\cdots+a_{k+1}=n+k+1} L_{m(a_1+1)} \cdot L_{m(a_2+1)} \cdots L_{m(a_{k+1}+1)}$$

$$= (-i)^{m(n+k+1)} \frac{2}{k!} \sum_{h=0}^{k+1} \left(\frac{i^{m+2} L_m}{2}\right)^h \frac{(k+1)!}{h!(k+1-h)!} U_{n+2k+1-h}^{(k)}\left(\frac{i^m L_m}{2}\right),$$

where k, m are any positive integers, $a_1, a_2, \ldots, a_{k+1}$ are nonnegative integers and i is the square root of -1. Falcón and Plaza [2, 3] presented many formulas about Fibonacci polynomials. This fact allowed them to present a family of integer sequences in a new and direct way. Zhang [1] also used the Chebyshev polynomials to solve some calculating problems of the general summations. Wu and Yang [4] studied Chebyshev polynomials and got a lot of properties.

In this paper, we combine Sergio Falcón and Wenpeng Zhang's ideas. Then we obtain the following theorems and corollaries. These results strengthen the connections of two kinds of polynomials. They are also helpful in dealing with some calculating problems of the general summations or studying some integer sequences.

Theorem 1 *For any positive integer n, we have the identities*

$$T_{2n}(x) = \sum_{s=1}^{n+1} \sum_{k=0}^{n} \frac{2^{2k} \cdot (2sn-n)(n+k-1)! F_{2s-1}(x)}{(-1)^{s+n-1}(n-k)!(k+s)!(k-s+1)!},$$

$$T_{2n-1}(x) = \sum_{s=1}^{n} \sum_{k=1}^{n} \frac{2^{2k-1}(2ns-s)(n+k-2)! F_{2s}(x)}{(-1)^{s+n}(n-k)!(k+s)!(k-s)!}.$$

Theorem 2 *For any positive integer n, we have*

$$U_{2n}(x) = \sum_{s=1}^{n+1} \sum_{k=0}^{n} \frac{2^{2k-1}(1-2s)(n+k)! F_{2s-1}(x)}{(-1)^{s+n}(n-k)!(k+s)!(k-s+1)!},$$

$$U_{2n-1}(x) = \sum_{s=1}^{n} \sum_{k=0}^{n-1} \frac{2^{2k+2} s(n+k-1)!(-1)^{s+n} F_{2s}(x)}{(n-k+1)!(k+s+1)!(k-s+1)!}.$$

Theorem 3 *For any positive integer n, we have the following forms:*

$$F_{2n-1}(x) = \sum_{k=0}^{n-1} \sum_{j=1}^{n-1} \frac{2^{2j+2-2n}(2k+1)(n-j-1)!}{j!(n-k-j-1)!(n+k-j)!} U_{2k}(x),$$

$$F_{2n}(x) = \sum_{k=1}^{n-1} \sum_{j=1}^{n-1} \frac{2^{2j-2n+2} k(2n-j-1)!}{j!(n-k-j)!(n+k-j)!} U_{2k}(x).$$

Theorem 4 *For any positive integer n, we have the following forms:*

$$F_{2n}(x) = \sum_{k=1}^{n}\sum_{j=1}^{n-1} \frac{2^{2j+2-2n}(2n-j-1)!T_{2k-1}(x)}{j!(2n+2k-2j-2)!(2n-2k-2j)!},$$

$$F_{2n-1}(x) = \sum_{k=1}^{n-1}\sum_{j=1}^{n-1} \frac{2^{2j+3-2n}(2n-j-2)!T_{2k}(x)}{j!(n+k-j-1)!(n-k-j-1)!} + \sum_{j=1}^{n-1} \frac{2^{2j+2-2n}(2n-j-2)!}{j!(n-j-1)!(n-j-1)!}.$$

Corollary 1 *For any positive integer n, we have the following identities:*

$$F_{2n-1}\left(\frac{i^m L_m}{2}\right) = \sum_{k=0}^{n-1}\sum_{j=1}^{n-1} \frac{2^{2j+2-2n}(2k+1)(n-j-1)!F_{m(2k+1)}}{j!(n-k-j-1)!(n+k-j)!(-1)^{km}F_m},$$

$$F_{2n}\left(\frac{i^m L_m}{2}\right) = \sum_{k=1}^{n-1}\sum_{j=1}^{n-1} \frac{2^{2j-2n+2}k(2n-j-1)!(-i)F_{2km}}{(-1)^{km}j!(n-k-j)!(n+k-j)!F_m}.$$

Corollary 2 *For any positive integer n, we have the following identities:*

$$F_{2n}\left(\frac{i^m L_m}{2}\right) = \sum_{k=1}^{n}\sum_{j=1}^{n-1} \frac{2^{2j+2-2n}(2n-j-1)!(-1)^{km}(-i)L_{2km-m}}{j!(2n+2k-2j-2)!(2n-2k-2j)!},$$

$$F_{2n-1}\left(\frac{i^m L_m}{2}\right) = \sum_{k=1}^{n-1}\sum_{j=1}^{n-1} \frac{2^{2j+3-2n}(2n-j-2)!(-1)^{km}L_{2km}}{j!(n+k-j-1)!(n-k-j-1)!}$$

$$+ \sum_{j=1}^{n-1} \frac{2^{2j+2-2n}(2n-j-2)!}{j!(n-j-1)!(n-j-1)!}.$$

2 Some lemmas

Lemma 1 *For any nonnegative integers m and n, we have these identities*

$$\int_{-1}^{1} \frac{T_m(x)T_n(x)}{\sqrt{1-x^2}}\,dx = \begin{cases} 0, & m \neq n, \\ \frac{\pi}{2}, & m = n > 0, \\ \pi, & m = n = 0; \end{cases} \tag{1}$$

$$\int_{-1}^{1} U_m(x)U_n(x)\sqrt{1-x^2}\,dx = \begin{cases} 0, & m \neq n, \\ \frac{\pi}{2}, & m = n; \end{cases} \tag{2}$$

$$T_n(\cos\theta) = \cos n\theta; \tag{3}$$

$$U_n(\cos\theta) = \frac{\sin(n+1)\theta}{\sin\theta}. \tag{4}$$

Proof See reference [5]. □

Lemma 2 *For any positive integers m and n, we have these identities*

$$U_n\left(\frac{i}{2}\right) = i^n F_{n+1},$$

$$T_n\left(\frac{i}{2}\right) = \frac{i^n}{2} L_{n+1},$$

$$T_n\left(T_m\left(\frac{i}{2}\right)\right) = i^{mn}L_{mn},$$
$$U_n\left(T_m\left(\frac{i}{2}\right)\right) = i^{mn}\frac{F_{m(n+1)}}{F_m}.$$

Proof See reference [1]. □

Lemma 3 *For any positive integer n, we have*

$$T_n(T_m(x)) = T_{nm}(x),$$
$$U_n(T_m(x)) = \frac{U_{m(n+1)-1}(x)}{U_{m-1}(x)}.$$

Proof See reference [1]. □

Lemma 4 *For any positive integer n, let*

$$F_n(x) = \sum_{k=0}^{+\infty} a_{nk}U_k(x) \tag{5}$$

and

$$F_n(x) = \frac{1}{2}b_{n0}T_0(x) + \sum_{k=1}^{+\infty} b_{nk}T_k(x), \tag{6}$$

then we can get

$$a_{nk} = \begin{cases} \sum_{j=1}^{\lfloor \frac{n-1}{2} \rfloor} \frac{2(k+1)(n-j-1)!}{j!(n-k-2j-1)!!(n+k+1-2j)!!}, & k+n \text{ is odd,} \\ 0, & \text{otherwise,} \end{cases} \tag{7}$$

$$b_{nk} = \begin{cases} \sum_{j=1}^{\lfloor \frac{n-1}{2} \rfloor} \frac{2(n-j-1)!}{j!(n+k-2j-1)!!(n-k-2j-1)!!}, & k+n \text{ is odd,} \\ 0, & \text{otherwise.} \end{cases} \tag{8}$$

Proof To begin with, we multiply $\sqrt{1-x^2}U_m(x)$ to both sides of (5), then integrate it from −1 to 1, we can get the following identity by applying property (2):

$$\int_{-1}^{1} \sqrt{1-x^2}F_n(x)U_m(x)\,dx = \sum_{k=1}^{\infty}\int_{-1}^{1} a_{nk}\sqrt{1-x^2}U_m(x)U_k(x)\,dx = \frac{\pi}{2}a_{nm},$$

and then we have

$$a_{nm} = \frac{2}{\pi}\int_{-1}^{1} \sqrt{1-x^2}F_n(x)U_m(x)\,dx.$$

From reference [2] we know

$$F_n(x) = \sum_{j=0}^{\lfloor \frac{n-1}{2} \rfloor} \binom{n-j-1}{j} x^{n-2j-1}, \tag{9}$$

where $n \geq 1$. We define

$$w_{nk} = \frac{2}{\pi}\int_0^{\pi} \cos^n\theta \sin(k+1)\theta \sin\theta \, d\theta.$$

From reference [6] we know

$$w_{nk} = \begin{cases} \frac{2(k+1)n!}{(n+k+2)!!(n-k)!!}, & k+n \text{ is even and } n \geq k, \\ 0, & \text{otherwise}, \end{cases} \tag{10}$$

where n and k are any nonnegative integers. Let $x = \cos\theta$, then we can get the following identity by applying property (2) and property (9):

$$\begin{aligned} a_{nm} &= \frac{2}{\pi}\int_0^{\pi} F_n(\cos\theta)\sin(m+1)\theta \sin\theta \, d\theta \\ &= \frac{2}{\pi}\sum_{j=1}^{\lfloor \frac{n-1}{2} \rfloor} \binom{n-j-1}{j} \int_0^{\pi} \cos^{n-2j-1}\theta \sin(m+1)\theta \sin\theta \, d\theta \\ &= \sum_{j=1}^{\lfloor \frac{n-1}{2} \rfloor} \binom{n-j-1}{j} w_{n-2j-1,m}, \end{aligned}$$

and then we have $a_{nm} = 0$ if $n+m$ is even. If $n+m$ is odd, we have

$$\begin{aligned} a_{nm} &= \sum_{j=1}^{\lfloor \frac{n-1}{2} \rfloor} \binom{n-j-1}{j} \cdot \frac{2(m+1)(n-2j-1)!}{(n-m-2j-1)!!(n+m+1-2j)!!} \\ &= \sum_{j=1}^{\lfloor \frac{n-1}{2} \rfloor} \frac{(n-j-1)!}{j!(n-2j-1)!} \cdot \frac{2(m+1)(n-2j-1)!}{(n-m-2j-1)!!(n+m+1-2j)!!} \\ &= \sum_{j=1}^{\lfloor \frac{n-1}{2} \rfloor} \frac{2(m+1)(n-j-1)!}{j!(n-m-2j-1)!!(n+m+1-2j)!!}. \end{aligned}$$

We finish proving property (7).

In order to prove property (8), we must multiply $\frac{T_m(x)}{\sqrt{1-x^2}}$ to both sides of (6), then integrate it from −1 to 1, we can get the following identity by applying property (1):

$$\int_{-1}^{1} \frac{F_n(x)T_m(x)}{\sqrt{1-x^2}}\,dx = \int_{-1}^{1} \frac{b_{n0}T_0(x)T_m(x)}{2\sqrt{1-x^2}}\,dx + \sum_{k=1}^{\infty}\int_{-1}^{1} \frac{b_{nk}T_m(x)T_n(x)}{\sqrt{1-x^2}}\,dx = \frac{\pi}{2}b_{nm},$$

and then we have

$$b_{nm} = \frac{2}{\pi}\int_{-1}^{1} \frac{F_n(x)T_m(x)}{\sqrt{1-x^2}}\,dx.$$

We define

$$q_{nk} = \frac{2}{\pi}\int_0^{\pi} \cos^n\theta \cos k\theta \, d\theta.$$

From reference [6] we know

$$q_{nk} = \begin{cases} \frac{2n!}{(n+k)!!(n-k)!!}, & k+n \text{ is even and } n \geq k, \\ 0, & \text{otherwise,} \end{cases} \tag{11}$$

where n and k are any nonnegative integers. Let $x = \cos\theta$, then we can get the following identity by applying property (2) and property (11):

$$\begin{aligned} b_{nm} &= \frac{2}{\pi}\int_0^{\pi} \frac{T_m(\cos\theta)F_n(\cos\theta)}{\sin\theta}\sin\theta\, d\theta \\ &= \frac{2}{\pi}\sum_{j=1}^{\lfloor\frac{n-1}{2}\rfloor}\binom{n-j-1}{j}\int_0^{\pi}\cos^{n-2j-1}\theta\cos m\theta\, d\theta \\ &= \sum_{j=1}^{\lfloor\frac{n-1}{2}\rfloor}\binom{n-j-1}{j} q_{n-2j-1,m}, \end{aligned}$$

and then we have $b_{nm} = 0$ if $n+m$ is even. If $n+m$ is odd, we have

$$\begin{aligned} b_{nm} &= \sum_{j=1}^{\lfloor\frac{n-1}{2}\rfloor}\binom{n-j-1}{j}\cdot\frac{2(n-2j-1)!}{(n+m-2j-1)!!(n-m-2j-1)!!} \\ &= \sum_{j=1}^{\lfloor\frac{n-1}{2}\rfloor}\frac{(n-j-1)!}{j!(n-2j-1)!}\cdot\frac{2(n-2j-1)!}{(n+m-2j-1)!!(n-m-2j-1)!!} \\ &= \sum_{j=1}^{\lfloor\frac{n-1}{2}\rfloor}\frac{2(n-j-1)!}{j!(n+m-2j-1)!!(n-m-2j-1)!!}. \end{aligned}$$

This proves Lemma 4. □

Lemma 5 *For any positive integers m and n, we have the following identities*:

$$F_n(2i\cos\theta) = \frac{i^{n+3}\sin n\theta}{\sin\theta}, \tag{12}$$

$$\int_{-2i}^{2i}\sqrt{x^2+4}F_m(x)F_n(x)\,dx = \begin{cases} 2i^{2m-1}\pi, & m = n > 0, \\ 0, & \textit{otherwise.} \end{cases} \tag{13}$$

Proof As we know,

$$F_n(x) = \frac{1}{2^n\sqrt{x^2+4}}\left[\left(x+\sqrt{x^2+4}\right)^n - \left(x-\sqrt{x^2+4}\right)^n\right].$$

Let $x = 2i\cos\theta$, then we have

$$\begin{aligned} F_n(2i\cos\theta) &= \frac{1}{2\sin\theta}\left[(i\cos\theta+\sin\theta)^n - (i\cos\theta-\sin\theta)^n\right] \\ &= \frac{1}{2\sin\theta}\left(i^n e^{-in\theta} - i^n e^{in\theta}\right) \end{aligned}$$

$$= \frac{i^n}{2\sin\theta}(\cos n\theta - i\sin n\theta - \cos n\theta - i\sin n\theta)$$
$$= \frac{i^{n+3}\sin n\theta}{\sin\theta}.$$

This proves property (12). Let $x = 2i\cos\theta$ in the following identity:

$$A = \int_{-2i}^{2i} \sqrt{x^2+4}F_m(x)F_n(x)\,dx,$$

then we can get

$$\begin{aligned} A &= \int_0^{\pi} 2\sin\theta F_n(2i\cos\theta)F_m(2i\cos\theta)2i\sin\theta\,d\theta \\ &= \int_0^{\pi} 4i\sin^2\theta\frac{i^{n+3}\sin n\theta}{\sin\theta}\frac{i^{m+3}\sin m\theta}{\sin\theta}\,d\theta \\ &= 4i^{n+m-1}\int_0^{\pi}\sin n\theta\sin m\theta\,d\theta \\ &= 2i^{n+m-1}\int_0^{\pi}\cos(n-m)\theta - \cos(m+n)\theta\,d\theta. \end{aligned}$$

Then we can get property (13). This proves Lemma 5. □

Lemma 6 *For any positive integer n, we have*

$$T_{2n}(x) = \sum_{k=0}^{n}\frac{(-1)^{n-k}\cdot 2^{2k}\cdot n}{n+k}\binom{n+k}{2k}x^{2k},$$
$$T_{2n+1}(x) = \sum_{k=0}^{n}\frac{(-1)^{n-k}\cdot 2^{2k}\cdot(2n+1)}{n+k+1}\binom{n+k+1}{2k+1}x^{2k+1},$$
$$U_{2n}(x) = \sum_{k=0}^{n}\frac{(-1)^{n-k}\cdot 2^{2k}\cdot(2k+1)}{n+k+1}\binom{n+k+1}{2k+1}x^{2k},$$
$$U_{2n+1}(x) = \sum_{k=0}^{n}\frac{(-1)^{n-k}\cdot 2^{2k+2}\cdot(k+1)}{n+k+2}\binom{n+k+2}{2k+2}x^{2k+1}.$$

Proof From Theorem 2 of reference [4], we can get the following result easily:

$$\begin{aligned} T_{2n}(x) &= \sum_{k=0}^{n}(-1)^{n-k}\left[2^{2k}\binom{n+k}{n-k} - 2^{2k-1}\binom{n+k-1}{2k-1}\right]x^{2k} \\ &= \sum_{k=0}^{n}(-1)^{n-k}\cdot 2^{2k-1}\left[\frac{2(n+k)!}{(2k)!(n-k)!} - \frac{(n+k-1)!}{(2k-1)!(n-k)!}\right]x^{2k} \\ &= \sum_{k=0}^{n}(-1)^{n-k}\cdot 2^{2k}\frac{n(n+k-1)!}{(2k)!(n-k)!}x^{2k} \\ &= \sum_{k=0}^{n}\frac{(-1)^{n-k}\cdot 2^{2k}\cdot n}{n+k}\binom{n+k}{2k}x^{2k}. \end{aligned}$$

From Theorem 2 of reference [4], we know

$$T_{2n+1}(x) = \sum_{k=0}^{n} (-1)^{n-k} \left[2^{2k+1} \binom{n+k+1}{n-k} - 2^{2k} \binom{n+k}{2k} \right] x^{2k+1}.$$

In a similar way, we can get

$$T_{2n+1}(x) = \sum_{k=0}^{n} \frac{(-1)^{n-k} \cdot 2^{2k} \cdot (2n+1)}{n+k+1} \binom{n+k+1}{2k+1} x^{2k+1}.$$

We can get $T'_n(x) = nU_{n-1}(x)$ easily from the definition of the Chebyshev polynomials. If we derive both sides of the above properties, we will get

$$U_{2n}(x) = \sum_{k=0}^{n} \frac{(-1)^{n-k} \cdot 2^{2k} \cdot (2k+1)}{n+k+1} \binom{n+k+1}{2k+1} x^{2k},$$

$$U_{2n+1}(x) = \sum_{k=0}^{n} \frac{(-1)^{n-k} \cdot 2^{2k+2} \cdot (k+1)}{n+k+2} \binom{n+k+2}{2k+2} x^{2k+1}.$$

This proves Lemma 6. □

Lemma 7 *For any positive integer n, let*

$$T_{2n}(x) = \sum_{s=1}^{+\infty} c_{2n,s} F_s(x) \tag{14}$$

and

$$T_{2n-1}(x) = \sum_{s=1}^{+\infty} c_{2n-1,s} F_s(x), \tag{15}$$

then we can get

$$c_{2n,s} = \begin{cases} \sum_{k=0}^{n} \frac{2^{4k+1} sn \cdot i^{3s+2n+1} (n+k-1)!}{(n-k)!(2k+s+1)!!(2k-s+1)!!}, & s \text{ is odd}, \\ 0, & \text{otherwise}, \end{cases} \tag{16}$$

$$c_{2n-1,s} = \begin{cases} \sum_{k=1}^{n} \frac{(2ns-s) i^{3s+2n} (n+k-2)!}{2^{2-4k} (n-k)!(2k+s)!!(2k-s)!!}, & s \text{ is even}, \\ 0, & \text{otherwise}. \end{cases} \tag{17}$$

Proof At first, we multiply $\sqrt{x^2+4}F_m(x)$ to both sides of (14), then integrate it from $-2i$ to $2i$. We can get the following identity by applying Lemma 5, where m is any positive integer:

$$\int_{-2i}^{2i} \sqrt{x^2+4} F_m(x) T_{2n}(x)\, dx = \sum_{s=1}^{\infty} \int_{-2i}^{2i} c_{2n,s} \sqrt{x^2+4} F_s(x) F_m(x)\, dx = 2i^{2m-1} \pi\, c_{2n,m},$$

and then we have

$$c_{2n,m} = \frac{(-i)^{2m-1}}{2\pi} \int_{-2i}^{2i} \sqrt{x^2+4} F_m(x) T_{2n}(x)\, dx.$$

Let $x = 2i\cos\theta$, then we can get the following identity by applying Lemma 5 and property (10):

$$\begin{aligned} c_{2n,m} &= \frac{(-i)^{2m-1}}{2\pi}\int_0^{\pi} T_{2n}(2i\cos\theta)F_m(2i\cos\theta)4i\sin^2\theta\, d\theta \\ &= \frac{-i^{2m}}{2\pi}\int_0^{\pi} T_{2n}(2i\cos\theta)\cdot\frac{i^{m+3}\sin m\theta}{\sin\theta}\cdot(2\sin\theta)^2\, d\theta \\ &= \frac{2i^{3m+1}}{\pi}\int_0^{\pi} T_{2n}(2i\cos\theta)\sin m\theta\sin\theta\, d\theta \\ &= \sum_{k=0}^{n}\frac{2^{4k}\cdot n\cdot i^{3m+2n+1}}{n+k}\binom{n+k}{2k}w_{2k,m-1}, \end{aligned}$$

so when m is even, we have $c_{2n,m} = 0$. When m is odd, we have

$$\begin{aligned} c_{2n,m} &= \sum_{k=0}^{n}\frac{2^{4k}\cdot n\cdot i^{3m+2n+1}}{n+k}\cdot\binom{n+k}{2k}\cdot\frac{2m\cdot(2k)!}{(2k+m+1)!!(2k-m+1)!!} \\ &= \sum_{k=0}^{n}\frac{2^{4k}\cdot n\cdot i^{3m+2n+1}}{n+k}\cdot\frac{(n+k)!}{(2k)!(n-k)!}\cdot\frac{2m\cdot(2k)!}{(2k+m+1)!!(2k-m+1)!!} \\ &= \sum_{k=0}^{n}\frac{2^{4k+1}\cdot mn\cdot i^{3m+2n+1}(n+k-1)!}{(n-k)!(2k+m+1)!!(2k-m+1)!!}. \end{aligned}$$

In a similar way, we can get the following result easily:

$$\begin{aligned} c_{2n-1,m} &= \frac{(-i)^{2m-1}}{2\pi}\int_0^{\pi} T_{2n-1}(2i\cos\theta)F_m(2i\cos\theta)4i\sin^2\theta\, d\theta \\ &= \sum_{k=1}^{n}\frac{2^{4k-3}\cdot(2n-1)\cdot i^{3m+2n}}{n+k-1}\binom{n+k-1}{2k-1}w_{2k-1,m-1}, \end{aligned}$$

and we have $c_{2n-1,m} = 0$ if m is odd. If m is even, we have

$$\begin{aligned} c_{2n-1,m} &= \sum_{k=1}^{n}\frac{2^{4k-3}\cdot(2n-1)\cdot i^{3m+2n}}{n+k-1}\binom{n+k-1}{2k-1}\frac{2m(2k-1)!}{(2k+m)!!(2k-m)!!} \\ &= \sum_{k=1}^{n}\frac{(2nm-m)i^{3m+2n}(n+k-2)!}{2^{2-4k}(n-k)!(2k+m)!!(2k-m)!!}. \end{aligned}$$

This proves Lemma 7. □

Lemma 8 *For any positive integer n, let*

$$U_{2n}(x) = \sum_{s=1}^{+\infty} d_{2n,s}F_s(x) \tag{18}$$

and

$$U_{2n-1}(x) = \sum_{s=1}^{+\infty} d_{2n-1,s}F_s(x), \tag{19}$$

then we can get

$$d_{2n,s} = \begin{cases} \sum_{k=0}^{n} \frac{2^{4k+1} \cdot i^{3s+2n+1} s(n+k)!}{(n-k)!(2k+s+1)!!(2k-s+1)!!}, & s \text{ is odd}, \\ 0, & \text{otherwise}, \end{cases} \tag{20}$$

$$d_{2n-1,s} = \begin{cases} \sum_{k=0}^{n-1} \frac{2^{4k+3} \cdot i^{3s+2n} s(n+k-1)!}{(n-k-1)!(2k+s+2)!!(2k+2-s)!!}, & s \text{ is even}, \\ 0, & \text{otherwise}. \end{cases} \tag{21}$$

Proof At first, we multiply $\sqrt{x^2+4}F_m(x)$ to both sides of (18), then integrate it from $-2i$ to $2i$, we can get the following identities by applying Lemma 5, where m is any positive integer:

$$\int_{-2i}^{2i} \sqrt{x^2+4}F_m(x)U_{2n}(x)\,dx = \sum_{k=1}^{\infty} \int_{-2i}^{2i} d_{ns}\sqrt{x^2+4}F_s(x)F_m(x)\,dx = 2i^{2m-1}\pi\, d_{2n,m},$$

then we have

$$d_{2n,m} = \frac{(-i)^{2m-1}}{2\pi} \int_{-2i}^{2i} \sqrt{x^2+4}F_m(x)U_{2n}(x)\,dx.$$

Let $x = 2i\cos\theta$, then we can get the following identity by applying Lemma 5 and property (10):

$$\begin{aligned} d_{2n,m} &= \frac{i^{2m+1}}{2\pi} \int_0^{\pi} U_{2n}(2i\cos\theta)F_m(2i\cos\theta)4i\sin^2\theta\,d\theta \\ &= \frac{i^{2m+2}}{2\pi} \int_0^{\pi} U_{2n}(2i\cos\theta)\frac{i^{m+3}\sin m\theta}{\sin\theta}(2\sin\theta)^2\,d\theta \\ &= \frac{2i^{3m+1}}{\pi} \int_0^{\pi} U_{2n}(2i\cos\theta)\sin m\theta\sin\theta\,d\theta \\ &= \sum_{k=0}^{n} \frac{2^{4k}\cdot(2k+1)\cdot i^{3m+1+2n}}{n+k+1}\binom{n+k+1}{2k+1} w_{2k,m-1}, \end{aligned}$$

so we have $d_{2n,m} = 0$ if m is even. If m is odd, we have

$$\begin{aligned} d_{2n,m} &= \sum_{k=0}^{n} \frac{2^{4k}\cdot(2k+1)\cdot i^{3m+1+2n}}{n+k+1}\binom{n+k+1}{2k+1}\cdot\frac{2m\cdot(2k)!}{(2k+m+1)!!(2k-m+1)!!} \\ &= \sum_{k=0}^{n} \frac{2^{4k}\cdot(2k+1)\cdot i^{3m+1+2n}}{n+k+1}\cdot\frac{(n+k+1)!}{(2k+1)!(n-k)!}\cdot\frac{2m\cdot(2k)!}{(2k+m+1)!!(2k-m+1)!!} \\ &= \sum_{k=0}^{n} \frac{2^{4k+1}\cdot i^{3m+2n+1}m(n+k)!}{(n-k)!(2k+m+1)!!(2k-m+1)!!}. \end{aligned}$$

In a similar way, we have

$$\begin{aligned} d_{2n-1,m} &= \frac{(-i)^{2m-1}}{2\pi} \int_0^{\pi} U_{2n-1}(2i\cos\theta)F_m(2i\cos\theta)4i\sin^2\theta\,d\theta \\ &= \sum_{k=0}^{n-1} \frac{2^{4k+3}\cdot(k+1)\cdot i^{3m+2n}}{n+k+1}\binom{n+k+1}{2k+2} w_{2k+1,m-1}. \end{aligned}$$

When m is odd, we have $d_{2n-1,m} = 0$. When m is even, we have

$$d_{2n-1,m} = \sum_{k=0}^{n-1} \frac{2^{4k+3} \cdot (k+1) \cdot i^{3m+2n}}{n+k+1} \binom{n+k+1}{2k+2} \frac{2m(2k+1)!}{(2k+m+2)!!(2k+2-m)!!}$$
$$= \sum_{k=0}^{n-1} \frac{2^{4k+3} \cdot i^{3m+2n} m(n+k-1)!}{(n-k-1)!(2k+m+2)!!(2k+2-m)!!}.$$

This proves Lemma 8. □

3 Proof of the theorems and corollaries

In this section, we will prove our theorems and corollaries. First of all, we can prove all the theorems from Lemma 4, Lemma 7 and Lemma 8 easily.

Proof of Corollary 1 We can get the following properties from Lemma 2 and Lemma 3 by letting $x = T_m(x)$ in Theorem 3:

$$F_{2n-1}(T_m(x)) = \sum_{k=0}^{n-1} \sum_{j=1}^{n-1} \frac{2^{2j+2-2n}(2k+1)(n-j-1)!}{j!(n-k-j-1)!(n+k-j)!} U_{2k}(T_m(x)),$$
$$F_{2n}(T_m(x)) = \sum_{k=1}^{n-1} \sum_{j=1}^{n-1} \frac{2^{2j-2n+2}k(2n-j-1)!}{j!(n-k-j)!(n+k-j)!} U_{2k}(T_m(x)).$$

Then, taking $x = \frac{i}{2}$ in the above identities, according to Lemma 2, we can get Corollary 1. □

Proof of Corollary 2 We can get the following properties from Lemma 2 and Lemma 3 by letting $x = T_m(x)$ in Theorem 4:

$$F_{2n}(T_m(x)) = \sum_{k=1}^{n} \sum_{j=1}^{n-1} \frac{2^{2j+2-2n}(2n-j-1)!T_{2k-1}(T_m(x))}{j!(2n+2k-2j-2)!(2n-2k-2j)!},$$
$$F_{2n-1}(T_m(x)) = \sum_{k=1}^{n-1} \sum_{j=1}^{n-1} \frac{2^{2j+3-2n}(2n-j-2)!T_{2k}(T_m(x))}{j!(n+k-j-1)!(n-k-j-1)!}$$
$$+ \sum_{j=1}^{n-1} \frac{2^{2j+2-2n}(2n-j-2)!}{j!(n-j-1)!(n-j-1)!} T_0(T_m(x)).$$

Then, taking $x = \frac{i}{2}$ in the above identities, according to Lemma 2, we can get Corollary 2. □

Competing interests
The author declares that there is no conflict of interests regarding the publication of the article.

Acknowledgements
The author would like to thank the referees for their very helpful and detailed comments which have significantly improved the presentation of this paper.

References

1. Zhang, W: Some identities involving the Fibonacci numbers and Lucas numbers. Fibonacci Q. **42**, 149-154 (2004)
2. Falcón, S, Plaza, Á: On *k*-Fibonacci sequences and polynomials and their derivatives. Chaos Solitons Fractals **39**, 1005-1019 (2009)
3. Falcón, S, Plaza, Á: The *k*-Fibonacci sequences and the Pascal 2-triangle. Chaos Solitons Fractals **38**, 38-49 (2007)
4. Wu, X, Yang, G: The general formula of Chebyshev polynomials. J. Wuhan Transp. Univ. **24**, 573-576 (2000)
5. Abramowitz, M, Stegun, IA: Handbook of Mathematical Functions: With Formulas, Graphs and Mathematical Tables, chap. 22, pp. 771-802. Dover New York (1965)
6. Ma, R, Zhang, W: Several identities involving the Fibonacci numbers and Lucas numbers. Fibonacci Q. **5**, 164-171 (2007)

2

A coupled system of fractional q-integro-difference equations with nonlocal fractional q-integral boundary conditions

Suthep Suantai[1], Sotiris K Ntouyas[2,3], Suphawat Asawasamrit[4] and Jessada Tariboon[4*]

*Correspondence:
jessadat@kmutnb.ac.th
[4]Nonlinear Dynamic Analysis Research Center, Department of Mathematics, Faculty of Applied Science, King Mongkut's University of Technology North Bangkok, Bangkok, 10800, Thailand
Full list of author information is available at the end of the article

Abstract

In this paper, we investigate the existence and the uniqueness of solutions for coupled and uncoupled systems of fractional q-integro-difference equations with nonlocal fractional q-integral boundary conditions. The existence and the uniqueness of the solutions are established by using the Banach contraction principle, while the existence of solutions is derived by applying Leray-Schauder's alternative. Examples illustrating our results are also presented.

MSC: 34A08; 34A12; 34B15; 93A10

Keywords: fractional q-difference equations; existence; uniqueness; fixed point theorems

1 Introduction

In this paper, we investigate a coupled system of fractional q-integro-difference equations with nonlocal fractional q-integral boundary conditions given by

$$\begin{cases} D_q^{\alpha}x(t) = f(t, x(t), I_r^{\delta}y(t)), & t \in [0,T], 1 < \alpha \leq 2, \\ D_p^{\beta}y(t) = g(t, y(t), I_z^{\varepsilon}x(t)), & t \in [0,T], 1 < \beta \leq 2, \\ x(0) = 0, & \lambda_1 I_m^{\gamma}x(\eta) = I_n^{\kappa}y(\xi), \\ y(0) = 0, & \lambda_2 I_h^{\mu}y(\theta) = I_k^{\nu}x(\tau), \end{cases} \tag{1.1}$$

where $0 < p, q, r, z, m, n, h, k < 1$ are quantum numbers, $\eta, \xi, \theta, \tau \in (0, T)$ are fixed points, $\delta, \varepsilon, \gamma, \kappa, \mu, \nu > 0$, and $\lambda_1, \lambda_2 \in \mathbb{R}$ are given constants, D_{ω}^{ρ} is the fractional ω-derivative of Riemann-Liouville type of order ρ, when $\rho \in \{\alpha, \beta\}$ and $\omega \in \{p, q\}$, I_{ϕ}^{ψ} is the fractional ϕ-integral of order ψ with $\phi \in \{r, z, m, n, h, k\}$ and $\psi \in \{\delta, \varepsilon, \gamma, \kappa, \mu, \nu\}$ and $f, g : [0, T] \times \mathbb{R} \times \mathbb{R} \to \mathbb{R}$ are continuous functions.

The early work on q-difference calculus or *quantum calculus* dates back to Jackson's paper [1]. Basic definitions and properties of quantum calculus can be found in the book [2]. The fractional q-difference calculus had its origin in the works by Al-Salam [3] and Agarwal [4]. Motivated by recent interest in the study of fractional-order differential equations, the topic of q-fractional equations has attracted the attention of many researchers. The details of some recent development of the subject can be found in [5–18], and the references cited therein, whereas the background material on q-fractional calculus can be found in a recent book [19].

Recently in [20], we have studied the existence and the uniqueness of solutions of a class of boundary value problems for fractional q-integro-difference equations with nonlocal fractional q-integral conditions which have different quantum numbers. Here we extend the results of [20] to a coupled system of fractional q-integro-difference equations with nonlocal fractional q-integral boundary conditions.

The paper is organized as follows: In Section 2 we will present some useful preliminaries and lemmas. Some auxiliary lemmas are presented in Section 3. In Section 4, we establish an existence and a uniqueness result via the Banach contraction principle, and an existence result by applying Leray-Schauder's alternative. Results on the uncoupled integral boundary conditions case are contained in Section 5. Examples illustrating our results are also presented.

2 Preliminaries

To make this paper self-contained, below we recall some well-known facts on fractional q-calculus. The presentation here can be found in, for example, [6, 19].

For $q \in (0,1)$, define

$$[a]_q = \frac{1-q^a}{1-q}, \quad a \in \mathbb{R}. \tag{2.1}$$

The q-analog of the power function $(a-b)^k$ with $k \in \mathbb{N}_0 := \{0,1,2,\ldots\}$ is

$$(a-b)^{(0)} = 1, \qquad (a-b)^{(k)} = \prod_{i=0}^{k-1}\left(a-bq^i\right), \quad k \in \mathbb{N}, a, b \in \mathbb{R}. \tag{2.2}$$

More generally, if $\gamma \in \mathbb{R}$, then

$$(a-b)^{(\gamma)} = a^\gamma \prod_{i=0}^{\infty} \frac{1-(b/a)q^i}{1-(b/a)q^{\gamma+i}}, \quad a \neq 0. \tag{2.3}$$

Note if $b = 0$, then $a^{(\gamma)} = a^\gamma$. We also use the notation $0^{(\gamma)} = 0$ for $\gamma > 0$. The q-gamma function is defined by

$$\Gamma_q(t) = \frac{(1-q)^{(t-1)}}{(1-q)^{t-1}}, \quad t \in \mathbb{R} \setminus \{0,-1,-2,\ldots\}. \tag{2.4}$$

Obviously, $\Gamma_q(t+1) = [t]_q \Gamma_q(t)$.

The q-derivative of a function h is defined by

$$(D_q h)(t) = \frac{h(t)-h(qt)}{(1-q)t} \quad \text{for } t \neq 0 \quad \text{and} \quad (D_q h)(0) = \lim_{t\to 0}(D_q h)(t), \tag{2.5}$$

and q-derivatives of higher order are given by

$$\left(D_q^0 h\right)(t) = h(t) \quad \text{and} \quad \left(D_q^k h\right)(t) = D_q\left(D_q^{k-1} h\right)(t), \quad k \in \mathbb{N}. \tag{2.6}$$

The q-integral of a function h defined on the interval $[0,b]$ is given by

$$(I_q h)(t) = \int_0^t h(s)\, d_q s = t(1-q)\sum_{i=0}^{\infty} h\left(tq^i\right)q^i, \quad t \in [0,b]. \tag{2.7}$$

If $a \in [0, b]$ and h is defined in the interval $[0, b]$, then its integral from a to b is defined by

$$\int_a^b h(s)\, d_q s = \int_0^b h(s)\, d_q s - \int_0^a h(s)\, d_q s. \tag{2.8}$$

Similar to derivatives, an operator I_q^k is given by

$$\big(I_q^0 h\big)(t) = h(t) \quad \text{and} \quad \big(I_q^k h\big)(t) = I_q\big(I_q^{k-1} h\big)(t), \quad k \in \mathbb{N}. \tag{2.9}$$

The fundamental theorem of calculus applies to these operators D_q and I_q, *i.e.*,

$$(D_q I_q h)(t) = h(t), \tag{2.10}$$

and if h is continuous at $t = 0$, then

$$(I_q D_q h)(t) = h(t) - h(0). \tag{2.11}$$

Definition 2.1 Let $\nu \geq 0$ and h be a function defined on $[0, T]$. The fractional q-integral of Riemann-Liouville type is given by $(I_q^0 h)(t) = h(t)$ and

$$\big(I_q^\nu h\big)(t) = \frac{1}{\Gamma_q(\nu)} \int_0^t (t - qs)^{(\nu-1)} h(s)\, d_q s, \quad \nu > 0, t \in [0, T]. \tag{2.12}$$

Definition 2.2 The fractional q-derivative of Riemann-Liouville type of order $\nu \geq 0$ is defined by $(D_q^0 h)(t) = h(t)$ and

$$\big(D_q^\nu h\big)(t) = \big(D_q^l I_q^{l-\nu} h\big)(t), \quad \nu > 0, \tag{2.13}$$

where l is the smallest integer greater than or equal to ν.

Definition 2.3 For any $t, s > 0$,

$$B_q(t, s) = \int_0^1 u^{(t-1)} (1 - qu)^{(s-1)}\, d_q u \tag{2.14}$$

is called the q-beta function.

The expression of q-beta function in terms of the q-gamma function can be written as

$$B_q(t, s) = \frac{\Gamma_q(t)\Gamma_q(s)}{\Gamma_q(t + s)}.$$

Lemma 2.4 [4] *Let $\alpha, \beta \geq 0$ and f be a function defined in $[0, T]$. Then the following formulas hold:*

(1) $(I_q^\beta I_q^\alpha f)(t) = (I_q^{\alpha+\beta} f)(t)$,

(2) $(D_q^\alpha I_q^\alpha f)(t) = f(t)$.

Lemma 2.5 [6] *Let $\alpha > 0$ and n be a positive integer. Then the following equality holds:*

$$\left(I_q^\alpha D_q^n f\right)(t) = \left(D_q^n I_q^\alpha f\right)(t) - \sum_{i=0}^{n-1} \frac{t^{\alpha-n+i}}{\Gamma_q(\alpha+i-n+1)} \left(D_q^i f\right)(0). \tag{2.15}$$

3 Some auxiliary lemmas

The following formulas have been modified from Lemmas 3.2 and 7 in [21] and [20], respectively.

Lemma 3.1 *Let $x, y, z > 0$ and $0 < u, v, w < 1$. Then, for $\phi \in \mathbb{R}_+$, we have*

(i) $I_u^x I_v^y(1)(\phi) = \frac{\Gamma_u(y+1)}{\Gamma_u(x+y+1)\Gamma_v(y+1)} \phi^{x+y}$;

(ii) $I_u^x I_v^y I_w^z(1)(\phi) = \frac{\Gamma_u(y+z+1)\Gamma_v(z+1)}{\Gamma_u(x+y+z+1)\Gamma_v(y+z+1)\Gamma_w(z+1)} \phi^{x+y+z}$.

Lemma 3.2 *Given $u, v \in C([0,T], \mathbb{R})$, the unique solution of the problem*

$$\begin{cases} D_q^\alpha x(t) = u(t), & t \in [0,T], 1 < \alpha \le 2, \\ D_p^\beta y(t) = v(t), & t \in [0,T], 1 < \beta \le 2, \\ x(0) = 0, & \lambda_1 I_m^\gamma x(\eta) = I_n^\kappa y(\xi), \\ y(0) = 0, & \lambda_2 I_h^\mu y(\theta) = I_k^\nu x(\tau), \end{cases} \tag{3.1}$$

is

$$\begin{aligned} x(t) = {} & \frac{1}{\Gamma_q(\alpha)} \int_0^t (t-qs)^{(\alpha-1)} u(s)\, d_q s + \frac{\lambda_2 \Omega_1}{\Omega} t^{\alpha-1} I_h^\mu I_p^\beta v(\theta) \\ & - \frac{\Omega_1}{\Omega} t^{\alpha-1} I_k^\nu I_q^\alpha u(\tau) + \frac{\lambda_1 \lambda_2 \Omega_4}{\Omega} t^{\alpha-1} I_m^\gamma I_q^\alpha u(\eta) \\ & - \frac{\lambda_2 \Omega_4}{\Omega} t^{\alpha-1} I_n^\kappa I_p^\beta v(\xi) \end{aligned} \tag{3.2}$$

and

$$\begin{aligned} y(t) = {} & \frac{1}{\Gamma_p(\beta)} \int_0^t (t-ps)^{(\beta-1)} v(s)\, d_p s + \frac{\lambda_1 \Omega_2}{\Omega} t^{\beta-1} I_m^\gamma I_q^\alpha u(\eta) \\ & - \frac{\Omega_2}{\Omega} t^{\beta-1} I_n^\kappa I_p^\beta v(\xi) + \frac{\lambda_1 \lambda_2 \Omega_3}{\Omega} t^{\beta-1} I_h^\mu I_p^\beta v(\theta) \\ & - \frac{\lambda_1 \Omega_3}{\Omega} t^{\beta-1} I_k^\nu I_q^\alpha u(\tau), \end{aligned} \tag{3.3}$$

where

$$\Omega_1 = \frac{\Gamma_n(\beta)}{\Gamma_n(\beta+\kappa)} \xi^{\beta+\kappa-1},$$

$$\Omega_2 = \frac{\Gamma_k(\alpha)}{\Gamma_k(\alpha+\nu)} \tau^{\alpha+\nu-1},$$

$$\Omega_3 = \frac{\Gamma_m(\alpha)}{\Gamma_m(\alpha+\gamma)} \eta^{\alpha+\gamma-1},$$

$$\Omega_4 = \frac{\Gamma_h(\beta)}{\Gamma_h(\beta+\mu)} \theta^{\beta+\mu-1},$$

$$\Omega = \Omega_1 \Omega_2 - \lambda_1 \lambda_2 \Omega_3 \Omega_4 \neq 0.$$

Proof From $1 < \alpha \leq 2$, we let $n = 2$. Applying Lemma 2.5, the equations in (3.1) can be expressed as equivalent integral equations

$$x(t) = c_1 t^{\alpha-1} + c_2 t^{\alpha-2} + \frac{1}{\Gamma_q(\alpha)} \int_0^t (t-qs)^{(\alpha-1)} u(s)\, d_q s, \tag{3.4}$$

$$y(t) = d_1 t^{\beta-1} + d_2 t^{\beta-2} + \frac{1}{\Gamma_p(\beta)} \int_0^t (t-ps)^{(\beta-1)} v(s)\, d_p s \tag{3.5}$$

for $c_1, c_2, d_1, d_2 \in \mathbb{R}$. The conditions $x(0) = 0$ and $y(0) = 0$ imply that $c_2 = 0$ and $d_2 = 0$, respectively. Taking the Riemann-Liouville fractional ϕ-integral of order $\psi > 0$ for (3.4) and (3.5), we have the system

$$\begin{aligned} I_\phi^\psi x(t) &= c_1 \frac{\Gamma_\phi(\alpha)}{\Gamma_\phi(\alpha+\psi)} t^{\alpha+\psi-1} \\ &\quad + \frac{1}{\Gamma_\phi(\psi)\Gamma_q(\alpha)} \int_0^t \int_0^s (t-\phi s)^{(\psi-1)} (s-qw)^{(\alpha-1)} u(w)\, d_q w\, d_\phi s, \end{aligned} \tag{3.6}$$

$$\begin{aligned} I_\phi^\psi y(t) &= d_1 \frac{\Gamma_\phi(\beta)}{\Gamma_\phi(\beta+\psi)} t^{\beta+\psi-1} \\ &\quad + \frac{1}{\Gamma_\phi(\psi)\Gamma_p(\beta)} \int_0^t \int_0^s (t-\phi s)^{(\psi-1)} (s-pw)^{(\beta-1)} v(w)\, d_p w\, d_\phi s. \end{aligned} \tag{3.7}$$

Substituting (ψ, ϕ, t) by (γ, m, η), (ν, k, τ) in (3.6), and (κ, n, ξ), (μ, h, θ) in (3.7) and using Lemma 2.4 with nonlocal conditions in (3.1), we have

$$\begin{aligned} c_1 &= \frac{\lambda_2 \Omega_1}{\Omega} I_h^\mu I_p^\beta v(\theta) - \frac{\Omega_1}{\Omega} I_k^\nu I_q^\alpha u(\tau) \\ &\quad + \frac{\lambda_1 \lambda_2 \Omega_4}{\Omega} I_m^\gamma I_q^\alpha u(\eta) - \frac{\lambda_2 \Omega_4}{\Omega} I_n^\kappa I_p^\beta v(\xi) \end{aligned}$$

and

$$\begin{aligned} d_1 &= \frac{\lambda_1 \Omega_2}{\Omega} I_m^\gamma I_q^\alpha u(\eta) - \frac{\Omega_2}{\Omega} I_n^\kappa I_p^\beta v(\xi) \\ &\quad + \frac{\lambda_1 \lambda_2 \Omega_3}{\Omega} I_h^\mu I_p^\beta v(\theta) - \frac{\lambda_1 \Omega_3}{\Omega} I_k^\nu I_q^\alpha u(\tau). \end{aligned}$$

Substituting the values of c_1, c_2, d_1, and d_2 in (3.4) and (3.5), we obtain the solutions (3.2) and (3.3) as required. □

4 Main results

Let $\mathcal{C} = C([0,T], \mathbb{R})$ denotes the Banach space of all continuous functions from $[0,T]$ to $\mathbb{R}$. Let us introduce the space $X = \{x(t) | x(t) \in C([0,T], \mathbb{R})\}$ endowed with the norm $\|x\| = \sup\{|x(t)|, t \in [0,T]\}$. Obviously $(X, \|\cdot\|)$ is a Banach space. Also let $Y = \{y(t) | y(t) \in C([0,T], \mathbb{R})\}$ be endowed with the norm $\|y\| = \sup\{|y(t)|, t \in [0,T]\}$. Obviously the product space $(X \times Y, \|(x,y)\|)$ is a Banach space with norm $\|(x,y)\| = \|x\| + \|y\|$.

In view of Lemma 3.2, we define an operator $\mathcal{K} : X \times Y \to X \times Y$ by

$$\mathcal{K}(x,y)(t) = \begin{pmatrix} \mathcal{K}_1(x,y)(t) \\ \mathcal{K}_2(x,y)(t) \end{pmatrix},$$

where

$$
\begin{aligned}
\mathcal{K}_1(x,y)(t) = {} & I_q^\alpha f\big(s,x(s),I_r^\delta y(s)\big)(t) + \frac{\lambda_2\Omega_1}{\Omega}t^{\alpha-1}I_h^\mu I_p^\beta g\big(s,y(s),I_z^\varepsilon x(s)\big)(\theta) \\
& - \frac{\Omega_1}{\Omega}t^{\alpha-1}I_k^\nu I_q^\alpha f\big(s,x(s),I_r^\delta y(s)\big)(\tau) \\
& + \frac{\lambda_1\lambda_2\Omega_4}{\Omega}t^{\alpha-1}I_m^\gamma I_q^\alpha f\big(s,x(s),I_r^\delta y(s)\big)(\eta) \\
& - \frac{\lambda_2\Omega_4}{\Omega}t^{\alpha-1}I_n^\kappa I_p^\beta g\big(s,y(s),I_z^\varepsilon x(s)\big)(\xi)
\end{aligned} \tag{4.1}
$$

and

$$
\begin{aligned}
\mathcal{K}_2(x,y)(t) = {} & I_p^\beta g\big(s,y(s),I_z^\varepsilon x(s)\big)(t) + \frac{\lambda_1\Omega_2}{\Omega}t^{\beta-1}I_m^\gamma I_q^\alpha f\big(s,x(s),I_r^\delta y(s)\big)(\eta) \\
& - \frac{\Omega_2}{\Omega}t^{\beta-1}I_n^\kappa I_p^\beta g\big(s,y(s),I_z^\varepsilon x(s)\big)(\xi) \\
& + \frac{\lambda_1\lambda_2\Omega_3}{\Omega}t^{\beta-1}I_h^\mu I_p^\beta g\big(s,y(s),I_z^\varepsilon x(s)\big)(\theta) \\
& - \frac{\lambda_1\Omega_3}{\Omega}t^{\beta-1}I_k^\nu I_q^\alpha f\big(s,x(s),I_r^\delta y(s)\big)(\tau).
\end{aligned} \tag{4.2}
$$

For the sake of convenience, we set

$$
\begin{aligned}
A_1 &= \frac{T^\alpha}{\Gamma_q(\alpha+1)}, \qquad A_2 = \frac{T^\beta}{\Gamma_p(\beta+1)}, \\
A_3 &= \frac{\Gamma_q(\delta+1)T^{\alpha+\delta}}{\Gamma_q(\alpha+\delta+1)\Gamma_r(\delta+1)}, \qquad A_4 = \frac{\Gamma_p(\varepsilon+1)T^{\beta+\varepsilon}}{\Gamma_p(\beta+\varepsilon+1)\Gamma_z(\varepsilon+1)}, \\
A_5 &= \frac{\Gamma_m(\alpha+1)\eta^{\gamma+\alpha}}{\Gamma_m(\gamma+\alpha+1)\Gamma_q(\alpha+1)}, \qquad A_6 = \frac{\Gamma_n(\beta+1)\xi^{\kappa+\beta}}{\Gamma_n(\kappa+\beta+1)\Gamma_p(\beta+1)}, \\
A_7 &= \frac{\Gamma_h(\beta+1)\theta^{\mu+\beta}}{\Gamma_h(\mu+\beta+1)\Gamma_p(\beta+1)}, \qquad A_8 = \frac{\Gamma_k(\alpha+1)\tau^{\nu+\alpha}}{\Gamma_k(\nu+\alpha+1)\Gamma_q(\alpha+1)}, \\
A_9 &= \frac{\Gamma_m(\alpha+\delta+1)\Gamma_q(\delta+1)\eta^{\gamma+\alpha+\delta}}{\Gamma_m(\gamma+\alpha+\delta+1)\Gamma_q(\alpha+\delta+1)\Gamma_r(\delta+1)}, \\
A_{10} &= \frac{\Gamma_n(\beta+\varepsilon+1)\Gamma_p(\varepsilon+1)\xi^{\kappa+\beta+\varepsilon}}{\Gamma_n(\kappa+\beta+\varepsilon+1)\Gamma_p(\beta+\varepsilon+1)\Gamma_z(\varepsilon+1)}, \\
A_{11} &= \frac{\Gamma_h(\beta+\varepsilon+1)\Gamma_p(\varepsilon+1)\theta^{\mu+\beta+\varepsilon}}{\Gamma_h(\mu+\beta+\varepsilon+1)\Gamma_p(\beta+\varepsilon+1)\Gamma_z(\varepsilon+1)}, \\
A_{12} &= \frac{\Gamma_k(\alpha+\delta+1)\Gamma_q(\delta+1)\tau^{\nu+\alpha+\delta}}{\Gamma_k(\nu+\alpha+\delta+1)\Gamma_q(\alpha+\delta+1)\Gamma_r(\delta+1)}.
\end{aligned}
$$

Theorem 4.1 *Assume that $f, g : [0,T] \times \mathbb{R}^2 \to \mathbb{R}$ are continuous functions and there exist positive constants M_i, N_i, $i = 1, 2$, such that for all $t \in [0,T]$ and $u_i, v_i \in \mathbb{R}$, $i = 1, 2$,*

$$
\big|f(t,u_1,u_2) - f(t,v_1,v_2)\big| \le M_1|u_1 - v_1| + M_2|u_2 - v_2|
$$

and

$$
\big|g(t,u_1,u_2) - g(t,v_1,v_2)\big| \le N_1|u_1 - v_1| + N_2|u_2 - v_2|.
$$

In addition, we suppose that

$$B_1 + B_2 + C_1 + C_2 < 1,$$

where

$$B_1 = M_1A_1 + \frac{|\lambda_2|T^{\alpha-1}}{|\Omega|}\left(\Omega_1 N_2 A_{11} + |\lambda_1|\Omega_4 M_1 A_5 + \Omega_4 N_2 A_{10}\right) + \frac{\Omega_1}{|\Omega|}T^{\alpha-1}M_1A_8,$$

$$B_2 = M_2A_3 + \frac{|\lambda_2|T^{\alpha-1}}{|\Omega|}\left(\Omega_1 N_1 A_7 + |\lambda_1|\Omega_4 M_2 A_9 + \Omega_4 N_1 A_6\right) + \frac{\Omega_1}{|\Omega|}T^{\alpha-1}M_2A_{12},$$

$$C_1 = N_2A_4 + \frac{|\lambda_1|T^{\beta-1}}{|\Omega|}\left(\Omega_2 M_1 A_5 + |\lambda_2|\Omega_3 N_2 A_{11} + \Omega_3 M_1 A_8\right) + \frac{\Omega_2}{|\Omega|}T^{\beta-1}N_2A_{10},$$

$$C_2 = N_1A_2 + \frac{|\lambda_1|T^{\beta-1}}{|\Omega|}\left(\Omega_2 M_2 A_9 + |\lambda_2|\Omega_3 N_1 A_7 + \Omega_3 M_2 A_{12}\right) + \frac{\Omega_2}{|\Omega|}T^{\beta-1}N_1A_6.$$

Then the system (1.1) *has a unique solution on* $[0, T]$.

Proof Firstly, we define $\sup_{t\in[0,T]}|f(t,0,0)| = G_1 < \infty$ and $\sup_{t\in[0,T]}|g(t,0,0)| = G_2 < \infty$ such that

$$r \geq \max\left\{\frac{B_3}{1-(B_1+B_2)}, \frac{C_3}{1-(C_1+C_2)}\right\},$$

where

$$B_3 = G_1A_1 + \frac{|\lambda_2|T^{\alpha-1}}{|\Omega|}\left(\Omega_1 G_2 A_7 + |\lambda_1|\Omega_4 G_1 A_5 + \Omega_4 G_2 A_6\right) + \frac{\Omega_1}{|\Omega|}T^{\alpha-1}G_1A_8,$$

$$C_3 = G_2A_2 + \frac{|\lambda_1|T^{\beta-1}}{|\Omega|}\left(\Omega_2 G_1 A_5 + |\lambda_2|\Omega_3 G_2 A_7 + \Omega_3 G_1 A_8\right) + \frac{\Omega_2}{|\Omega|}T^{\beta-1}G_2A_6.$$

We will show that $\mathcal{K}B_r \subset B_r$, where $B_r = \{(x,y) \in X \times Y : \|(x,y)\| \leq r\}$.

For $(x,y) \in B_r$, taking into account Lemma 3.1, we have

$$\begin{aligned}
&\left|\mathcal{K}_1(x,y)(t)\right| \\
&\leq \sup_{t\in T}\Bigg\{ I_q^\alpha\left|f\left(s,x(s),I_r^\delta y(s)\right)\right|(t) + \frac{|\lambda_2|\Omega_1}{|\Omega|}t^{\alpha-1}I_h^\mu I_p^\beta\left|g\left(s,y(s),I_z^\varepsilon x(s)\right)\right|(\theta) \\
&\quad + \frac{\Omega_1}{|\Omega|}t^{\alpha-1}I_k^\nu I_q^\alpha\left|f\left(s,x(s),I_r^\delta y(s)\right)\right|(\tau) + \frac{|\lambda_1||\lambda_2|\Omega_4}{|\Omega|}t^{\alpha-1}I_m^\gamma I_q^\alpha\left|f\left(s,x(s),I_r^\delta y(s)\right)\right|(\eta) \\
&\quad + \frac{|\lambda_2|\Omega_4}{|\Omega|}t^{\alpha-1}I_n^\kappa I_p^\beta\left|g\left(s,y(s),I_z^\varepsilon x(s)\right)\right|(\xi)\Bigg\} \\
&\leq I_q^\alpha\left(\left|f\left(s,x(s),I_r^\delta y(s)\right) - f(s,0,0)\right| + \left|f(s,0,0)\right|\right)(t) \\
&\quad + \frac{|\lambda_2|\Omega_1}{|\Omega|}T^{\alpha-1}I_h^\mu I_p^\beta\left(\left|g\left(s,y(s),I_z^\varepsilon x(s)\right) - g(s,0,0)\right| + \left|g(s,0,0)\right|\right)(\theta) \\
&\quad + \frac{\Omega_1}{|\Omega|}T^{\alpha-1}I_k^\nu I_q^\alpha\left(\left|f\left(s,x(s),I_r^\delta y(s)\right) - f(s,0,0)\right| + \left|f(s,0,0)\right|\right)(\tau) \\
&\quad + \frac{|\lambda_1||\lambda_2|\Omega_4}{|\Omega|}T^{\alpha-1}I_m^\gamma I_q^\alpha\left(\left|f\left(s,x(s),I_r^\delta y(s)\right) - f(s,0,0)\right| + \left|f(s,0,0)\right|\right)(\eta)
\end{aligned}$$

$$
\begin{aligned}
&+\frac{|\lambda_2|\Omega_4}{|\Omega|}T^{\alpha-1}I_n^{\kappa}I_p^{\beta}\left(\left|g\left(s,y(s),I_z^{\varepsilon}x(s)\right)-g(t,0,0)\right|+\left|g(t,0,0)\right|\right)(\xi)\\
&\le M_1\|x\|A_1+M_2\|y\|A_3+G_1A_1\\
&+\frac{|\lambda_2|\Omega_1}{|\Omega|}T^{\alpha-1}\left(N_1\|y\|A_7+N_2\|x\|A_{11}+G_2A_7\right)\\
&+\frac{\Omega_1}{|\Omega|}T^{\alpha-1}\left(M_1\|x\|A_8+M_2\|y\|A_{12}+G_1A_8\right)\\
&+\frac{|\lambda_1||\lambda_2|\Omega_4}{|\Omega|}T^{\alpha-1}\left(M_1\|x\|A_5+M_2\|y\|A_9+G_1A_5\right)\\
&+\frac{|\lambda_2|\Omega_4}{|\Omega|}T^{\alpha-1}\left(N_1\|y\|A_6+N_2\|x\|A_{10}+G_2A_6\right)\\
&=B_1\|x\|+B_2\|y\|+B_3\\
&\le (B_1+B_2)r+B_3\le r.
\end{aligned}
$$

In a similar way, we get

$$
\begin{aligned}
&\left|\mathcal{K}_2(x,y)(t)\right|\\
&\le I_p^{\beta}\left(\left|g\left(s,y(s),I_z^{\varepsilon}x(s)\right)-g(s,0,0)\right|+\left|g(s,0,0)\right|\right)(t)\\
&+\frac{|\lambda_1|\Omega_2}{|\Omega|}T^{\beta-1}I_m^{\gamma}I_q^{\alpha}\left(\left|f\left(s,x(s),I_r^{\delta}y(s)\right)-f(s,0,0)\right|+\left|f(s,0,0)\right|\right)(\eta)\\
&+\frac{\Omega_2}{|\Omega|}T^{\beta-1}I_n^{\kappa}I_p^{\beta}\left(\left|g\left(s,y(s),I_z^{\varepsilon}x(s)\right)-g(s,0,0)\right|+\left|g(s,0,0)\right|\right)(\xi)\\
&+\frac{|\lambda_1||\lambda_2|\Omega_3}{|\Omega|}T^{\beta-1}I_h^{\mu}I_p^{\beta}\left(\left|g\left(s,y(s),I_z^{\varepsilon}x(s)\right)-g(s,0,0)\right|+\left|g(s,0,0)\right|\right)(\theta)\\
&+\frac{|\lambda_1|\Omega_3}{|\Omega|}T^{\beta-1}I_k^{\upsilon}I_q^{\alpha}\left(\left|f\left(s,x(s),I_r^{\delta}y(s)\right)-f(s,0,0)\right|+\left|f(s,0,0)\right|\right)(\tau)\\
&\le N_1\|y\|A_2+N_2\|x\|A_4+G_2A_2\\
&+\frac{|\lambda_1|\Omega_2}{|\Omega|}T^{\beta-1}\left(M_1\|x\|A_5+M_2\|y\|A_9+G_1A_5\right)\\
&+\frac{\Omega_2}{|\Omega|}T^{\beta-1}\left(N_1\|y\|A_6+N_2\|x\|A_{10}+G_2A_6\right)\\
&+\frac{|\lambda_1||\lambda_2|\Omega_3}{|\Omega|}T^{\beta-1}\left(N_1\|y\|A_7+N_2\|x\|A_{11}+G_2A_7\right)\\
&+\frac{|\lambda_1|\Omega_3}{|\Omega|}T^{\beta-1}\left(M_1\|x\|A_8+M_2\|y\|A_{12}+G_1A_8\right)\\
&=C_1\|x\|+C_2\|y\|+C_3\\
&\le (C_1+C_2)r+B_3\le r.
\end{aligned}
$$

Consequently, $\|\mathcal{K}(x,y)(t)\|\le r$.

Next, for $(x_2,y_2),(x_1,y_1)\in X\times Y$, and for any $t\in[0,T]$, we have

$$
\begin{aligned}
&\left|\mathcal{K}_1(x_2,y_2)(t)-\mathcal{K}_1(x_1,y_1)(t)\right|\\
&\le I_q^{\alpha}\left(\left|f\left(s,x_2(s),I_r^{\delta}y_2(s)\right)-f\left(s,x_1(s),I_r^{\delta}y_1(s)\right)\right|\right)(t)
\end{aligned}
$$

$$
\begin{aligned}
&+\frac{|\lambda_2|\Omega_1}{|\Omega|}T^{\alpha-1}I_h^{\mu}I_p^{\beta}\left(\left|g\left(s,y_2(s),I_z^{\varepsilon}x_2(s)\right)-g\left(s,y_1(s),I_z^{\varepsilon}x_1(s)\right)\right|\right)(\theta)\\
&+\frac{\Omega_1}{|\Omega|}T^{\alpha-1}I_k^{\nu}I_q^{\alpha}\left(\left|f\left(s,x_2(s),I_r^{\delta}y_2(s)\right)-f\left(s,x_1(s),I_r^{\delta}y_1(s)\right)\right|\right)(\tau)\\
&+\frac{|\lambda_1||\lambda_2|\Omega_4}{|\Omega|}T^{\alpha-1}I_m^{\gamma}I_q^{\alpha}\left(\left|f\left(s,x_2(s),I_r^{\delta}y_2(s)\right)-f\left(s,x_1(s),I_r^{\delta}y_1(s)\right)\right|\right)(\eta)\\
&+\frac{|\lambda_2|\Omega_4}{|\Omega|}T^{\alpha-1}I_n^{\kappa}I_p^{\beta}\left(\left|g\left(s,y_2(s),I_z^{\varepsilon}x_2(s)\right)-g\left(s,y_1(s),I_z^{\varepsilon}x_1(s)\right)\right|\right)(\xi)\\
&\le M_1\|x_2-x_1\|I_q^{\alpha}(1)(T)+M_2\|y_2-y_1\|I_q^{\alpha}I_r^{\delta}(1)(T)\\
&+\frac{|\lambda_2|\Omega_1}{|\Omega|}T^{\alpha-1}\left(N_1\|y_2-y_1\|I_h^{\mu}I_p^{\beta}(1)(\theta)+N_2\|x_2-x_1\|I_h^{\mu}I_p^{\beta}I_z^{\varepsilon}(1)(\theta)\right)\\
&+\frac{\Omega_1}{|\Omega|}T^{\alpha-1}\left(M_1\|x_2-x_1\|I_k^{\nu}I_q^{\alpha}(1)(\tau)+M_2\|y_2-y_1\|I_k^{\nu}I_q^{\alpha}I_r^{\delta}(1)(\tau)\right)\\
&+\frac{|\lambda_1||\lambda_2|\Omega_4}{|\Omega|}T^{\alpha-1}\left(M_1\|x_2-x_1\|I_m^{\gamma}I_q^{\alpha}(1)(\eta)+M_2\|y_2-y_1\|I_m^{\gamma}I_q^{\alpha}I_r^{\delta}(1)(\eta)\right)\\
&+\frac{|\lambda_2|\Omega_4}{|\Omega|}T^{\alpha-1}\left(N_1\|y_2-y_1\|I_n^{\kappa}I_p^{\beta}(1)(\xi)+N_2\|x_2-x_1\|I_n^{\kappa}I_p^{\beta}I_z^{\varepsilon}(1)(\xi)\right)\\
&=B_1\|x_2-x_1\|+B_2\|y_2-y_1\|.
\end{aligned}
$$

Therefore, we have

$$
\left\|\mathcal{K}_1(x_2,y_2)(t)-\mathcal{K}_1(x_1,y_1)(t)\right\|\le(B_1+B_2)\left(\|x_2-x_1\|+\|y_2-y_1\|\right). \tag{4.3}
$$

In the same way, we have

$$
\begin{aligned}
&\left|\mathcal{K}_2(x_2,y_2)(t)-\mathcal{K}_2(x_1,y_1)(t)\right|\\
&\le I_p^{\beta}\left(\left|g\left(s,y_2(s),I_z^{\varepsilon}x_2(s)\right)-g\left(s,y_1(s),I_z^{\varepsilon}x_1(s)\right)\right|\right)(t)\\
&+\frac{|\lambda_1|\Omega_2}{|\Omega|}T^{\beta-1}I_m^{\gamma}I_q^{\alpha}\left(\left|f\left(s,x_2(s),I_r^{\delta}y_2(s)\right)-f\left(s,x_1(s),I_r^{\delta}y_1(s)\right)\right|\right)(\eta)\\
&+\frac{\Omega_2}{|\Omega|}T^{\beta-1}I_n^{\kappa}I_p^{\beta}\left(\left|g\left(s,y_2(s),I_z^{\varepsilon}x_2(s)\right)-g\left(s,y_1(s),I_z^{\varepsilon}x_1(s)\right)\right|\right)(\xi)\\
&+\frac{|\lambda_1||\lambda_2|\Omega_3}{|\Omega|}T^{\beta-1}I_h^{\mu}I_p^{\beta}\left(\left|g\left(s,y_2(s),I_z^{\varepsilon}x_2(s)\right)-g\left(s,y_1(s),I_z^{\varepsilon}x_1(s)\right)\right|\right)(\theta)\\
&+\frac{|\lambda_1|\Omega_3}{|\Omega|}T^{\beta-1}I_k^{\nu}I_q^{\alpha}\left(\left|f\left(s,x_2(s),I_r^{\delta}y_2(s)\right)-f\left(s,x_1(s),I_r^{\delta}y_1(s)\right)\right|\right)(\tau)\\
&\le N_1\|y_2-y_1\|A_2+N_2\|x_2-x_1\|A_4\\
&+\frac{|\lambda_1|\Omega_2}{|\Omega|}T^{\beta-1}\left(M_1\|x_2-x_1\|A_5+M_2\|y_2-y_1\|A_9\right)\\
&+\frac{\Omega_2}{|\Omega|}T^{\beta-1}\left(N_1\|y_2-y_1\|A_6+N_2\|x_2-x_1\|A_{10}\right)\\
&+\frac{|\lambda_1||\lambda_2|\Omega_3}{|\Omega|}T^{\beta-1}\left(N_1\|y_2-y_1\|A_7+N_2\|x_2-x_1\|A_{11}\right)\\
&+\frac{|\lambda_1|\Omega_3}{|\Omega|}T^{\beta-1}\left(M_1\|x_2-x_1\|A_8+M_2\|y_2-y_1\|A_{12}\right)\\
&=C_1\|x_2-x_1\|+C_2\|y_2-y_1\|,
\end{aligned}
$$

which implies

$$\left\|\mathcal{K}_2(x_2,y_2)(t)-\mathcal{K}_2(x_1,y_1)(t)\right\| \leq (C_1+C_2)\left(\|x_2-x_1\|+\|y_2-y_1\|\right). \tag{4.4}$$

It follows from (4.3) and (4.4) that

$$\left\|\mathcal{K}(x_2,y_2)(t)-\mathcal{K}(x_1,y_1)(t)\right\| \leq (B_1+B_2+C_1+C_2)\left(\|x_2-x_1\|+\|y_2-y_1\|\right).$$

Since $B_1 + B_2 + C_1 + C_2 < 1$, therefore, $\mathcal{K}$ is a contraction operator. So, by Banach's fixed point theorem, the operator $\mathcal{K}$ has a unique fixed point, which is the unique solution of problem (1.1). The proof is completed. □

In the next result, we prove the existence of solutions for the problem (1.1) by applying the Leray-Schauder alternative.

Lemma 4.2 (Leray-Schauder alternative, see [22], p.4) *Let $F : E \to E$ be a completely continuous operator (i.e., a map that restricted to any bounded set in E is compact). Let*

$$\mathcal{E}(F) = \left\{x \in E : x = \lambda F(x) \text{ for some } 0 < \lambda < 1\right\}.$$

Then either the set $\mathcal{E}(F)$ is unbounded, or F has at least one fixed point.

For convenience, we set constants

$$\begin{aligned}
E_0 &= P_0A_1 + \frac{|\lambda_2|T^{\alpha-1}}{|\Omega|}\left(\Omega_1Q_0A_7 + |\lambda_1|\Omega_4P_0A_5 + \Omega_4Q_0A_6\right) + \frac{\Omega_1}{|\Omega|}T^{\alpha-1}P_0A_8,\\
E_1 &= P_1A_1 + \frac{|\lambda_2|T^{\alpha-1}}{|\Omega|}\left(\Omega_1Q_2A_{11} + |\lambda_1|\Omega_4P_1A_5 + \Omega_4Q_2A_{10}\right) + \frac{\Omega_1}{|\Omega|}T^{\alpha-1}P_1A_8,\\
E_3 &= P_2A_3 + \frac{|\lambda_2|T^{\alpha-1}}{|\Omega|}\left(\Omega_1Q_1A_7 + |\lambda_1|\Omega_4P_2A_9 + \Omega_4Q_1A_6\right) + \frac{\Omega_1}{|\Omega|}T^{\alpha-1}P_2A_{12},\\
F_0 &= Q_0A_2 + \frac{|\lambda_1|T^{\beta-1}}{|\Omega|}\left(\Omega_2P_0A_5 + |\lambda_2|\Omega_3Q_0A_7 + \Omega_3P_0A_8\right) + \frac{\Omega_2}{|\Omega|}T^{\beta-1}Q_0A_6,\\
F_1 &= Q_2A_4 + \frac{|\lambda_1|T^{\beta-1}}{|\Omega|}\left(\Omega_2P_1A_5 + |\lambda_2|\Omega_3Q_2A_{11} + \Omega_3P_1A_8\right) + \frac{\Omega_2}{|\Omega|}T^{\beta-1}Q_2A_{10},\\
F_2 &= Q_1A_2 + \frac{|\lambda_1|T^{\beta-1}}{|\Omega|}\left(\Omega_2P_2A_9 + |\lambda_2|\Omega_3Q_1A_7 + \Omega_3P_2A_{12}\right) + \frac{\Omega_2}{|\Omega|}T^{\beta-1}Q_1A_6
\end{aligned}$$

and

$$G^* = \max\left\{1-(E_1+F_1), 1-(E_2+F_2)\right\}.$$

Theorem 4.3 *Assume that there exist real constants $P_i, Q_i \geq 0$ $(i = 1,2)$, and $P_0 > 0$, $Q_0 > 0$ such that for all $u_i, v_i \in \mathbb{R}$ $(i = 1,2)$ we have*

$$\begin{aligned}
\left|f(t,u_1,u_2)\right| &\leq P_0 + P_1|u_1| + P_2|u_2|,\\
\left|g(t,v_1,v_2)\right| &\leq Q_0 + Q_1|v_1| + Q_2|v_2|.
\end{aligned}$$

In addition it is assumed that

$$E_1 + F_1 < 1 \quad and \quad E_2 + F_2 < 1.$$

Then there exists at least one solution for the system (1.1).

Proof We first prove that the operator $\mathcal{K}: X \times Y \to X \times Y$ is completely continuous. The continuity of functions f and g imply that the operator $\mathcal{K}$ is continuous. Let $\Phi \subset X \times Y$ be a bounded set. Then there exist positive constants D_1 and D_2 such that

$$\left|f(t, u_1(t), u_2(t))\right| \leq D_1, \qquad \left|g(t, v_1(t), v_2(t))\right| \leq D_2, \quad \forall (u_1, u_2), (v_1, v_2) \in \Phi.$$

Then for any $(u_1, u_2), (v_1, v_2) \in \Phi$, and using Lemma 3.1, we have

$$\begin{aligned}
\|\mathcal{K}_1(x,y)\| \leq{}& I_q^{\alpha}\left|f\left(s, x(s), I_r^{\delta} y(s)\right)\right|(t) + \frac{|\lambda_2|\Omega_1}{|\Omega|} T^{\alpha-1} I_h^{\mu} I_p^{\beta}\left|g\left(s, y(s), I_z^{\varepsilon} x(s)\right)\right|(\theta) \\
&+ \frac{\Omega_1}{|\Omega|} T^{\alpha-1} I_k^{\nu} I_q^{\alpha}\left|f\left(s, x(s), I_r^{\delta} y(s)\right)\right|(\tau) \\
&+ \frac{|\lambda_1||\lambda_2|\Omega_4}{|\Omega|} T^{\alpha-1} I_m^{\gamma} I_q^{\alpha}\left|f\left(s, x(s), I_r^{\delta} y(s)\right)\right|(\eta) \\
&+ \frac{|\lambda_2|\Omega_4}{|\Omega|} T^{\alpha-1} I_n^{\kappa} I_p^{\beta}\left|g\left(s, y(s), I_z^{\varepsilon} x(s)\right)\right|(\xi) \\
\leq{}& D_1 A_1 + \frac{D_2|\lambda_2|\Omega_1}{|\Omega|} T^{\alpha-1} A_7 + \frac{D_1 \Omega_1}{|\Omega|} T^{\alpha-1} A_8 \\
&+ \frac{D_1|\lambda_1||\lambda_2|\Omega_4}{|\Omega|} T^{\alpha-1} A_5 + \frac{D_2|\lambda_2|\Omega_4}{|\Omega|} T^{\alpha-1} A_6.
\end{aligned}$$

In the same way, we deduce that

$$\begin{aligned}
\|\mathcal{K}_2(x,y)\| \leq{}& I_p^{\beta}\left(\left|g\left(s, y(s), I_z^{\varepsilon} x(s)\right)\right|\right)(t) + \frac{|\lambda_1|\Omega_2}{|\Omega|} T^{\beta-1} I_m^{\gamma} I_q^{\alpha}\left(\left|f\left(s, x(s), I_r^{\delta} y(s)\right)\right|\right)(\eta) \\
&+ \frac{\Omega_2}{|\Omega|} T^{\beta-1} I_n^{\kappa} I_p^{\beta}\left(\left|g\left(s, y(s), I_z^{\varepsilon} x(s)\right)\right|\right)(\xi) \\
&+ \frac{|\lambda_1||\lambda_2|\Omega_3}{|\Omega|} T^{\beta-1} I_h^{\mu} I_p^{\beta}\left(\left|g\left(s, y(s), I_z^{\varepsilon} x(s)\right)\right|\right)(\theta) \\
&+ \frac{|\lambda_1|\Omega_3}{|\Omega|} T^{\beta-1} I_k^{\nu} I_q^{\alpha}\left(\left|f\left(s, x(s), I_r^{\delta} y(s)\right)\right|\right)(\tau) \\
\leq{}& D_2 A_2 + \frac{D_1|\lambda_1|\Omega_2}{|\Omega|} T^{\beta-1} A_5 + \frac{D_2 \Omega_2}{|\Omega|} T^{\beta-1} A_6 \\
&+ \frac{D_2|\lambda_1||\lambda_2|\Omega_3}{|\Omega|} T^{\beta-1} A_7 + \frac{D_1|\lambda_1|\Omega_3}{|\Omega|} T^{\beta-1} A_8.
\end{aligned}$$

Thus, it follows from the above inequalities that the operator $\mathcal{K}$ is uniformly bounded.

Next, we show that $\mathcal{K}$ is equicontinuous. Let $t_1, t_2 \in [0, T]$ with $t_1 < t_2$. Then we have

$$\begin{aligned}
&\left|\mathcal{K}_1(x,y)(t_2) - \mathcal{K}_1(x,y)(t_1)\right| \\
&\quad \leq \left|I_q^{\alpha} f\left(s, x(s), I_r^{\delta} y(s)\right)(t_2) - I_q^{\alpha} f\left(s, x(s), I_r^{\delta} y(s)\right)(t_1)\right|
\end{aligned}$$

$$
\begin{aligned}
&+\frac{|\lambda_2|\Omega_1}{|\Omega|}\left|t_2^{\alpha-1}-t_1^{\alpha-1}\right|I_h^{\mu}I_p^{\beta}\left|g\left(s,y_2(s),I_z^{\varepsilon}x_2(s)\right)\right|(\theta)\\
&+\frac{\Omega_1}{|\Omega|}\left|t_2^{\alpha-1}-t_1^{\alpha-1}\right|I_k^{\nu}I_q^{\alpha}\left|f\left(s,x(s),I_r^{\delta}y(s)\right)\right|(\tau)\\
&+\frac{|\lambda_1||\lambda_2|\Omega_4}{|\Omega|}\left|t_2^{\alpha-1}-t_1^{\alpha-1}\right|I_m^{\gamma}I_q^{\alpha}\left|f\left(s,x(s),I_r^{\delta}y(s)\right)\right|(\eta)\\
&+\frac{|\lambda_2|\Omega_4}{|\Omega|}\left|t_2^{\alpha-1}-t_1^{\alpha-1}\right|I_n^{\kappa}I_p^{\beta}\left|g\left(s,y(s),I_z^{\varepsilon}x(s)\right)\right|(\xi)\\
\le{}&\frac{D_1}{\Gamma_q(\alpha)}\int_0^{t_1}\left[(t_2-qs)^{(\alpha-1)}-(t_1-qs)^{(\alpha-1)}\right]d_qs\\
&+\frac{D_1}{\Gamma_q(\alpha)}\int_{t_1}^{t_2}(t_2-qs)^{(\alpha-1)}\,d_qs\\
&+\frac{|\lambda_2|\Omega_1D_2}{|\Omega|}\left|t_2^{\alpha-1}-t_1^{\alpha-1}\right|A_7+\frac{\Omega_1D_1}{|\Omega|}\left|t_2^{\alpha-1}-t_1^{\alpha-1}\right|A_8\\
&+\frac{|\lambda_1||\lambda_2|\Omega_4D_1}{|\Omega|}\left|t_2^{\alpha-1}-t_1^{\alpha-1}\right|A_5+\frac{|\lambda_2|\Omega_4D_2}{|\Omega|}\left|t_2^{\alpha-1}-t_1^{\alpha-1}\right|A_6.
\end{aligned}
$$

Analogously, we can get

$$
\begin{aligned}
&\left|\mathcal{K}_2(x,y)(t_2)-\mathcal{K}_2(x,y)(t_1)\right|\\
&\quad\le\left|I_p^{\beta}g\left(s,y(s),I_z^{\varepsilon}x(s)\right)(t_2)-I_p^{\beta}g\left(s,y(s),I_z^{\varepsilon}x(s)\right)(t_1)\right|\\
&\quad+\frac{|\lambda_1|\Omega_2}{|\Omega|}\left|t_2^{\beta-1}-t_1^{\beta-1}\right|I_m^{\gamma}I_q^{\alpha}\left|f\left(s,x_2(s),I_r^{\delta}y_2(s)\right)\right|(\eta)\\
&\quad+\frac{\Omega_2}{|\Omega|}\left|t_2^{\beta-1}-t_1^{\beta-1}\right|I_n^{\kappa}I_p^{\beta}\left|g\left(s,y_2(s),I_z^{\varepsilon}x_2(s)\right)\right|(\xi)\\
&\quad+\frac{|\lambda_1||\lambda_2|\Omega_3}{|\Omega|}\left|t_2^{\beta-1}-t_1^{\beta-1}\right|I_h^{\mu}I_p^{\beta}\left|g\left(s,y_2(s),I_z^{\varepsilon}x_2(s)\right)\right|(\theta)\\
&\quad+\frac{|\lambda_1|\Omega_3}{|\Omega|}\left|t_2^{\beta-1}-t_1^{\beta-1}\right|I_k^{\nu}I_q^{\alpha}\left|f\left(s,x_2(s),I_r^{\delta}y_2(s)\right)\right|(\tau)\\
&\quad\le\frac{D_2}{\Gamma_p(\beta)}\int_0^{t_1}\left[(t_2-ps)^{(\beta-1)}-(t_1-ps)^{(\beta-1)}\right]d_ps\\
&\quad+\frac{D_2}{\Gamma_p(\beta)}\int_{t_1}^{t_2}(t_2-ps)^{(\beta-1)}\,d_ps\\
&\quad+\frac{|\lambda_1|\Omega_2D_1}{|\Omega|}\left|t_2^{\beta-1}-t_1^{\beta-1}\right|A_5+\frac{\Omega_2D_2}{|\Omega|}\left|t_2^{\beta-1}-t_1^{\beta-1}\right|A_6\\
&\quad+\frac{|\lambda_1||\lambda_2|\Omega_3D_2}{|\Omega|}\left|t_2^{\beta-1}-t_1^{\beta-1}\right|A_7+\frac{|\lambda_1|\Omega_3D_1}{|\Omega|}\left|t_2^{\beta-1}-t_1^{\beta-1}\right|A_8.
\end{aligned}
$$

Therefore, the operator $\mathcal{K}(x,y)$ is equicontinuous, and thus the operator $\mathcal{K}(x,y)$ is completely continuous.

Finally, it will be verified that the set $\mathcal{E}=\{(x,y)\in X\times Y:(x,y)=\lambda\mathcal{K}(x,y),0\le\lambda\le1\}$ is bounded. Let $(x,y)\in\mathcal{E}$, then $(x,y)=\lambda\mathcal{K}(x,y)$. For any $t\in[0,T]$, we have

$$
x(t)=\lambda\mathcal{K}_1(x,y)(t),\qquad y(t)=\lambda\mathcal{K}_2(x,y)(t).
$$

Then we have

$$\begin{aligned}
|x(t)| &\le P_0 I_q^{\alpha}(1)(t) + P_1\|x\| I_q^{\alpha}(1)(t) + P_2\|y\| I_q^{\alpha} I_r^{\delta}(1)(t) \\
&\quad + \frac{|\lambda_2|\Omega_1}{|\Omega|} T^{\alpha-1}\big(Q_0 I_h^{\mu} I_p^{\beta}(1)(\theta) + Q_1\|y\| I_h^{\mu} I_p^{\beta}(1)(\theta) + Q_2\|x\| I_h^{\mu} I_p^{\beta} I_z^{\varepsilon}(1)(\theta)\big) \\
&\quad + \frac{\Omega_1}{|\Omega|} T^{\alpha-1}\big(P_0 I_k^{\nu} I_q^{\alpha}(1)(\tau) + P_1\|x\| I_k^{\nu} I_q^{\alpha}(1)(\tau) + P_2\|y\| I_k^{\nu} I_q^{\alpha} I_r^{\delta}(1)(\tau)\big) \\
&\quad + \frac{|\lambda_1||\lambda_2|\Omega_4}{|\Omega|} T^{\alpha-1}\big(P_0 I_m^{\gamma} I_q^{\alpha}(1)(\eta) + P_1\|x\| I_m^{\gamma} I_q^{\alpha}(1)(\eta) + P_2\|y\| I_m^{\gamma} I_q^{\alpha} I_r^{\delta}(1)(\eta)\big) \\
&\quad + \frac{|\lambda_2|\Omega_4}{|\Omega|} T^{\alpha-1}\big(Q_0 I_n^{\kappa} I_p^{\beta}(1)(\xi) + Q_1\|y\| I_n^{\kappa} I_p^{\beta}(1)(\xi) + Q_2\|x\| I_n^{\kappa} I_p^{\beta} I_z^{\varepsilon}(1)(\xi)\big) \\
&\le E_0 + E_1\|x\| + E_2\|y\|
\end{aligned}$$

and

$$\begin{aligned}
|y(t)| &\le Q_0 I_p^{\beta}(1)(t) + Q_1\|y\| I_p^{\beta}(1)(t) + Q_1\|x\| I_p^{\beta} I_z^{\varepsilon}(1)(t) \\
&\quad + \frac{|\lambda_1|\Omega_2}{|\Omega|} T^{\beta-1}\big(P_0 I_m^{\gamma} I_q^{\alpha}(1)(\eta) + P_1\|x\| I_m^{\gamma} I_q^{\alpha}(1)(\eta) + P_2\|y\| I_m^{\gamma} I_q^{\alpha} I_r^{\delta}(1)(\eta)\big) \\
&\quad + \frac{\Omega_2}{|\Omega|} T^{\beta-1}\big(Q_0 I_n^{\kappa} I_p^{\beta}(1)(\xi) + Q_1\|y\| I_n^{\kappa} I_p^{\beta}(1)(\xi) + Q_2\|x\| I_n^{\kappa} I_p^{\beta} I_z^{\varepsilon}(1)(\xi)\big) \\
&\quad + \frac{|\lambda_1||\lambda_2|\Omega_3}{|\Omega|} T^{\beta-1}\big(Q_0 I_h^{\mu} I_p^{\beta}(1)(\theta) + Q_1\|y\| I_h^{\mu} I_p^{\beta}(1)(\theta) + Q_2\|x\| I_h^{\mu} I_p^{\beta} I_z^{\varepsilon}(1)(\theta)\big) \\
&\quad + \frac{|\lambda_1|\Omega_3}{|\Omega|} T^{\beta-1}\big(P_0 I_k^{\nu} I_q^{\alpha}(1)(\tau) + P_1\|x\| I_k^{\nu} I_q^{\alpha}(1)(\tau) + P_2\|y\| I_k^{\nu} I_q^{\alpha} I_r^{\delta}(1)(\tau)\big) \\
&\le F_0 + F_1\|x\| + F_2\|y\|,
\end{aligned}$$

which yields

$$\|x\| \le E_0 + E_1\|x\| + E_2\|y\|$$

and

$$\|y\| \le F_0 + F_1\|x\| + F_2\|y\|.$$

Therefore, we have

$$\|x\| + \|y\| \le (E_0 + F_0) + (E_1 + F_1)\|x\| + (E_2 + F_2)\|y\|,$$

and, consequently,

$$\|(x, y)\| \le \frac{E_0 + F_0}{G^*}$$

for any $t \in [0, T]$, which proves that $\mathcal{E}$ is bounded. Thus, by Lemma 4.2, the operator $\mathcal{K}$ has at least one fixed point. Hence the system (1.1) has at least one solution. The proof is complete. □

4.1 Examples

In this subsection, we present some examples to illustrate our results.

Example 4.4 Consider the following coupled system of fractional q-integro-difference equations:

$$\begin{cases} D_{1/2}^{3/2}x(t) = \frac{\cos^2 \pi t}{(e^t+4)^2} \cdot \frac{|x(t)|}{4+|x(t)|} + \frac{e^{-t^2}}{(t+8)^2} \cdot I_{1/4}^{\pi} y(t) + \frac{\sqrt{2}}{2}, & 0 < t < 2, \\ D_{1/3}^{4/3}y(t) = \frac{\sin^2 \pi t}{(11+t)^2} \cdot \frac{|y(t)|}{1+|y(t)|} + \frac{1}{(e^t+8)^2} \cdot I_{1/5}^{\pi/2} x(t) + \sqrt{3}, \\ x(0) = 0, \qquad \sqrt{2} I_{1/8}^{7/6} x(\frac{3}{2}) = I_{1/6}^{\sqrt{2}} y(\frac{1}{2}), \\ y(0) = 0, \qquad \frac{\sqrt{3}}{2} I_{1/9}^{e} y(\frac{1}{3}) = I_{1/7}^{\sqrt{3}} x(\frac{5}{3}). \end{cases} \tag{4.5}$$

Here $\alpha = 3/2$, $\delta = \pi$, $\beta = 4/3$, $\varepsilon = \pi/2$, $\gamma = 7/6$, $\kappa = \sqrt{2}$, $\mu = e$, $\nu = \sqrt{3}$, $q = 1/2$, $r = 1/4$, $p = 1/3$, $z = 1/5$, $m = 1/8$, $n = 1/6$, $h = 1/9$, $k = 1/7$, $\eta = 3/2$, $\xi = 1/2$, $\theta = 1/3$, $\tau = 5/3$, $\lambda_1 = \sqrt{2}$, $\lambda_2 = \sqrt{3}/2$, $T = 2$, $f(t, x, I_r^{\delta} y) = (|x| \cos^2 \pi t)/((e^t + 4)^2(4 + |x|)) + (e^{-t^2}/((t+8)^2))I_{1/4}^{\pi} y + \sqrt{2}/2$, and $g(t, y, I_r^{\delta} x) = (|y| \sin^2 \pi t)/((11 + t)^2(1 + |y|)) + (1/(e^t + 8)^2)I_{1/5}^{\pi/2} x + \sqrt{3}$. Since

$$|f(t, u_1, u_2) - f(t, v_1, v_2)| \le \frac{1}{100}|u_1 - v_1| + \frac{1}{64}|u_2 - v_2|$$

and

$$|g(t, u_1, u_2) - g(t, v_1, v_2)| \le \frac{1}{121}|u_1 - v_1| + \frac{1}{81}|u_2 - v_2|,$$

then the assumptions of Theorem 4.1 are satisfied with $M_1 = 1/100$, $M_2 = 1/64$, $N_1 = 1/121$, and $N_2 = 1/81$. By using the Maple program, we can find that

$$\Omega = \Omega_1\Omega_2 - \lambda_1\lambda_2\Omega_3\Omega_4 \approx 0.61154471 \neq 0$$

and

$$B_1 = M_1A_1 + \frac{|\lambda_2|T^{\alpha-1}}{|\Omega|}\left(\Omega_1 N_2 A_{11} + |\lambda_1|\Omega_4 M_1 A_5 + \Omega_4 N_2 A_{10}\right) + \frac{\Omega_1}{|\Omega|}T^{\alpha-1}M_1A_8$$
$$\approx 0.0577709,$$

$$B_2 = M_2A_3 + \frac{|\lambda_2|T^{\alpha-1}}{|\Omega|}\left(\Omega_1 N_1 A_7 + |\lambda_1|\Omega_4 M_2 A_9 + \Omega_4 N_1 A_6\right) + \frac{\Omega_1}{|\Omega|}T^{\alpha-1}M_2A_{12}$$
$$\approx 0.1489994,$$

$$C_1 = N_2A_4 + \frac{|\lambda_1|T^{\beta-1}}{|\Omega|}\left(\Omega_2 M_1 A_5 + |\lambda_2|\Omega_3 N_2 A_{11} + \Omega_3 M_1 A_8\right) + \frac{\Omega_2}{|\Omega|}T^{\beta-1}N_2A_{10}$$
$$\approx 0.22629179,$$

$$C_2 = N_1A_2 + \frac{|\lambda_1|T^{\beta-1}}{|\Omega|}\left(\Omega_2 M_2 A_9 + |\lambda_2|\Omega_3 N_1 A_7 + \Omega_3 M_2 A_{12}\right) + \frac{\Omega_2}{|\Omega|}T^{\beta-1}N_1A_6$$
$$\approx 0.28656994.$$

Therefore, we get

$$B_1 + B_2 + C_1 + C_2 \approx 0.71963204 < 1.$$

Hence, by Theorem 4.1, the problem (4.5) has a unique solution on $[0, 2]$.

Example 4.5 Consider the following coupled system of fractional q-integro-difference equations with fractional q-integral conditions:

$$\begin{cases} D_{1/3}^{4/3}x(t) = \frac{4e^{-t}}{(t+11)^2}\cdot\frac{|x(t)|}{2+|x(t)|} + \frac{1}{(e^{-t^2}+8)^2}\cdot I_{1/5}^{\sqrt{3}}y(t) + \sqrt{2}, \quad 0<t<\pi, \\ D_{1/4}^{7/5}y(t) = \frac{\cos^2 2\pi t}{(10+t)^2}\cdot\frac{|y(t)|}{1+|y(t)|} + \frac{1}{(e^t+7)^2}\cdot I_{1/6}^{\sqrt{2}}x(t) + \frac{1}{2}, \\ x(0)=0, \qquad I_{1/3}^{1/3}y(\frac{2\pi}{5}) + \sqrt{5}I_{1/2}^{\sqrt{5}}x(\frac{\pi}{5}) = 0, \\ y(0)=0, \qquad I_{1/8}^{\sqrt{2}/2}x(\frac{4\pi}{5}) + \sqrt{2}I_{1/7}^{\sqrt{\pi}}y(\frac{3\pi}{5}) = 0. \end{cases} \tag{4.6}$$

Here $\alpha = 4/3$, $\delta = \sqrt{3}$, $\beta = 7/5$, $\varepsilon = \sqrt{2}$, $\gamma = \sqrt{5}$, $\kappa = 1/3$, $\mu = \sqrt{\pi}$, $\nu = \sqrt{2}/2$, $q = 1/3$, $r = 1/5$, $p = 1/4$, $z = 1/6$, $m = 1/2$, $n = 1/3$, $h = 1/7$, $k = 1/8$, $\eta = \pi/5$, $\xi = 2\pi/5$, $\theta = 3\pi/5$, $\tau = 4\pi/5$, $\lambda_1 = -\sqrt{5}$, $\lambda_2 = -\sqrt{2}$, $T = \pi$, $f(t,x,I_r^\delta y) = (4e^{-t}/(t+11)^2)(|x|/(2+|x|)) + (1/(e^{-t^2}+8)^2)I_{1/5}^{\sqrt{3}}y + \sqrt{2}$, and $g(t,y,I_r^\delta x) = (\cos^2 2\pi t/(10+t)^2)(|y|/(1+|y|)) + (1/(e^t+7)^2)I_{1/6}^{\sqrt{2}}x + (1/2)$. Since

$$|f(t,u_1,u_2)| \le \sqrt{2} + \frac{1}{72}|u_1| + \frac{1}{81}|u_2|$$

and

$$|g(t,v_1,v_2)| \le \frac{1}{2} + \frac{1}{100}|v_1| + \frac{1}{81}|v_2|,$$

then the assumptions of Theorem 4.3 are satisfied with $P_0 = \sqrt{2}$, $P_1 = 1/72$, $P_2 = 1/81$, $Q_0 = 1/2$, $Q_1 = 1/100$, and $Q_2 = 1/81$. By using the Maple program, we can find that

$$\Omega = \Omega_1\Omega_2 - \lambda_1\lambda_2\Omega_3\Omega_4 \approx -4.18385985 \neq 0$$

and

$$\begin{aligned} E_0 &= P_0A_1 + \frac{|\lambda_2|T^{\alpha-1}}{|\Omega|}\left(\Omega_1Q_0A_7 + |\lambda_1|\Omega_4P_0A_5 + \Omega_4Q_0A_6\right) + \frac{\Omega_1}{|\Omega|}T^{\alpha-1}P_0A_8 \\ &\approx 11.08984581, \\ E_1 &= P_1A_1 + \frac{|\lambda_2|T^{\alpha-1}}{|\Omega|}\left(\Omega_1Q_2A_{11} + |\lambda_1|\Omega_4P_1A_5 + \Omega_4Q_2A_{10}\right) + \frac{\Omega_1}{|\Omega|}T^{\alpha-1}P_1A_8 \\ &\approx 0.1629615871, \\ E_2 &= P_2A_3 + \frac{|\lambda_2|T^{\alpha-1}}{|\Omega|}\left(\Omega_1Q_1A_7 + |\lambda_1|\Omega_4P_2A_9 + \Omega_4Q_1A_6\right) + \frac{\Omega_1}{|\Omega|}T^{\alpha-1}P_2A_{12} \\ &\approx 0.34091157, \\ F_0 &= Q_0A_2 + \frac{|\lambda_1|T^{\beta-1}}{|\Omega|}\left(\Omega_2P_0A_5 + |\lambda_2|\Omega_3Q_0A_7 + \Omega_3P_0A_8\right) + \frac{\Omega_2}{|\Omega|}T^{\beta-1}Q_0A_6 \\ &\approx 25.68580671, \\ F_1 &= Q_2A_4 + \frac{|\lambda_1|T^{\beta-1}}{|\Omega|}\left(\Omega_2P_1A_5 + |\lambda_2|\Omega_3Q_2A_{11} + \Omega_3P_1A_8\right) + \frac{\Omega_2}{|\Omega|}T^{\beta-1}Q_2A_{10} \\ &\approx 0.68153261, \end{aligned}$$

$$F_2 = Q_1A_2 + \frac{|\lambda_1|T^{\beta-1}}{|\Omega|}\left(\Omega_2P_2A_9 + |\lambda_2|\Omega_3Q_1A_7 + \Omega_3P_2A_{12}\right) + \frac{\Omega_2}{|\Omega|}T^{\beta-1}Q_1A_6$$

$$\approx 0.5902944$$

and

$$G^* = \max\{1-(E_1+F_1), 1-(E_2+F_2)\} = \max\{0.1555058, 0.06879403\} = 0.1555058.$$

Therefore, we get

$$E_1 + F_1 \approx 0.8444942 < 1 \quad \text{and} \quad E_2 + F_2 \approx 0.93120597 < 1.$$

Hence, by Theorem 4.3, the problem (4.6) has at least one solution on $[0, \pi]$.

5 Uncoupled integral boundary conditions case

In this section we consider the following system:

$$\begin{cases} D_q^{\alpha}x(t) = f(t, x(t), I_r^{\delta}y(t)), & t \in [0,T], 1 < \alpha \le 2, \\ D_p^{\beta}y(t) = g(t, y(t), I_z^{\varepsilon}x(t)), & t \in [0,T], 1 < \beta \le 2, \\ x(0) = 0, & \lambda_1 I_m^{\gamma}x(\eta) = I_n^{\kappa}x(\xi), \\ y(0) = 0, & \lambda_2 I_h^{\mu}y(\theta) = I_k^{\nu}y(\tau). \end{cases} \tag{5.1}$$

Lemma 5.1 (Auxiliary lemma, see [20]) *For $h \in C([0,T], \mathbb{R})$, the unique solution of the problem*

$$\begin{cases} D_q^{\alpha}x(t) = h(t), & t \in [0,T], 1 < \alpha \le 2, \\ x(0) = 0, & \lambda_1 I_m^{\gamma}x(\eta) = I_n^{\kappa}x(\xi), \end{cases} \tag{5.2}$$

is given by

$$x(t) = I_q^{\alpha}h(t) + \frac{\lambda_1 t^{\alpha-1}}{\Lambda}I_m^{\gamma}I_q^{\alpha}h(\eta) - \frac{t^{\alpha-1}}{\Lambda}I_n^{\kappa}I_q^{\alpha}h(\xi), \tag{5.3}$$

where

$$\Lambda = \frac{\Gamma_n(\alpha)}{\Gamma_n(\kappa+\alpha)}\xi^{\kappa+\alpha-1} - \lambda_1\frac{\Gamma_m(\alpha)}{\Gamma_m(\gamma+\alpha)}\eta^{\gamma+\alpha-1} \neq 0.$$

5.1 Existence results for uncoupled case

In view of Lemma 5.1, we define an operator $\mathcal{T}: X \times Y \to X \times Y$ by

$$\mathcal{T}(x,y)(t) = \begin{pmatrix} \mathcal{T}_1(x,y)(t) \\ \mathcal{T}_2(x,y)(t) \end{pmatrix},$$

where

$$\mathcal{T}_1(x,y)(t) = I_q^{\alpha}f\left(s, x(s), I_r^{\delta}y(s)\right)(t) + \frac{\lambda_1 t^{\alpha-1}}{\Lambda}I_m^{\gamma}I_q^{\alpha}f\left(s, x(s), I_r^{\delta}y(s)\right)(\eta)$$
$$- \frac{t^{\alpha-1}}{\Lambda}I_n^{\kappa}I_q^{\alpha}f\left(s, x(s), I_r^{\delta}y(s)\right)(\xi)$$

and

$$\mathcal{T}_2(x,y)(t) = I_p^\beta g\big(s,y(s),I_z^\varepsilon x(s)\big)(t) + \frac{\lambda_2 t^{\beta-1}}{\Psi} I_h^\mu I_p^\beta g\big(s,y(s),I_z^\varepsilon x(s)\big)(\theta)$$
$$- \frac{t^{\beta-1}}{\Psi} I_k^\nu I_p^\beta g\big(s,y(s),I_z^\varepsilon x(s)\big)(\tau),$$

where

$$\Psi = \frac{\Gamma_k(\beta)}{\Gamma_k(\nu+\beta)}\tau^{\nu+\beta-1} - \lambda_2 \frac{\Gamma_h(\beta)}{\Gamma_h(\mu+\beta)}\theta^{\mu+\beta-1} \neq 0.$$

In the sequel, we set constants

$$A_{13} = \frac{\Gamma_n(\alpha+1)\xi^{\kappa+\alpha}}{\Gamma_n(\kappa+\alpha+1)\Gamma_q(\alpha+1)},$$
$$A_{14} = \frac{\Gamma_n(\alpha+\delta+1)\Gamma_q(\delta+1)\xi^{\kappa+\alpha+\delta}}{\Gamma_n(\kappa+\alpha+\delta+1)\Gamma_q(\alpha+\delta+1)\Gamma_r(\delta+1)},$$
$$A_{15} = \frac{\Gamma_k(\beta+1)\tau^{\nu+\beta}}{\Gamma_k(\nu+\beta+1)\Gamma_p(\beta+1)},$$
$$A_{16} = \frac{\Gamma_k(\beta+\varepsilon+1)\Gamma_p(\varepsilon+1)\tau^{\nu+\beta+\varepsilon}}{\Gamma_k(\nu+\beta+\varepsilon+1)\Gamma_p(\beta+\varepsilon+1)\Gamma_z(\varepsilon+1)},$$
$$H_1 = \overline{M}_1 A_1 + \frac{|\lambda_1|T^{\alpha-1}}{|\Lambda|}\overline{M}_1 A_5 + \frac{T^{\alpha-1}}{|\Lambda|}\overline{M}_1 A_{13},$$
$$H_2 = \overline{M}_2 A_3 + \frac{|\lambda_1|T^{\alpha-1}}{|\Lambda|}\overline{M}_2 A_9 + \frac{T^{\alpha-1}}{|\Lambda|}\overline{M}_2 A_{14},$$
$$L_1 = \overline{N}_2 A_4 + \frac{|\lambda_2|T^{\beta-1}}{|\Psi|}\overline{N}_2 A_{11} + \frac{T^{\beta-1}}{|\Psi|}\overline{N}_2 A_{16},$$
$$L_2 = \overline{N}_1 A_2 + \frac{|\lambda_2|T^{\beta-1}}{|\Psi|}\overline{N}_1 A_7 + \frac{T^{\beta-1}}{|\Psi|}\overline{N}_1 A_{15}.$$

Now we present the existence and the uniqueness result for the problem (5.1). We do not provide the proof of this result as it is similar to the one for Theorem 4.1.

Theorem 5.2 *Assume that $f,g:[0,T]\times\mathbb{R}^2\to\mathbb{R}$ are continuous functions and there exist constants $\overline{K}_i, \overline{L}_i$, $i=1,2$ such that for all $t\in[0,T]$ and $u_i, v_i\in\mathbb{R}$, $i=1,2$,*

$$\big|f(t,u_1,u_2)-f(t,v_1,v_2)\big| \le \overline{M}_1|u_1-v_1| + \overline{M}_2|u_2-v_2|$$

and

$$\big|g(t,u_1,u_2)-g(t,v_1,v_2)\big| \le \overline{N}_1|u_1-v_1| + \overline{N}_2|u_2-v_2|.$$

In addition, assume that

$$H_1+H_2+L_1+L_2<1.$$

Then the boundary value problem (5.1) has a unique solution on $[0,T]$.

The second result deals with the existence of solutions for the problem (5.1), is analogous to Theorem 4.3 and is given below.

Theorem 5.3 *Assume that there exist real constants* $\bar{m}_i, \bar{n}_i \geq 0$ $(i = 1, 2)$, *and* $\bar{m}_0 > 0, \bar{n}_0 > 0$ *such that* $\forall x_i \in \mathbb{R}$ $(i = 1, 2)$ *we have*

$$|f(t, x_1, x_2)| \leq \bar{m}_0 + \bar{m}_1|x_1| + \bar{m}_2|x_2|,$$
$$|g(t, x_1, x_2)| \leq \bar{n}_0 + \bar{n}_1|x_1| + \bar{n}_2|x_2|.$$

In addition it is assumed that

$$U_1 + V_1 < 1 \quad \text{and} \quad U_2 + V_2 < 1,$$

where U_i, V_i, $i = 1, 2$, *are given by*

$$U_1 = \bar{m}_1 A_1 + \frac{|\lambda_1| T^{\alpha-1}}{|\Lambda|}\bar{m}_1 A_5 + \frac{T^{\alpha-1}}{|\Lambda|}\bar{m}_1 A_{13},$$
$$U_2 = \bar{m}_2 A_3 + \frac{|\lambda_1| T^{\alpha-1}}{|\Lambda|}\bar{m}_2 A_9 + \frac{T^{\alpha-1}}{|\Lambda|}\bar{m}_2 A_{14},$$
$$V_1 = \bar{n}_2 A_4 + \frac{|\lambda_2| T^{\beta-1}}{|\Psi|}\bar{n}_2 A_{11} + \frac{T^{\beta-1}}{|\Psi|}\bar{n}_2 A_{16},$$
$$V_2 = \bar{n}_1 A_2 + \frac{|\lambda_2| T^{\beta-1}}{|\Psi|}\bar{n}_1 A_7 + \frac{T^{\beta-1}}{|\Psi|}\bar{n}_1 A_{15}.$$

Then the boundary value problem (5.1) *has at least one solution on* $[0, T]$.

Proof By setting

$$G^* = \min\{1 - (U_1 + V_1), 1 - (U_2 + V_2)\},$$

the proof is similar to that of Theorem 4.3. So we omit it. □

5.2 Examples

In this subsection, we present two examples of uncoupled case of nonlocal conditions.

Example 5.4 Consider the following system of fractional q-integro-difference equations with q-integral conditions:

$$\begin{cases} D_{1/9}^{3/2}x(t) = \frac{e^{-t\sin\pi t}}{(t+3)^2} \cdot \frac{|x(t)|}{9+|x(t)|} + \frac{\cos^2\pi t}{\pi(t+7)^2} \cdot I_{1/8}^{\sqrt{2}}y(t) - \frac{1}{3}, & 0 < t < 3, \\ D_{1/7}^{5/4}y(t) = \frac{2\pi e^{-2t}}{(7\pi+t)^2} \cdot \frac{|y(t)|}{2+|y(t)|} + \frac{\sin 2\pi t}{(3e^t+5)^2} \cdot I_{1/6}^{\sqrt{3}}x(t) + \frac{1}{3}, & \\ x(0) = 0, \qquad \frac{1}{2} I_{1/3}^{\frac{\sqrt{2}}{2}} x(\frac{3}{4}) = I_{1/4}^{\sqrt{3}} x(\frac{9}{4}), & \\ y(0) = 0, \qquad I_{1/6}^{\sqrt{\pi}} y(3) + \frac{1}{3} I_{1/5}^{\pi} y(\frac{3}{2}) = 0. & \end{cases} \tag{5.4}$$

Here $\alpha = 3/2$, $\delta = \sqrt{2}$, $\beta = 5/4$, $\varepsilon = \sqrt{3}$, $\gamma = \sqrt{2}/2$, $\kappa = \sqrt{3}/2$, $\mu = \pi$, $\nu = \sqrt{\pi}$, $q = 1/9$, $r = 1/8$, $p = 1/7$, $z = 1/6$, $m = 1/3$, $n = 1/4$, $h = 1/5$, $k = 1/6$, $\eta = 3/4$, $\xi = 9/4$, $\theta = 3/2$, $\tau = 3$, $\lambda_1 =$

$1/2$, $\lambda_2 = -1/3$, $T = 3$, $f(t,x,I_r^\delta y) = (e^{-t\sin\pi t}/(t+3)^2)(|x|/(9+|x|)) + (\cos^2\pi t/\pi(t+7)^2)I_{1/8}^{\sqrt{2}}y - (1/3)$, and $g(t,y,I_r^\delta x) = (2\pi e^{-2t}/(7\pi+t)^2)(|y|/(2+|y|)) + (\sin 2\pi t/(3e^t+5)^2)I_{1/6}^{\sqrt{3}}x + (1/3)$. Since

$$\left|f(t,u_1,u_2) - f(t,v_1,v_2)\right| \le \frac{1}{81}|u_1 - v_1| + \frac{1}{49\pi}|u_2 - v_2|$$

and

$$\left|g(t,u_1,u_2) - g(t,v_1,v_2)\right| \le \frac{1}{36\pi}|u_1 - v_1| + \frac{1}{64}|u_2 - v_2|,$$

then the assumptions of Theorem 5.2 are satisfied with $\overline{M}_1 = 1/81$, $\overline{M}_2 = 1/49\pi$, $\overline{N}_1 = 1/36\pi$, and $\overline{N}_2 = 1/64$. By using the Maple program, we can find that

$$\Lambda = \frac{\Gamma_n(\alpha)}{\Gamma_n(\kappa+\alpha)}\xi^{\kappa+\alpha-1} - \lambda_1\frac{\Gamma_m(\alpha)}{\Gamma_m(\gamma+\alpha)}\eta^{\gamma+\alpha-1} \approx 1.9245172 \neq 0,$$

$$\Psi = \frac{\Gamma_k(\beta)}{\Gamma_k(\nu+\beta)}\tau^{\nu+\beta-1} - \lambda_2\frac{\Gamma_h(\beta)}{\Gamma_h(\mu+\beta)}\theta^{\mu+\beta-1} \approx 8.37494759 \neq 0$$

and

$$H_1 = \overline{M}_1 A_1 + \frac{|\lambda_1|T^{\alpha-1}}{|\Lambda|}\overline{M}_1 A_5 + \frac{T^{\alpha-1}}{|\Lambda|}\overline{M}_1 A_{13} \approx 0.11268247,$$

$$H_2 = \overline{M}_2 A_3 + \frac{|\lambda_1|T^{\alpha-1}}{|\Lambda|}\overline{M}_2 A_9 + \frac{T^{\alpha-1}}{|\Lambda|}\overline{M}_2 A_{14} \approx 0.19713212,$$

$$L_1 = \overline{N}_2 A_4 + \frac{|\lambda_2|T^{\beta-1}}{|\Psi|}\overline{N}_2 A_{11} + \frac{T^{\beta-1}}{|\Psi|}\overline{N}_2 A_{16} \approx 0.58490031,$$

$$L_2 = \overline{N}_1 A_2 + \frac{|\lambda_2|T^{\beta-1}}{|\Psi|}\overline{N}_1 A_7 + \frac{T^{\beta-1}}{|\Psi|}\overline{N}_1 A_{15} \approx 0.06286768.$$

Therefore, we get

$$H_1 + H_2 + L_1 + L_2 \approx 0.95758259 < 1.$$

Hence, by Theorem 5.2, the problem (5.4) has a unique solution on $[0,3]$.

Example 5.5 Consider the following system of fractional q-integro-difference equations:

$$\begin{cases} D_{\sqrt{2}/2}^{\sqrt{\pi}}x(t) = \frac{25e^{-t}}{(e^{-t}+4)^2}\cdot\frac{|x(t)|}{5+|x(t)|} + \frac{3\pi^2}{(t+3\pi)^2}\cdot I_{\sqrt{3}/2}^{1/2}y(t) + \frac{1}{\sqrt{5}}, \quad 0<t<1, \\ D_{\pi/4}^{\pi/2}y(t) = \frac{9e^{-t\cos^2\pi t}}{(t+6)^2}\cdot\frac{|y(t)|}{1+|y(t)|} + \frac{6}{(t+6)^2}\cdot I_{\pi/5}^{3/2}x(t) + \frac{\sqrt{2}}{3}, \\ x(0) = 0, \qquad \frac{\sqrt{3}}{2}I_{\pi/6}^{4/5}x(1) + I_{\pi/7}^{2/3}x(\frac{3}{4}) = 0, \\ y(0) = 0, \qquad 5I_{\pi/8}^{\sqrt{3}}y(\frac{1}{2}) = I_{\pi/9}^{1/3}y(\frac{1}{4}). \end{cases} \tag{5.5}$$

Here $\alpha = \sqrt{\pi}$, $\delta = 1/2$, $\beta = \pi/2$, $\varepsilon = 3/2$, $\gamma = 4/5$, $\kappa = 2/3$, $\mu = \sqrt{3}$, $\nu = 1/3$, $q = \sqrt{2}/2$, $r = \sqrt{3}/2$, $p = \pi/4$, $z = \pi/5$, $m = \pi/6$, $n = \pi/7$, $h = \pi/8$, $k = \pi/9$, $\eta = 1$, $\xi = 3/4$, $\theta = 1/2$, $\tau = 1/4$, $\lambda_1 = -\sqrt{3}/2$, $\lambda_2 = 5$, $T = 1$, $f(t,x,I_r^\delta y) = (25e^{-t}/(e^{-t}+4)^2)(|x|/(5+|x|)) + (3\pi^2/(t+3\pi)^2)I_{\sqrt{3}/2}^{1/2}y + (1/\sqrt{5})$, and $g(t,y,I_r^\delta x) = (9e^{-t\cos^2\pi t}/(t+6)^2)(|y|/(1+|y|)) + (6/(t+6)^2)I_{\pi/5}^{3/2}x +$

$(\sqrt{2}/3)$. Since

$$|f(t,x_1,x_2)| \le \frac{1}{\sqrt{5}} + \frac{1}{5}|x_1| + \frac{1}{3}|x_2|$$

and

$$|g(t,x_1,x_2)| \le \frac{\sqrt{2}}{3} + \frac{1}{4}|x_1| + \frac{1}{6}|x_2|,$$

then the assumptions of Theorem 5.3 are satisfied with $\bar{m}_0 = 1/\sqrt{5}$, $\bar{m}_1 = 1/5$, $\bar{m}_2 = 1/3$, $\bar{n}_0 = \sqrt{2}/3$, $\bar{n}_1 = 1/4$, and $\bar{n}_2 = 1/6$. By using the Maple program, we can find that

$$\Lambda = \frac{\Gamma_n(\alpha)}{\Gamma_n(\kappa+\alpha)}\xi^{\kappa+\alpha-1} - \lambda_1\frac{\Gamma_m(\alpha)}{\Gamma_m(\gamma+\alpha)}\eta^{\gamma+\alpha-1} \approx 1.21235918 \neq 0,$$

$$\Psi = \frac{\Gamma_k(\beta)}{\Gamma_k(\nu+\beta)}\tau^{\nu+\beta-1} - \lambda_2\frac{\Gamma_h(\beta)}{\Gamma_h(\mu+\beta)}\theta^{\mu+\beta-1} \approx -0.32647283 \neq 0$$

and

$$U_1 = \bar{m}_1A_1 + \frac{|\lambda_1|T^{\alpha-1}}{|\Lambda|}\bar{m}_1A_5 + \frac{T^{\alpha-1}}{|\Lambda|}\bar{m}_1A_{13} \approx 0.23965603,$$

$$U_2 = \bar{m}_2A_3 + \frac{|\lambda_1|T^{\alpha-1}}{|\Lambda|}\bar{m}_2A_9 + \frac{T^{\alpha-1}}{|\Lambda|}\bar{m}_2A_{14} \approx 0.27471434,$$

$$V_1 = \bar{n}_2A_4 + \frac{|\lambda_2|T^{\beta-1}}{|\Psi|}\bar{n}_2A_{11} + \frac{T^{\beta-1}}{|\Psi|}\bar{n}_2A_{16} \approx 0.04758258,$$

$$V_2 = \bar{n}_1A_2 + \frac{|\lambda_2|T^{\beta-1}}{|\Psi|}\bar{n}_1A_7 + \frac{T^{\beta-1}}{|\Psi|}\bar{n}_1A_{15} \approx 0.36424461.$$

Therefore, we get

$$U_1 + V_1 \approx 0.28723861 < 1 \quad \text{and} \quad U_2 + V_2 \approx 0.63895895 < 1.$$

Hence, by Theorem 5.3, the problem (5.5) has at least one solution on $[0,1]$.

Competing interests

The authors declare that they have no competing interests.

Authors' contributions

All authors contributed equally to the writing of this paper. All authors read and approved the final manuscript.

Author details

[1]Department of Mathematics, Faculty of Science, Chiang Mai University, Chiang Mai, 50200, Thailand. [2]Department of Mathematics, University of Ioannina, Ioannina, 451 10, Greece. [3]Nonlinear Analysis and Applied Mathematics (NAAM) Research Group, Department of Mathematics, Faculty of Science, King Abdulaziz University, P.O. Box 80203, Jeddah, 21589, Saudi Arabia. [4]Nonlinear Dynamic Analysis Research Center, Department of Mathematics, Faculty of Applied Science, King Mongkut's University of Technology North Bangkok, Bangkok, 10800, Thailand.

Acknowledgements

This paper was supported by the Thailand Research Fund under the project RTA5780007.

References

1. Jackson, FH: q-Difference equations. Am. J. Math. **32**, 305-314 (1910)
2. Kac, V, Cheung, P: Quantum Calculus. Springer, New York (2002)
3. Al-Salam, WA: Some fractional q-integrals and q-derivatives. Proc. Edinb. Math. Soc. **15**(2), 135-140 (1966/1967)
4. Agarwal, RP: Certain fractional q-integrals and q-derivatives. Proc. Camb. Philos. Soc. **66**, 365-370 (1969)
5. Ernst, T: The history of q-calculus and a new method. UUDM Report 2000:16, Department of Mathematics, Uppsala University (2000)
6. Ferreira, RAC: Nontrivial solutions for fractional q-difference boundary value problems. Electron. J. Qual. Theory Differ. Equ. **2010**, 70 (2010)
7. Goodrich, CS: Existence and uniqueness of solutions to a fractional difference equation with nonlocal conditions. Comput. Math. Appl. **61**, 191-202 (2011)
8. Kilbas, AA, Srivastava, HM, Trujillo, JJ: Theory and Applications of Fractional Differential Equations. North-Holland Mathematics Studies, vol. 204. Elsevier, Amsterdam (2006)
9. Srivastava, HM: Some generalizations and basic (or q-) extensions of the Bernoulli, Euler and Genocchi polynomials. Appl. Math. Inf. Sci. **5**, 390-444 (2011)
10. Ma, J, Yang, J: Existence of solutions for multi-point boundary value problem of fractional q-difference equation. Electron. J. Qual. Theory Differ. Equ. **2011**, 92 (2011)
11. Graef, JR, Kong, L: Positive solutions for a class of higher order boundary value problems with fractional q-derivatives. Appl. Math. Comput. **218**, 9682-9689 (2012)
12. Ahmad, B, Ntouyas, SK, Purnaras, IK: Existence results for nonlocal boundary value problems of nonlinear fractional q-difference equations. Adv. Differ. Equ. **2012**, 140 (2012)
13. Ahmad, B, Ntouyas, SK: Existence of solutions for nonlinear fractional q-difference inclusions with nonlocal Robin (separated) conditions. Mediterr. J. Math. **10**, 1333-1351 (2013)
14. Ahmad, B, Ntouyas, SK, Alsaedi, A, Al-Hutami, H: Nonlinear q-fractional differential equations with nonlocal and sub-strip type boundary conditions. Electron. J. Qual. Theory Differ. Equ. **2014**, 26 (2014)
15. Ahmad, B, Nieto, JJ, Alsaedi, A, Al-Hutami, H: Existence of solutions for nonlinear fractional q-difference integral equations with two fractional orders and nonlocal four-point boundary conditions. J. Franklin Inst. **351**, 2890-2909 (2014)
16. Alsaedi, A, Ahmad, B, Al-Hutami, H: A study of nonlinear fractional q-difference equations with nonlocal integral boundary conditions. Abstr. Appl. Anal. **2013**, Article ID 410505 (2013)
17. Li, X, Han, Z, Sun, S: Existence of positive solutions of nonlinear fractional q-difference equation with parameter. Adv. Differ. Equ. **2013**, 260 (2013)
18. Ahmad, B, Nieto, JJ, Alsaedi, A, Al-Hutami, H: Boundary value problems of nonlinear fractional q-difference (integral) equations with two fractional orders and four-point nonlocal integral boundary conditions. Filomat **28**(8), 1719-1736 (2014)
19. Annaby, MH, Mansour, ZS: q-Fractional Calculus and Equations. Lecture Notes in Mathematics, vol. 2056. Springer, Berlin (2012)
20. Asawasamrit, S, Tariboon, J, Ntouyas, SK: Existence of solutions for fractional q-integrodifference equations with nonlocal fractional q-integral conditions. Abstr. Appl. Anal. **2014**, Article ID 474138 (2014)
21. Pongarm, N, Asawasamrit, S, Tariboon, J, Ntouyas, SK: Multi-strip fractional q-integral boundary value problems for nonlinear fractional q-difference equations. Adv. Differ. Equ. **2014**, 193 (2014)
22. Granas, A, Dugundji, J: Fixed Point Theory. Springer, New York (2003)

3

Multi-point boundary value problems for a coupled system of nonlinear fractional differential equations

Chengbo Zhai* and Mengru Hao

*Correspondence: cbzhai@sxu.edu.cn
School of Mathematical Sciences, Shanxi University, Taiyuan, Shanxi 030006, P.R. China

Abstract

In this paper, we investigate the existence and uniqueness of solutions to the coupled system of nonlinear fractional differential equations

$$\begin{cases} -D_{0+}^{\nu_1}y_1(t) = \lambda_1 a_1(t)f(y_1(t), y_2(t)), \\ -D_{0+}^{\nu_2}y_2(t) = \lambda_2 a_2(t)g(y_1(t), y_2(t)), \end{cases}$$

where D_{0+}^{ν} is the standard Riemann-Liouville fractional derivative of order ν, $t \in (0,1)$, $\nu_1, \nu_2 \in (n-1, n]$ for $n > 3$ and $n \in \mathbf{N}$, and $\lambda_1, \lambda_2 > 0$, with the multi-point boundary value conditions: $y_1^{(i)}(0) = 0 = y_2^{(i)}(0)$, $0 \le i \le n-2$; $D_{0+}^{\beta}y_1(1) = \sum_{i=1}^{m-2} b_i D_{0+}^{\beta}y_1(\xi_i)$; $D_{0+}^{\beta}y_2(1) = \sum_{i=1}^{m-2} b_i D_{0+}^{\beta}y_2(\xi_i)$, where $n-2 < \beta < n-1$, $0 < \xi_1 < \xi_2 < \cdots < \xi_{m-2} < 1$, $b_i \ge 0$ $(i = 1, 2, \ldots, m-2)$ with $\rho_1 := \sum_{i=1}^{m-2} b_i \xi_i^{\nu_1-\beta-1} < 1$, and $\rho_2 := \sum_{i=1}^{m-2} b_i \xi_i^{\nu_2-\beta-1} < 1$. Our analysis relies on the Banach contraction principle and Krasnoselskii's fixed point theorem.

MSC: 26A33; 34B18; 34B27

Keywords: existence and uniqueness; solutions; Riemann-Liouville fractional derivative; multi-point boundary value problems; Banach contraction principle; Krasnoselskii's fixed point theorem

1 Introduction

Fractional calculus is a generalization of ordinary differentiation and integration to arbitrary non-integer order. The first definition of fractional derivative was introduced at the end of the nineteenth century by Liouville and Riemann, but the concept of non-integer derivative and integral, as a generalization of the traditional integer order differential and integral calculus, was mentioned already in 1695 by Leibniz and L'Hospital. With the help of fractional calculus, we can describe natural phenomena and mathematical models more accurately. The fractional differential equations play an important role in various fields of engineering, physics, economics and biological sciences, *etc.* (see [1–4] for example). In consequence, the subject of fractional differential equations is gaining much importance and attention. For more details on basic theory of fractional differential equations, one can see the monographs of Diethelm [1], Kilbas *et al.* [2], Miller and Ross [3], Podlubny [4] and Tarasov [5], and the papers [6–13] and the references therein.

As is known to all, the initial and boundary value problems for nonlinear fractional differential equations arise from the study of models of control, porous media, electrochemistry, viscoelasticity, electromagnetics, *etc.* Recently, the existence and uniqueness of solutions of initial and boundary value problems for nonlinear fractional equations have been extensively studied (see [14–20]), and some are coupled systems of nonlinear fractional differential equations (see [8, 14, 17, 21, 22]).

In [15], Mophou studied the mild solutions to impulsive fractional differential equations

$$\begin{cases} D_t^\alpha x(t) = Ax(t) + f(t, x(t)), & t \in I = [0, T], t \neq t_k, \\ x(0) = x_0 \in X, \\ \Delta x|_{t=t_k} = I_k(x(t_k^-)), & k = 1, 2, \ldots, m, \end{cases}$$

where $0 < \alpha < 1$, the operator $A : D(A) \subset X \to X$ is a generator of C_0-semigroup $(T(t))_{t \geq 0}$ on a Banach space X, D_t^α is the Caputo fractional derivative, $f : I \times X \to X$ is a given continuous function, $I_k : X \to X$, $0 = t_0 < t_1 < \cdots < t_m < t_{m+1} = T$, $\Delta x|_{t=t_k} = x(t_k^+) - x(t_k^-)$. Some existence and uniqueness results for the equations were established by means of Krasnoselskii's fixed point theorem.

By using the same fixed point theorem, Goodrich [23] considered the existence of a positive solution to the following system of differential equations of fractional order:

$$\begin{cases} -D_{0^+}^{\nu_1} y_1(t) = \lambda_1 a_1(t) f(y_1(t), y_2(t)), \\ -D_{0^+}^{\nu_2} y_2(t) = \lambda_2 a_2(t) g(y_1(t), y_2(t)), \end{cases} \tag{1}$$

where $D_{0^+}^\nu$ is the standard Riemann-Liouville fractional derivative of order ν, $t \in (0, 1)$, $\nu_1, \nu_2 \in (n-1, n]$ for $n > 3$ and $n \in \mathbf{N}$, and $\lambda_1, \lambda_2 > 0$, with the following boundary value conditions:

$$y_1^{(i)}(0) = 0 = y_2^{(i)}(0), \quad 0 \leq i \leq n-2,$$
$$\left[D_{0+}^\alpha y_1(t)\right]_{t=1} = 0 = \left[D_{0+}^\alpha y_2(t)\right]_{t=1}, \quad 1 \leq \alpha \leq n-2,$$

under the assumptions that a_1, a_2, f, g are nonnegative and continuous.

Very recently, Sun *et al.* [14] considered the coupled system of multi-term nonlinear fractional differential equations

$$\begin{cases} D^\alpha u(t) = f(t, v(t), D^{\beta_1} v(t), \ldots, D^{\beta_N} v(t)), & D^{\alpha - i} u(0) = 0, \quad i = 1, 2, \ldots, n_1, \\ D^\sigma v(t) = g(t, u(t), D^{\rho_1} u(t), \ldots, D^{\rho_N} u(t)), & D^{\sigma - j} v(0) = 0, \quad j = 1, 2, \ldots, n_2, \end{cases}$$

where $t \in (0, 1]$, $\alpha > \beta_1 > \beta_2 > \cdots > \beta_N > 0$, $\sigma > \rho_1 > \rho_2 \cdots > \rho_N > 0$, $n_1 = [\alpha] + 1$, $n_2 = [\sigma] + 1$, $\beta_q, \rho_q < 1$ and $q \in \{1, 2, \ldots, N\}$. By using the Schauder fixed point theorem and the Banach contraction principle, some results of existence and uniqueness of solutions for the coupled system are obtained.

However, to our knowledge, there are few works that deal with multi-point boundary value problems for a coupled system of nonlinear fractional differential equations. The purpose of this article is to investigate the solutions for the coupled system of nonlinear fractional differential equations (1) with the multi-point boundary conditions:

$$y_1^{(i)}(0) = 0 = y_2^{(i)}(0), \quad 0 \leq i \leq n-2, \tag{2}$$

$$D_{0^+}^{\beta}y_1(1)=\sum_{i=1}^{m-2}b_iD_{0^+}^{\beta}y_1(\xi_i), \tag{3}$$

$$D_{0^+}^{\beta}y_2(1)=\sum_{i=1}^{m-2}b_iD_{0^+}^{\beta}y_2(\xi_i), \tag{4}$$

where $n-2<\beta<n-1$, $0<\xi_1<\xi_2<\cdots<\xi_{m-2}<1$, $b_i\geq 0$ $(i=1,2,\ldots,m-2)$ with $\rho_1:=\sum_{i=1}^{m-2}b_i\xi_i^{\nu_1-\beta-1}<1$, and $\rho_2:=\sum_{i=1}^{m-2}b_i\xi_i^{\nu_2-\beta-1}<1$.

Motivated by the above-mentioned works and recent works on coupled systems of fractional differential equations, we consider the existence and uniqueness of solutions of coupled system (1)-(4) by means of the Banach contraction principle and Krasnoselskii's fixed point theorem. In our paper, we do not suppose that a_1, a_2, f, g are nonnegative.

With this context in mind, the outline of this paper is as follows. In Section 2 we recall certain results from the theory of continuous fractional calculus. In Section 3 we provide some conditions under which problem (1)-(4) will have a unique solution or at least one solution.

2 Preliminaries

For the convenience of the reader, we present here some definitions, lemmas and basic results that will be used in the proofs of our theorems.

Definition 2.1 (see [4]) Let $\nu>0$ with $\nu\in\mathbf{R}$. Suppose that $y:[a,+\infty)\to\mathbf{R}$. Then the νth Riemann-Liouville fractional integral is defined to be

$$D_{a^+}^{-\nu}y(t):=\frac{1}{\Gamma(\nu)}\int_a^t y(s)(t-s)^{\nu-1}\,ds,$$

whenever the right-hand side is defined. Similarly, with $\nu>0$ and $\nu\in\mathbf{R}$, we define the νth Riemann-Liouville fractional derivative to be

$$D_{a^+}^{\nu}y(t):=\frac{1}{\Gamma(n-\nu)}\frac{d^n}{dt^n}\int_a^t\frac{y(s)}{(t-s)^{\nu+1-n}}\,ds,$$

where $n\in\mathbf{N}$ is the unique positive integer satisfying $n-1\leq\nu<n$ and $t>a$.

Lemma 2.2 (see [23]) *Let $\alpha\in\mathbf{R}$. Then $D^nD_{a^+}^{\alpha}y(t)=D_{a^+}^{n+\alpha}y(t)$, for each $n\in N_0$, where $y(t)$ is assumed to be sufficiently regular so that both sides of the equality are well defined. Moreover, if $\beta\in(-\infty,0]$ and $\gamma\in[0,+\infty)$, then $D_{a^+}^{\gamma}D_{a^+}^{\beta}y(t)=D_{a^+}^{\gamma+\beta}y(t)$.*

Lemma 2.3 (see [23]) *The general solution to $D_{a^+}^{\nu}y(t)=0$, where $n-1<\nu\leq n$ and $\nu>0$, is the function $y(t)=c_1t^{\nu-1}+c_2t^{\nu-2}+\cdots+c_nt^{\nu-n}$, where $c_i\in\mathbf{R}$ for each i.*

Lemma 2.4 *Let $h\in C^n([0,1])$ be given. Then the unique solution to problem $-D_{0^+}^{\nu}y(t)=h(t)$ together with the boundary conditions $y^{(i)}(0)=0$ and $D_{0^+}^{\beta}y(1)=\sum_{i=1}^{m-2}b_iD_{0^+}^{\beta}y(\xi_i)$, where $n-2<\beta<n-1$ and $0\leq i\leq n-2$, is*

$$y(t)=\int_0^1 G(t,s)h(s)\,ds, \tag{5}$$

where

$$G(t,s) = \begin{cases} -\frac{(t-s)^{\nu-1}}{\Gamma(\nu)} + \frac{t^{\nu-1}}{\Gamma(\nu)(1-\rho)}[(1-s)^{\nu-\beta-1} - \sum_{i=1}^{m-2} b_i(\xi_i - s)^{\nu-\beta-1}], \\ \quad s \le t, \xi_{i-1} < s \le \xi_i, i = 1,2,\dots m-1, \\ \frac{t^{\nu-1}}{\Gamma(\nu)(1-\rho)}[(1-s)^{\nu-\beta-1} - \sum_{i=1}^{m-2} b_i(\xi_i - s)^{\nu-\beta-1}], \\ \quad t \le s, \xi_{i-1} < s \le \xi_i, i = 1,2,\dots m-1, \end{cases} \tag{6}$$

is the Green function for this problem, where $\rho = \sum_{i=1}^{m-2} b_i \xi_i^{\nu-\beta-1} < 1$ *and* $\xi_0 = 0$, $\xi_{m-1} = 1$.

Proof We know that the general solution to our problem is

$$y(t) = c_1 t^{\nu-1} + c_2 t^{\nu-2} + \dots + c_n t^{\nu-n} - D_{0^+}^{-\nu} h(t),$$

we immediately observe that the boundary value condition $y^{(i)}(0) = 0, 0 \le i \le n-2$, implies that $c_2 = c_3 = \dots = c_n = 0$. On the other hand, $D_{0^+}^{\beta} y(1) = \sum_{i=1}^{m-2} b_i D_{0^+}^{\beta} y(\xi_i)$ implies that

$$-\frac{1}{\Gamma(\nu-\beta)} \int_0^1 (1-s)^{\nu-\beta-1} h(s)\, ds + c_1 \frac{\Gamma(\nu)}{\Gamma(\nu-\beta)}$$
$$= \sum_{i=1}^{m-2} b_i \left(-\frac{1}{\Gamma(\nu-\beta)} \int_0^{\xi_i} (\xi_i - s)^{\nu-\beta-1} h(s)\, ds + c_1 \frac{\Gamma(\nu)}{\Gamma(\nu-\beta)} \xi_i^{\nu-\beta-1} \right).$$

That is to say,

$$c_1 = \frac{1}{\Gamma(\nu)(1-\rho)} \left[\int_0^1 (1-s)^{\nu-\beta-1} h(s)\, ds - \sum_{i=1}^{m-2} b_i \int_0^{\xi_i} (\xi_i - s)^{\nu-\beta-1} h(s)\, ds \right].$$

Therefore, the unique solution is

$$\begin{aligned} y(t) &= -\int_0^t \frac{(t-s)^{\nu-1}}{\Gamma(\nu)} h(s)\, ds \\ &\quad + \frac{t^{\nu-1}}{\Gamma(\nu)(1-\rho)} \left[\int_0^1 (1-s)^{\nu-\beta-1} h(s)\, ds - \sum_{i=1}^{m-2} b_i \int_0^{\xi_i} (\xi_i - s)^{\nu-\beta-1} h(s)\, ds \right] \\ &= \int_0^1 G(t,s) h(s)\, ds. \end{aligned}$$

The proof is complete. □

3 Main results

This section deals with the existence and uniqueness of solutions to problem (1)-(4).

Let E represent the Banach space of $C[0,1]$ when equipped with the usual supremum norm $\|\cdot\|$. Then put $X := E \times E$, where X is equipped with the norm

$$\|(y_1, y_2)\| := \|y_1\| + \|y_2\|$$

for $(y_1, y_2) \in X$. Observe that X is also a Banach space (see [24]). In addition, define two operators $T_1, T_2 : X \to E$ by

$$\big(T_1(y_1, y_2)\big)(t) := \lambda_1 \int_0^1 G_1(t,s)a_1(s)f\big(s, y_1(s), y_2(s)\big)\,ds$$

and

$$\big(T_2(y_1, y_2)\big)(t) := \lambda_2 \int_0^1 G_2(t,s)a_2(s)g\big(s, y_1(s), y_2(s)\big)\,ds,$$

where $G_1(t,s)$ is the Green function of Lemma 2.4 with ν replaced by ν_1 and ρ replaced by ρ_1, and likewise, $G_2(t,s)$ is the Green function of Lemma 2.4 with ν replaced by ν_2 and ρ replaced by ρ_2. Now, we define an operator $S : X \to X$ by

$$\begin{aligned}\big(S(y_1, y_2)\big)(t) &:= \Big(\big(T_1(y_1, y_2)\big)(t), \big(T_2(y_1, y_2)\big)(t)\Big)\\ &= \Big(\lambda_1 \int_0^1 G_1(t,s)a_1(s)f\big(s, y_1(s), y_2(s)\big)\,ds,\\ &\quad \lambda_2 \int_0^1 G_2(t,s)a_2(s)g\big(s, y_1(s), y_2(s)\big)\,ds\Big).\end{aligned} \tag{7}$$

We claim that whenever $(y_1, y_2) \in X$ is a fixed point of the operator defined in (7), it follows that $y_1(t)$ and $y_2(t)$ solve problem (1)-(4). We shall look for fixed points of the operator S, seeing as these fixed points coincide with solutions of problem (1)-(4).

To establish the main results, we need the following assumptions:

(H_1) $f, g : [0,1] \times \mathbf{R} \times \mathbf{R} \to \mathbf{R}$ and $a_1, a_2 : [0,1] \to \mathbf{R}$;

(H_2) f, g, a_1, a_2 are continuous;

(H_3) there exist positive functions $L_1(t)$ and $L_2(t)$ such that

$$\begin{aligned}&|a_1(t)| \cdot \big|f\big(t, y_1(t), y_2(t)\big) - f\big(t, u_1(t), u_2(t)\big)\big| \le L_1(t)\big\|(y_1 - u_1, y_2 - u_2)\big\|,\\ &|a_2(t)| \cdot \big|g\big(t, y_1(t), y_2(t)\big) - g\big(t, u_1(t), u_2(t)\big)\big| \le L_2(t)\big\|(y_1 - u_1, y_2 - u_2)\big\|\end{aligned}$$

for all $t \in [0,1]$ and $(y_1, y_2), (u_1, u_2) \in X$.

Further, we set

$$\begin{aligned}D_L^{-\nu} &= \max\Big\{\sup_{t\in[0,1]} \big|D_{0^+}^{-\nu_1} L_1(t)\big|, \sup_{t\in[0,1]} \big|D_{0^+}^{-\nu_2} L_2(t)\big|\Big\},\\ D^{-(\nu-\beta)}L(1) &= \max\Big\{\big|D_{0^+}^{-(\nu_1-\beta)} L_1(1)\big|, \big|D_{0^+}^{-(\nu_2-\beta)} L_2(1)\big|\Big\},\\ D^{-(\nu-\beta)}L(\xi_i) &= \max\Big\{\big|D_{0^+}^{-(\nu_1-\beta)} L_1(\xi_i)\big|, \big|D_{0^+}^{-(\nu_2-\beta)} L_2(\xi_i)\big|\Big\}, \quad i = 1, 2, \ldots, m-2.\end{aligned}$$

(H_4) The parameters λ_1, λ_2 satisfy $\lambda_1, \lambda_2 < \Lambda$, where

$$\begin{aligned}\Lambda = \min\Bigg\{&\frac{1}{4}\left[D_L^{-\nu} + \frac{\Gamma(\nu_1-\beta)}{\Gamma(\nu_1)(1-\rho_1)}\left(\sum_{i=1}^{m-2} b_i D^{-(\nu-\beta)}L(\xi_i) + D^{-(\nu-\beta)}L(1)\right)\right]^{-1},\\ &\frac{1}{4}\left[D_L^{-\nu} + \frac{\Gamma(\nu_2-\beta)}{\Gamma(\nu_2)(1-\rho_2)}\left(\sum_{i=1}^{m-2} b_i D^{-(\nu-\beta)}L(\xi_i) + D^{-(\nu-\beta)}L(1)\right)\right]^{-1}\Bigg\}.\end{aligned}$$

(H_5) The parameters λ_1, λ_2 satisfy $\lambda_1, \lambda_2 < \Lambda$, where

$$\Lambda = \min\left\{\frac{1}{4}\left[\frac{\Gamma(\nu_1-\beta)}{\Gamma(\nu_1)(1-\rho_1)}\left(\sum_{i=1}^{m-2} b_i D^{-(\nu-\beta)}L(\xi_i) + D^{-(\nu-\beta)}L(1)\right)\right]^{-1},\right.$$
$$\left.\frac{1}{4}\left[\frac{\Gamma(\nu_2-\beta)}{\Gamma(\nu_2)(1-\rho_2)}\left(\sum_{i=1}^{m-2} b_i D^{-(\nu-\beta)}L(\xi_i) + D^{-(\nu-\beta)}L(1)\right)\right]^{-1}\right\}.$$

(H_6) There exists $\mu \in L^1([0,1], \mathbf{R}^+)$ such that

$$|a_1(t)f(t, y_1(t), y_2(t))| \le \mu(t), \qquad |a_2(t)g(t, y_1(t), y_2(t))| \le \mu(t),$$
$$\forall (t, y_1, y_2) \in [0,1] \times X.$$

We are ready to state the existence and uniqueness result.

Theorem 3.1 *Suppose that conditions* (H_1)-(H_4) *are satisfied. Then the boundary value problem* (1)-(4) *has a unique solution.*

Proof Let us set

$$M = \max\left\{\sup_{t\in[0,1]} |a_1(t)f(t,0,0)|, \sup_{t\in[0,1]} |a_2(t)g(t,0,0)|\right\}$$

and choose

$$r \ge \max\left\{4\lambda_1\left[\frac{M}{\Gamma(\nu_1+1)} + \frac{M}{\Gamma(\nu_1)(1-\rho_1)(\nu_1-\beta)}\left(\sum_{i=1}^{m-2} b_i\xi_i^{\nu_1-\beta} + 1\right)\right],\right.$$
$$\left.4\lambda_2\left[\frac{M}{\Gamma(\nu_2+1)} + \frac{M}{\Gamma(\nu_2)(1-\rho_2)(\nu_2-\beta)}\left(\sum_{i=1}^{m-2} b_i\xi_i^{\nu_2-\beta} + 1\right)\right]\right\}.$$

Now, we show that $S(\Omega_r) \subset \Omega_r$, where $\Omega_r = \{(y_1, y_2) \in X : \|(y_1, y_2)\| \le r\}$, and S is a contraction. In fact, for all $(y_1, y_2) \in \Omega_r$, we obtain

$$\begin{aligned}
&\|T_1(y_1, y_2)\| \\
&= \max_{t\in[0,1]}\Bigg| -\int_0^t \frac{(t-s)^{\nu_1-1}}{\Gamma(\nu_1)}\lambda_1 a_1(s)f(s, y_1(s), y_2(s))\,ds \\
&\quad + \frac{t^{\nu_1-1}}{\Gamma(\nu_1)(1-\rho_1)}\Bigg[\int_0^1 (1-s)^{\nu_1-\beta-1}\lambda_1 a_1(s)f(s, y_1(s), y_2(s))\,ds \\
&\quad - \sum_{i=1}^{m-2} b_i \int_0^{\xi_i} (\xi_i - s)^{\nu_1-\beta-1}\lambda_1 a_1(s)f(s, y_1(s), y_2(s))\,ds\Bigg]\Bigg| \\
&\le \max_{t\in[0,1]}\Bigg\{\int_0^t \frac{(t-s)^{\nu_1-1}}{\Gamma(\nu_1)}\lambda_1 |a_1(s)f(s, y_1(s), y_2(s))|\,ds \\
&\quad + \frac{t^{\nu_1-1}}{\Gamma(\nu_1)(1-\rho_1)}\Bigg[\int_0^1 (1-s)^{\nu_1-\beta-1}\lambda_1 |a_1(s)f(s, y_1(s), y_2(s))|\,ds
\end{aligned}$$

$$+\sum_{i=1}^{m-2} b_i \int_0^{\xi_i} (\xi_i - s)^{\nu_1-\beta-1} \lambda_1 \left| a_1(s) f\left(s, y_1(s), y_2(s)\right) \right| ds \Bigg] \Bigg\}$$

$$\le \max_{t\in[0,1]} \Bigg\{ \int_0^t \frac{(t-s)^{\nu_1-1}}{\Gamma(\nu_1)} \lambda_1 |a_1(s)| \left(\left| f\left(s, y_1(s), y_2(s)\right) - f(s,0,0) \right| + |f(s,0,0)| \right) ds$$

$$+ \frac{t^{\nu_1-1}}{\Gamma(\nu_1)(1-\rho_1)} \Bigg[\int_0^1 (1-s)^{\nu_1-\beta-1} \lambda_1 |a_1(s)| \left(\left| f\left(s, y_1(s), y_2(s)\right) - f(s,0,0) \right| \right.$$

$$\left. + |f(s,0,0)| \right) ds$$

$$+\sum_{i=1}^{m-2} b_i \int_0^{\xi_i} (\xi_i - s)^{\nu_1-\beta-1} \lambda_1 |a_1(s)| \left(\left| f\left(s, y_1(s), y_2(s)\right) - f(s,0,0) \right| + |f(s,0,0)| \right) ds \Bigg] \Bigg\}$$

$$\le \max_{t\in[0,1]} \Bigg\{ \int_0^t \frac{(t-s)^{\nu_1-1}}{\Gamma(\nu_1)} \lambda_1 \left(L_1(s) \|(y_1, y_2)\| + M \right) ds$$

$$+ \frac{t^{\nu_1-1}}{\Gamma(\nu_1)(1-\rho_1)} \Bigg[\int_0^1 (1-s)^{\nu_1-\beta-1} \lambda_1 \left(L_1(s) \|(y_1, y_2)\| + M \right) ds$$

$$+\sum_{i=1}^{m-2} b_i \int_0^{\xi_i} (\xi_i - s)^{\nu_1-\beta-1} \lambda_1 \left(L_1(s) \|(y_1, y_2)\| + M \right) ds \Bigg] \Bigg\}$$

$$\le \max_{t\in[0,1]} \Bigg\{ \lambda_1 \left(D_{0^+}^{-\nu_1} L_1(t) r + \frac{M t^{\nu_1}}{\Gamma(\nu_1+1)} \right) + \lambda_1 \frac{\Gamma(\nu_1-\beta) t^{\nu_1-1}}{\Gamma(\nu_1)(1-\rho_1)}$$

$$\times \left[\sum_{i=1}^{m-2} b_i \left(D_{0^+}^{-(\nu_1-\beta)} L_1(\xi_i) r + \frac{M \xi_i^{\nu_1-\beta}}{\Gamma(\nu_1-\beta+1)} \right) + D_{0^+}^{-(\nu_1-\beta)} L_1(1) r + \frac{M}{\Gamma(\nu_1-\beta+1)} \right] \Bigg\}$$

$$\le \lambda_1 \left(D_L^{-\nu} r + \frac{M}{\Gamma(\nu_1+1)} \right) + \lambda_1 \frac{\Gamma(\nu_1-\beta)}{\Gamma(\nu_1)(1-\rho_1)}$$

$$\times \left[\sum_{i=1}^{m-2} b_i \left(D^{-(\nu-\beta)} L(\xi_i) r + \frac{M \xi_i^{\nu_1-\beta}}{\Gamma(\nu_1-\beta+1)} \right) + D^{-(\nu-\beta)} L(1) r + \frac{M}{\Gamma(\nu_1-\beta+1)} \right]$$

$$\le \lambda_1 \left[D_L^{-\nu} + \frac{\Gamma(\nu_1-\beta)}{\Gamma(\nu_1)(1-\rho_1)} \left(\sum_{i=1}^{m-2} b_i D^{-(\nu-\beta)} L(\xi_i) + D^{-(\nu-\beta)} L(1) \right) \right] r$$

$$+ \lambda_1 M \left[\frac{1}{\Gamma(\nu_1+1)} + \frac{1}{\Gamma(\nu_1)(1-\rho_1)(\nu_1-\beta)} \left(\sum_{i=1}^{m-2} b_i \xi_i^{\nu_1-\beta} + 1 \right) \right]$$

$$\le \frac{1}{4} r + \frac{1}{4} r = \frac{r}{2},$$

that is to say, $\|T_1(y_1, y_2)\| \le \frac{r}{2}$.

Then, for $(y_1, y_2), (u_1, u_2) \in X$ and for each $t \in [0,1]$, we obtain

$$\|T_1(y_1, y_2) - T_1(u_1, u_2)\|$$

$$\le \max_{t\in[0,1]} \Bigg\{ \int_0^t \frac{(t-s)^{\nu_1-1}}{\Gamma(\nu_1)} \lambda_1 |a_1(s)| \cdot \left| f\left(s, y_1(s), y_2(s)\right) - f\left(s, u_1(s), u_2(s)\right) \right| ds$$

$$+ \frac{t^{\nu_1-1}}{\Gamma(\nu_1)(1-\rho_1)} \Bigg[\int_0^1 (1-s)^{\nu_1-\beta-1} \lambda_1 |a_1(s)|$$

$$\times \left|f\big(s,y_1(s),y_2(s)\big)-f\big(s,u_1(s),u_2(s)\big)\right| ds$$

$$\left.\left. + \sum_{i=1}^{m-2} b_i \int_0^{\xi_i} (\xi_i - s)^{\nu_1-\beta-1}\lambda_1 |a_1(s)| \cdot \left|f\big(s,y_1(s),y_2(s)\big)-f\big(s,u_1(s),u_2(s)\big)\right| ds\right]\right\}$$

$$\leq \lambda_1 D_L^{-\nu}\left\|(y_1-u_1,y_2-u_2)\right\| + \lambda_1 \frac{\Gamma(\nu_1-\beta)}{\Gamma(\nu_1)(1-\rho_1)}$$

$$\times \left[\sum_{i=1}^{m-2} b_i D^{-(\nu_1-\beta)}L(\xi_i) + D^{-(\nu_1-\beta)}L(1)\right]\left\|(y_1-u_1,y_2-u_2)\right\|$$

$$\leq \frac{1}{4}\left\|(y_1-u_1,y_2-u_2)\right\|.$$

That is to say, $\|T_1(y_1,y_2)-T_1(u_1,u_2)\| \leq \frac{1}{4}\|(y_1-u_1,y_2-u_2)\| = \frac{1}{4}\|(y_1,y_2)-(u_1,u_2)\|$.

Hence, we find that $T_1:\Omega_r \to B_{\frac{r}{2}}$ and T_1 is a contraction, where $B_{\frac{r}{2}} = \{y \in B : \|y\| \leq \frac{r}{2}\}$. Similarly, we have $T_2:\Omega_r \to B_{\frac{r}{2}}$ and T_2 is a contraction. Consequently, for any $(y_1,y_2) \in \Omega_r$,

$$\left\|S(y_1,y_2)\right\| = \left\|\big(T_1(y_1,y_2),T_2(y_1,y_2)\big)\right\| \leq \frac{r}{2}+\frac{r}{2} \leq r,$$

i.e., $S(\Omega_r) \subset \Omega_r$. And, for $(y_1,y_2),(u_1,u_2) \in \Omega_r$ and for each $t \in [0,1]$,

$$\left\|S(y_1,y_2)-S(u_1,u_2)\right\| \leq \frac{1}{2}\left\|(y_1-u_1,y_2-u_2)\right\| = \frac{1}{2}\left\|(y_1,y_2)-(u_1,u_2)\right\|.$$

So, $S:\Omega_r \to \Omega_r$ and S is a contraction. Thus, the conclusion of the theorem follows from the contraction mapping principle. □

Our next result is based on the following well-known fixed point theorem due to Krasnoselskii.

Lemma 3.2 (Krasnoselskii [25]) *Let K be a closed convex and nonempty subset of a Banach space E. Let T, S be the operators such that*:

(i) *$Tx + Sy \in K$ whenever $x,y \in K$;*
(ii) *T is compact and continuous;*
(iii) *S is a contraction mapping.*

Then there exists $z \in K$ such that $z = Tz + Sz$.

Now we are ready to state and prove the following existence result.

Theorem 3.3 *Suppose that conditions* (H_1)-(H_3), (H_5), (H_6) *are satisfied. Then there exists at least one solution of the boundary value problem* (1)-(4).

Proof Let us fix

$$r \geq \max\left\{\frac{1}{2}\lambda_1\|\mu\|_{L^1}\left[\frac{1}{\Gamma(\nu_1+1)}+\frac{1}{\Gamma(\nu_1)(1-\rho_1)(\nu_1-\beta)}\left(\sum_{i=1}^{m-2} b_i\xi_i^{\nu_1-\beta}+1\right)\right],\right.$$

$$\left.\frac{1}{2}\lambda_2\|\mu\|_{L^1}\left[\frac{1}{\Gamma(\nu_2+1)}+\frac{1}{\Gamma(\nu_2)(1-\rho_2)(\nu_2-\beta)}\left(\sum_{i=1}^{m-2} b_i\xi_i^{\nu_2-\beta}+1\right)\right]\right\},$$

and consider $\Omega_r = \{(y_1, y_2) \in X : \|(y_1, y_2)\| \leq r\}$. We define the operators Q_1 and Q_2 on Ω_r as

$$T_1(y_1, y_2) := Q_1(y_1, y_2) + Q_2(y_1, y_2),$$

where

$$\big(Q_1(y_1, y_2)\big)(t) = -\int_0^t \frac{(t-s)^{\nu_1 - 1}}{\Gamma(\nu_1)} \lambda_1 a_1(s) f\big(s, y_1(s), y_2(s)\big)\, ds,$$

$$\big(Q_2(y_1, y_2)\big)(t) = \frac{t^{\nu_1 - 1}}{\Gamma(\nu_1)(1-\rho_1)} \Bigg[\int_0^1 (1-s)^{\nu_1 - \beta - 1} \lambda_1 a_1(s) f\big(s, y_1(s), y_2(s)\big)\, ds$$
$$- \sum_{i=1}^{m-2} b_i \int_0^{\xi_i} (\xi_i - s)^{\nu_1 - \beta - 1} \lambda_1 a_1(s) f\big(s, y_1(s), y_2(s)\big)\, ds\Bigg].$$

For all $(y_1, y_2), (u_1, u_2) \in \Omega_r$, we find that

$$\big\|Q_1(y_1, y_2) + Q_2(u_1, u_2)\big\|$$
$$\leq \|\mu\|_{L^1} \lambda_1 \left[\frac{1}{\Gamma(\nu_1 + 1)} + \frac{1}{\Gamma(\nu_1)(1-\rho_1)(\nu_1 - \beta)} \left(\sum_{i=1}^{m-2} b_i \xi_i^{\nu_1 - \beta} + 1\right)\right]$$
$$\leq \frac{r}{2}.$$

Thus, $Q_1(y_1, y_2) + Q_2(u_1, u_2) \in \Omega_{\frac{r}{2}}$ for all $(y_1, y_2), (u_1, u_2) \in \Omega_r$. From assumption ($H_3$), we have

$$\big\|Q_2(y_1, y_2) - Q_2(u_1, u_2)\big\| \leq \lambda_1 \frac{\Gamma(\nu_1 - \beta)}{\Gamma(\nu_1)(1-\rho_1)} \left[\sum_{i=1}^{m-2} b_i D^{-(\nu_1 - \beta)} L(\xi_i) + D^{-(\nu_1 - \beta)} L(1)\right]$$
$$\times \big\|(y_1 - u_1, y_2 - u_2)\big\|$$
$$\leq \frac{1}{4} \big\|(y_1 - u_1, y_2 - u_2)\big\|$$
$$= \frac{1}{4} \big\|(y_1, y_2) - (u_1, u_2)\big\|,$$

where $(y_1, y_2), (u_1, u_2) \in X$, $t \in [0,1]$. So, Q_2 is a contraction mapping. We next consider the operator Q_1. Evidently, the continuity of f implies that the operator Q_1 is continuous. Also, Q_1 is uniformly bounded on Ω_r as

$$\big\|Q_1(y_1, y_2)\big\| \leq \lambda_1 \frac{\|\mu\|_{L^1}}{\Gamma(\nu_1 + 1)} \leq 2r.$$

Now, we show that $Q_1(y_1, y_2)(t)$ is equicontinuous. In fact, since a_1, f are bounded on the compact set $[0,1]$ and $[0,1] \times \Omega_r$, respectively, we can define

$$M_1 = \max\left\{\sup_{(t, y_1, y_2) \in [0,1] \times \Omega_r} \big|a_1(t) f(t, y_1, y_2)\big|, \sup_{(t, y_1, y_2) \in [0,1] \times \Omega_r} \big|a_2(t) g(t, y_1, y_2)\big|\right\},$$

and we have, for any $t_1, t_2 \in [0,1]$,

$$\begin{aligned}
&\left|Q_1(y_1,y_2)(t_2) - Q_1(y_1,y_2)(t_1)\right| \\
&= \left| -\int_0^{t_2} \frac{(t_2-s)^{\nu_1-1}}{\Gamma(\nu_1)} \lambda_1 a_1(s) f\big(s, y_1(s), y_2(s)\big)\, ds \right. \\
&\quad \left. + \int_0^{t_1} \frac{(t_1-s)^{\nu_1-1}}{\Gamma(\nu_1)} \lambda_1 a_1(s) f\big(s, y_1(s), y_2(s)\big)\, ds \right| \\
&= \left| \int_0^{t_1} \frac{(t_2-s)^{\nu_1-1} - (t_1-s)^{\nu_1-1}}{\Gamma(\nu_1)} \lambda_1 a_1(s) f\big(s, y_1(s), y_2(s)\big)\, ds \right. \\
&\quad \left. + \int_{t_1}^{t_2} \frac{(t_2-s)^{\nu_1-1}}{\Gamma(\nu_1)} \lambda_1 a_1(s) f\big(s, y_1(s), y_2(s)\big)\, ds \right| \\
&\le \lambda_1 \frac{M_1}{\Gamma(\nu_1)} \left| \int_0^{t_1} \left((t_2-s)^{\nu_1-1} - (t_1-s)^{\nu_1-1}\right) ds + \int_{t_1}^{t_2} (t_2-s)^{\nu_1-1}\, ds \right| \\
&= \lambda_1 \frac{M_1}{\Gamma(\nu_1+1)} \left| -\left((t_2-t_1)^{\nu_1} - t_2^{\nu_1}\right) - t_1^{\nu_1} + (t_2-t_1)^{\nu_1} \right| \\
&= \lambda_1 \frac{M_1}{\Gamma(\nu_1+1)} \left| t_2^{\nu_1} - t_1^{\nu_1} \right|,
\end{aligned}$$

which is independent of (y_1, y_2). Therefore, Q_1 is equicontinuous on Ω_r. Hence, by the Arzela-Ascoli theorem, Q_1 is compact on Ω_r. Similarly, we set

$$T_2(y_1,y_2) := P_1(y_1,y_2) + P_2(y_1,y_2),$$

where

$$\begin{aligned}
\big(P_1(y_1,y_2)\big)(t) &= -\int_0^t \frac{(t-s)^{\nu_2-1}}{\Gamma(\nu_2)} \lambda_2 a_2(s) g\big(s, y_1(s), y_2(s)\big)\, ds, \\
\big(P_2(y_1,y_2)\big)(t) &= \frac{t^{\nu_2-1}}{\Gamma(\nu_2)(1-\rho_2)} \left[\int_0^1 (1-s)^{\nu_2-\beta-1} \lambda_2 a_2(s) g\big(s, y_1(s), y_2(s)\big)\, ds \right. \\
&\quad \left. - \sum_{i=1}^{m-2} b_i \int_0^{\xi_i} (\xi_i - s)^{\nu_2-\beta-1} \lambda_2 a_2(s) g\big(s, y_1(s), y_2(s)\big)\, ds \right].
\end{aligned}$$

And we can obtain a similar conclusion to the operator T_1. Now, we let

$$A(y_1,y_2) := \big(Q_1(y_1,y_2), P_1(y_1,y_2)\big), \qquad B(y_1,y_2) := \big(Q_2(y_1,y_2), P_2(y_1,y_2)\big).$$

Therefore, we have

$$\begin{aligned}
S(y_1,y_2) &= \big(T_1(y_1,y_2), T_2(y_1,y_2)\big) \\
&= \big(Q_1(y_1,y_2) + Q_2(y_1,y_2), P_1(y_1,y_2) + P_2(y_1,y_2)\big) \\
&= \big(Q_1(y_1,y_2), P_1(y_1,y_2)\big) + \big(Q_2(y_1,y_2), P_2(y_1,y_2)\big) \\
&= A(y_1,y_2) + B(y_1,y_2).
\end{aligned}$$

In view of the proof above, we get the following.

(I) For all $(y_1, y_2), (u_1, u_2) \in \Omega_r$, we find that

$$\|A(y_1, y_2) + B(u_1, u_2)\| = \|Q_1(y_1, y_2) + Q_2(u_1, u_2)\| + \|P_1(y_1, y_2) + P_2(u_1, u_2)\| \leq r.$$

Thus, $A(y_1, y_2) + B(u_1, u_2) \in \Omega_r$ for all $(y_1, y_2), (u_1, u_2) \in \Omega_r$.

(II) For $(y_1, y_2), (u_1, u_2) \in X$, $t \in [0, 1]$,

$$\begin{aligned}\|B(y_1, y_2) - B(u_1, u_2)\| &= \|Q_2(y_1, y_2) - Q_2(u_1, u_2)\| + \|P_2(y_1, y_2) - P_2(u_1, u_2)\| \\ &\leq \frac{1}{2}\|(y_1 - u_1, y_2 - u_2)\| = \frac{1}{2}\|(y_1, y_2) - (u_1, u_2)\|.\end{aligned}$$

So, B is a contraction mapping.

(III) The continuity of Q_1 and P_1 implies that the operator A is continuous. Also, A is uniformly bounded on Ω_r as

$$\|A(y_1, y_2)\| = \|Q_1(y_1, y_2)\| + \|P_1(y_1, y_2)\| \leq 4r.$$

Moreover, for any $t_1, t_2 \in [0, 1]$,

$$\begin{aligned}&|A(y_1, y_2)(t_2) - A(y_1, y_2)(t_1)| \\ &\quad = |(Q_1(y_1, y_2)(t_2) - Q_1(y_1, y_2)(t_1), P_1(y_1, y_2)(t_2) - P_1(y_1, y_2)(t_1))| \\ &\quad \leq |Q_1(y_1, y_2)(t_2) - Q_1(y_1, y_2)(t_1)| + |P_1(y_1, y_2)(t_2) - P_1(y_1, y_2)(t_1)| \\ &\quad \leq \lambda_1 \frac{M_1}{\Gamma(\nu_1 + 1)}|t_2^{\nu_1} - t_1^{\nu_1}| + \lambda_2 \frac{M_1}{\Gamma(\nu_2 + 1)}|t_2^{\nu_2} - t_1^{\nu_2}|,\end{aligned}$$

which is independent of (y_1, y_2). Therefore, A is equicontinuous on Ω_r. Hence, by the Arzela-Ascoli theorem, A is compact on Ω_r.

Thus all the assumptions of Lemma 3.2 are satisfied, so $S(y_1, y_2) = A(y_1, y_2) + B(y_1, y_2)$ has at least one fixed point. Hence, we obtain that (1)-(4) has at least one solution. □

4 Conclusions

There are few works that deal with multi-point boundary value problems for a coupled system of nonlinear fractional differential equations. In this article, we study multi-point boundary value problems for a coupled system of nonlinear fractional differential equations (1)-(4). By using Green's function, the Banach contraction principle and Krasnoselskii's fixed point theorem, we establish some new existence, uniqueness theorems of solutions for multi-point boundary value problems for a coupled system of nonlinear fractional differential equations (1)-(4).

Competing interests

The authors declare that they have no competing interests.

Authors' contributions

The authors declare that the study was realized in collaboration with the same responsibility. All authors read and approved the final manuscript.

Acknowledgements

The research was supported by the Youth Science Foundation of China (11201272) and the Science Foundation of Shanxi Province (2010021002-1).

References

1. Diethelm, K: The Analysis of Fractional Differential Equations. Lecture Notes in Math. (2010)
2. Kilbas, AA, Srivastava, HM, Trujillo, JJ: Theory and Applications of Fractional Differential Equations. North-Holland Mathematics Studies, vol. 204. Elsevier, Amsterdam (2006)
3. Miller, KS, Ross, B: An Introduction to the Fractional Calculus and Differential Equations. Wiley, New York (1993)
4. Podlubny, I: Fractional Differential Equations. Academic Press, San Diego (1999)
5. Tarasov, VE: Fractional Dynamics: Application of Fractional Calculus to Dynamics of Particles, Fields and Media. Springer, Berlin (2010)
6. Zhang, S: Existence of positive solution for some class of nonlinear fractional differential equations. J. Math. Anal. Appl. **278**, 136-148 (2003)
7. Benchohra, M, Henderson, J, Ntouyas, SK, Ouahab, A: Existence results for fractional order functional differential equations with infinite delay. J. Math. Anal. Appl. **338**, 1340-1350 (2008)
8. Ahmad, B, Nieto, JJ: Existence results for a coupled system of nonlinear fractional differential equations with three-point boundary conditions. Comput. Math. Appl. **58**, 1838-1843 (2009)
9. Chang, YK, Nieto, JJ: Some new existence results for fractional differential inclusions with boundary conditions. Math. Comput. Model. **49**, 605-609 (2009)
10. Agarwal, RP, Benchohra, M, Hamani, S: A survey on existence results for boundary value problems of nonlinear fractional differential equations and inclusions. Acta Appl. Math. **109**, 973-1033 (2010)
11. Zhou, Y, Jiao, F: Nonlocal Cauchy problem for fractional evolution equations. Nonlinear Anal., Real World Appl. **11**, 4465-4475 (2010)
12. Wang, J, Zhou, Y: A class of fractional evolution equations and optimal controls. Nonlinear Anal., Real World Appl. **12**, 262-272 (2011)
13. Wang, G, Zhang, L, Song, G: Systems of first order impulsive functional differential equations with deviating arguments and nonlinear boundary conditions. Nonlinear Anal. **74**, 974-982 (2011)
14. Sun, SR, Li, QP, Li, Y: Existence and uniqueness of solutions for a coupled system of multi-term nonlinear fractional differential equations. Comput. Math. Appl. **64**, 3310-3320 (2012)
15. Mophou, GM: Existence and uniqueness of mild solutions to impulsive fractional differential equations. Nonlinear Anal. **72**, 1604-1615 (2010)
16. Benchohra, M, Slimani, BA: Existence and uniqueness of solutions to impulsive fractional differential equations. Electron. J. Differ. Equ. **2009**, 10 (2009)
17. Ahmad, B, Alsaedi, A: Existence and uniqueness of solutions for coupled systems of higher-order nonlinear fractional differential equations. Fixed Point Theory Appl. **2010**, Article ID 364560 (2010)
18. Zhai, CB, Hao, MR: Fixed point theorems for mixed monotone operators with perturbation and applications to fractional differential equation boundary value problems. Nonlinear Anal. **75**, 2542-2551 (2012)
19. Yang, C, Zhai, CB: Uniqueness of positive solutions for a fractional differential equation via a fixed point theorem of a sum operator. Electron. J. Differ. Equ. **2012**, 70 (2012)
20. Zhai, CB, Yan, WP, Yang, C: A sum operator method for the existence and uniqueness of positive solutions to Riemann-Liouville fractional differential equation boundary value problems. Commun. Nonlinear Sci. Numer. Simul. **18**, 858-866 (2013)
21. Su, X: Boundary value problem for a coupled system of nonlinear fractional differential equations. Appl. Math. Lett. **22**, 64-69 (2009)
22. Bai, CZ, Fang, JX: The existence of a positive solution for a singular coupled system of a nonlinear fractional differential equations. Appl. Math. Comput. **150**(3), 611-621 (2004)
23. Goodrich, CS: Existence of a positive solution to systems of differential equations of fractional order. Comput. Math. Appl. **62**, 1251-1268 (2011)
24. Dunninger, D, Wang, H: Existence and multiplicity of positive solutions for elliptic systems. Nonlinear Anal. **29**, 1051-1060 (1997)
25. Krasnoselskii, MA: Topological Methods in the Theory of Nonlinear Integral Equations. Pergamon, New York (1964)

4

The existence of positive mild solutions for fractional differential evolution equations with nonlocal conditions of order $1 < \alpha < 2$

Xiancun Wang and Xiaobao Shu*

*Correspondence:
sxb0221@163.com
Department of Mathematics, Hunan University, Changsha, 410082, P.R. China

Abstract

In this paper, we investigate the existence of positive mild solutions of fractional evolution equations with nonlocal conditions of order $1 < \alpha < 2$ by using Schauder's fixed point theorem and Krasnoselskii's fixed point theorem.

Keywords: positive mild solutions; fractional evolution equations; solution operators; accretive operator; fixed point theorem

1 Introduction

The differential equations involving fractional derivatives have recently been studied extensively, because they have proved to be valuable in various fields of science and engineering. Indeed, we can find numerous applications in electrochemistry, electromagnetism, biology, and hydrogeology. For example space-fractional diffusion equations have been used in groundwater hydrology to model the transport of passive-tracers carried by fluid flow in a porous medium [1, 2], or to model activator-inhibitor dynamics with anomalous diffusion [3]. In particular, there has been a significant development in fractional evolution equations. The existence of solutions for fractional evolution equations has been studied by many authors during recent years. Many excellent results were obtained in this field [3–11].

Shu and Wang [4] studied the existence of mild solutions for the fractional differential equations with nonlocal conditions in a Banach space X:

$$\begin{cases} {}^cD_t^\alpha u(t) = Au(t) + f(t, u(t)) + \int_0^t q(t-s)g(s, u(s))\,ds, & t \in [0, T], \\ u(0) + m(u) = u_0 \in X, \qquad u'(0) + n(u) = u_1 \in X, \end{cases}$$

where D_t^α is the Caputo fractional derivative of order $1 < \alpha < 2$. By using the contraction mapping principle and Krasnoselskii's fixed point theorem, they obtained the existence of solutions for the equation.

Mu [5] considered the existence of mild solutions to the following semilinear fractional evolution equations:

$$\begin{cases} D^\alpha u(t) + Au(t) = f(t, u(t)), & t \in I, \\ u(0) = x_0 \in X, \end{cases} \tag{1.1}$$

where D^{α} is the Caputo fractional derivative of order $0 < \alpha < 1$. $-A$ is the infinitesimal generator of an analytic semigroup $T(t) = e^{At}|_{t\geq 0}$, and $f : I \times X \to X$ is continuous.

As is well known, a mild solution to system (1.1) satisfies the following equation:

$$u(t) = S_{\alpha}(t)x_0 + \int_0^t T_{\alpha}(t-s)f\big(s, u(s)\big)\,ds.$$

Since $0 < \alpha < 1$, we can combine the probability density function and semigroup theory to describe the corresponding solution operators $S_{\alpha}(t)$, $T_{\alpha}(t)$ (see [12]), where

$$T_{\alpha}(t) = \alpha\int_0^{\infty} \theta\phi_{\alpha}(\theta)t^{\alpha-1}T\big(t^{\alpha}\theta\big)\,d\theta, \qquad S_{\alpha}(t) = \int_0^{\infty} \phi_{\alpha}(\theta)T\big(t^{\alpha}\theta\big)\,d\theta,$$

where $\phi_{\alpha}(\theta)$ is the probability density function defined on $(0, \infty)$ such that its Laplace transform is

$$\int_0^{\infty} e^{-\theta x}\phi_{\alpha}(\theta)\,d\theta = \sum_{j=0}^{\infty} \frac{(-x)^j}{\Gamma(1+\alpha j)}, \qquad x > 0.$$

Thus, it is obvious that whatever operator A is, $T(t) = e^{At}$, $T_{\alpha}(t)$, $S_{\alpha}(t)$ are always positive if $-A$ is the infinitesimal generator of an analytic semigroup $T(t)$.

In [6], Mu and Fan investigate the existence and uniqueness of positive mild solutions of the following periodic boundary value problem for the fractional evolution equations in an ordered Banach space in [6]:

$$\begin{cases} D^{\alpha}u(t) + Au(t) = f(t, u(t)), & t \in [0, \omega], \\ u(0) = u(\omega), \end{cases}$$

where D^{α} is the Caputo fractional derivative of order $0 < \alpha < 1$. First, using the probability density function and semigroup theory the authors established the mild solutions of the associated linear fractional evolution equations. Then they estimated the spectral radius of resolvent operators accurately. With the aid of the estimation, the existence and uniqueness results of positive mild solutions are obtained by using the monotone iterative technique.

However, for $1 < \alpha < 2$, the existence of a positive mild solution for a fractional differential evolution equation still is an untreated topic in the literature. On the one hand, whether $K_{\alpha}(t)$, $T_{\alpha}(t)$ is positive is still unknown, meanwhile $S_{\alpha}(t)$ is not positive; see Remark 2.3. On the other hand, we do not know if we can still use the probability density function together with semigroup theory to describe the corresponding solution operators $S_{\alpha}(t)$, $K_{\alpha}(t)$, and $T_{\alpha}(t)$. So, it is difficult for us to investigate the positive mild solutions of fractional differential evolution equation in the way as [5] did. In [11], Bai and Lü investigate the positive solutions for nonlinear fractional differential equation by means of some fixed point theorem on cone. Given these, using the fixed point theorem on a cone, we investigate the existence of positive mild solution of such a fractional order differential equation.

Motivated by the above, in this paper, we study the existence of positive mild solutions of the following fractional differential evolution equation in an ordered Banach space X

with positive cone P:

$$\begin{cases} D_t^\alpha u(t) = Au(t) + f(t,u(t)) + \int_0^t q(t-s)g(s,u(s))\,ds, & t \in J = [0,T], \\ u(0) + m(u) = u_0 \in X, \qquad u'(0) + n(u) = u_1 \in X, \end{cases} \tag{1.2}$$

where D^α is the Caputo fractional derivatives of $1 < \alpha < 2$, A is a sectorial operator of type (M,θ,α,μ) defined from domain $D(A) \subset X$ into X, the nonlinear map $f, g : J \times X \to P$ are continuous, $q : [0,T] \to R^+$ is an integrable function on $[0,T]$, and $q = \max_{t\in[0,T]} \int_0^t |q(t-s)|\,ds$. $m, n : X \to X$ are also continuous.

The rest of this paper is organized as follows. In Section 2, we first present some basic definitions and theorems to be used. Then we investigate the properties of solution operators by means of classical Mittag-Leffler function. The main results of this article are given in Section 3. In Section 4, an example is considered to illustrate the applications of the main results presented in Section 3. Finally, in Section 5, we make a conclusion of this paper.

2 Preliminaries

In this part, we will introduce some basic definitions and theorems that are used throughout this paper.

2.1 Definitions and theorems

If X is an ordered Banach space with the norm $\|\cdot\|$. Let P be a cone in which defined a partial ordering in X by $x \le y$ if and only if $y - x \in P$. P is said to be normal if there exists a positive constant N such that $\theta \le x \le y$ implies $\|x\| \le N\|y\|$, where θ denotes the zero element of X, and the smallest N is called the normal constant of P. P is called solid if its interior P is nonempty. If $x \le y$ and $x \ne y$, we write $x < y$. If P is solid and $y - x \in \dot{P}$, we write $x \ll y$. For details on cone theory, see [13].

Besides, if X is an ordered Banach space, then $C(J,X)$ is also an ordered Banach space with the partial order $\le$ induced by the positive cone $K = \{x \in C(J,X) : x(t) \ge \theta, \text{for all } t \in J\}$.

Throughout this paper, we assume that P is a positive cone of ordered Banach space X, then $K = \{x \in C(J,X) : x(t) \ge \theta, \text{for all } t \in J\}$ also is the positive cone of Banach space $C(J,X)$.

In general, the Mittag-Leffler function is defined as [14]:

$$E_{\alpha,\beta}(z) = \sum_{k=0}^{\infty} \frac{z^k}{\Gamma(\alpha k + \beta)} = \frac{1}{2\pi i}\int_{H_\alpha} e^\mu \frac{\mu^{\alpha-\beta}}{\mu^\alpha - z}\,d\mu, \quad \alpha,\beta > 0, z \in C,$$

where H_α denotes a Hankel path, a contour starting and ending at $-\infty$, and encircling the disc $|\mu| \le |z|^{\frac{1}{\alpha}}$ counterclockwise.

Theorem 2.1 ([15]) *Let A be a densely defined operator in X satisfying the following conditions:*

(1) *For some $0 < \theta < \pi/2$, $\mu + S_\theta = \{\mu + \lambda : \lambda \in C, |\operatorname{Arg}(-\lambda)| < \theta\}$.*

(2) *There exists a constant M such that*

$$\|(\lambda I - A)^{-1}\| \le \frac{M}{|\lambda - \mu|}, \quad \lambda \notin \mu + S_\theta.$$

Then A is the infinitesimal generator of a semigroup $T(t)$ satisfying $\|T(t)\| \leq C$. Moreover, $T(t) = \frac{1}{2\pi i}\int_c e^{\lambda t}R(\lambda, A)\, d\lambda$ with c being a suitable path $\lambda \notin \mu + S_\theta$ for $\lambda \in c$.

Definition 2.1 ([4]) Let $A : D(A) \subseteq X \to X$ be a closed linear operator. A is said to be a sectorial operator of type (M, θ, α, μ) if there exist $0 < \theta < \pi/2$, $M > 0$, and $\mu \in R$ such that the α-resolvent of A exists outside the sector

$$\mu + S_\theta = \left\{\mu + \lambda^\alpha : \alpha \in C, \left|\mathrm{Arg}\left(-\lambda^\alpha\right)\right| < \theta\right\}$$

and

$$\left\|\left(\lambda^\alpha I - A\right)^{-1}\right\| \leq \frac{M}{|\lambda^\alpha - \mu|}, \quad \lambda \notin \mu + S_\theta.$$

Remark 2.1 ([4]) If A is a sectorial operator of type (M, θ, α, μ), then it is not difficult to see that A is the infinitesimal generator of a α-resolvent family $\{T_\alpha(t)\}_{t \geq 0}$ in a Banach space, where $T_\alpha(t) = \frac{1}{2\pi i}\int_C e^{\lambda t}R(\lambda^\alpha, A)\, d\lambda$.

Definition 2.2 ([4]) A function $x \in C([0, T], X)$ is called a mild solution of problem (1.2) if it satisfies the operator equation

$$\begin{aligned} u(t) &= S_\alpha(t)\left[u_0 - m(u)\right] + K_\alpha(t)\left[u_1 - n(u)\right] \\ &\quad + \int_0^t T_\alpha(t-s)\left[f\left(s, u(s)\right) + \int_0^s q(s-\tau)g\left(\tau, u(\tau)\right) d\tau\right] ds. \end{aligned}$$

Definition 2.3 ([5]) Let $R(t)_{(t \geq 0)}$ be an α-resolvent solution operator in X. If $R(t)x \geq \theta$ for every $x \geq \theta$, $x \in X$, and $t \geq 0$, then $R(t)_{(t \geq 0)}$ is called to be positive.

Definition 2.4 ([7]) Let $A : D(A) \to X$ be a linear operator. Operator A is said to be non-negative if and only if it satisfies both of the following conditions:

(1) There exists $K \geq 0$ such that for every value of $\lambda > 0$ and every $u \in D(A)$, we have

$$\lambda\|u\| \leq K\|\lambda u + Au\|.$$

(2) $R(\lambda I + A) = X$ for every value of $\lambda > 0$.

Definition 2.5 ([7]) If A is a linear operator and satisfies condition (1) in Definition 2.4 for $K = 1$, then A is said to be accretive. In addition, A is said to be m-accretive if condition (2) is also satisfied.

Definition 2.6 Operator A is a sectorial accretive operator of type (M, θ, α, μ) if and only if A is a sectorial operator of type (M, θ, α, μ) and A is accretive.

Remark 2.2 ([7]) Assume that A is a Hilbert space with inner product $(\cdot;\cdot)$. Then the necessary and sufficient condition for A to be accretive is $\mathrm{Re}(Au, u) \geq 0$ for every $u \in D(A)$. Particularly, if X is a real Hilbert space and A is positive, then we obtain $(Au, u) \geq 0$ for every $u \in D(A)$. Note that an ordered Banach space is a real space, implying that if X is an ordered Banach space and A is accretive, then $(Au, u) \geq 0$ for every $u \in D(A)$.

Now, we state two well-known fixed point theorems, which are needed to prove our main results.

Theorem 2.2 (Schauder's fixed point theorem) *Let Y be a nonempty, closed, bounded, and convex subset of a Banach space X, and suppose that $T : Y \to Y$ is a compact operator. Then T has at least one fixed point in Y.*

Theorem 2.3 (Krasnoselskii's fixed point theorem) *Let M be a closed convex and nonempty subset of a Banach space X. Let A, B be the operators such that* (i) $Ax + By \in M$ *whenever $x, y \in M$;* (ii) *A is compact and continuous;* (iii) *B is a contraction mapping. Then there exists $z \in M$ such that $z = Az + Bz$.*

2.2 Properties of solution operators

Lemma 2.1 *If A is a sectorial operator of type (M, θ, α, μ), then we have*

$$S_\alpha(t) = \frac{1}{2\pi i}\int_c e^{\lambda t}\lambda^{\alpha-1}R(\lambda^\alpha, A)\,d\lambda = E_{\alpha,1}(At^\alpha) = \sum_{k=0}^{\infty}\frac{(At^\alpha)^k}{\Gamma(1+\alpha k)},$$

$$T_\alpha(t) = \frac{1}{2\pi i}\int_c e^{\lambda t}R(\lambda^\alpha, A)\,d\lambda = t^{\alpha-1}E_{\alpha,\alpha}(At^\alpha) = t^{\alpha-1}\sum_{k=0}^{\infty}\frac{(At^\alpha)^k}{\Gamma(\alpha+\alpha k)}$$

and

$$K_\alpha(t) = \frac{1}{2\pi i}\int_c e^{\lambda t}\lambda^{\alpha-2}R(\lambda^\alpha, A)\,d\lambda = tE_{\alpha,2}(At^\alpha) = t\sum_{k=0}^{\infty}\frac{(At^\alpha)^k}{\Gamma(2+\alpha k)}.$$

Proof We note that

$$\frac{1}{\Gamma(s)} = \frac{1}{2\pi i}\int_c e^{\zeta}\zeta^{-s}\,d\zeta, \quad \mathrm{Re}\, s > 0. \tag{2.1}$$

Applying the transformation $\zeta = \eta^{\frac{1}{\alpha}}$ to (2.1) gives

$$\frac{1}{\Gamma(s)} = \frac{1}{2\pi\alpha i}\int_c e^{\eta^{\frac{1}{\alpha}}}\eta^{-\frac{s}{\alpha}+\frac{1}{\alpha}-1}\,d\eta.$$

Since A is a sectorial operator of type (M, θ, α, μ), it follows from the inequality

$$\|(\lambda^\alpha I - A)^{-1}\| \le \frac{M}{|\lambda^\alpha - \mu|}$$

that A is the infinitesimal generator of α-resolvent families $\{S_\alpha(t)\}_{t\ge0}$, $\{K_\alpha(t)\}_{t\ge0}$, and $\{T_\alpha(t)\}_{t\ge0}$ (see [4]). Hence, using the transformation $t^{-\alpha}\eta = \lambda^\alpha$ (*i.e.*, $t^{-\alpha}\,d\eta = \alpha\lambda^{\alpha-1}\,d\lambda$ and $e^{\eta^{\frac{1}{\alpha}}} = e^{t\lambda}$), we obtain

$$E_{\alpha,1}(At^\alpha) = \sum_{k=0}^{\infty}\frac{(At^\alpha)^k}{\Gamma(1+\alpha k)}$$
$$= \frac{1}{2\pi\alpha i}\sum_{k=0}^{\infty}\left\{\int_c e^{\eta^{\frac{1}{\alpha}}}\eta^{-k-1}\,d\eta\right\}(At^\alpha)^k$$

$$= \frac{1}{2\pi\alpha i}\int_c e^{\eta^{\frac{1}{\alpha}}}\eta^{-1}\left\{\sum_{k=0}^{\infty}\left(At^{\alpha}\eta^{-1}\right)^k\right\}d\eta$$
$$= \frac{1}{2\pi\alpha i}\int_c e^{\eta^{\frac{1}{\alpha}}}t^{-\alpha}\left(t^{-\alpha}\eta I - A\right)^{-1}d\eta$$
$$= \frac{1}{2\pi i}\int_c e^{\lambda t}\lambda^{\alpha-1}R(\lambda^{\alpha},A)\,d\lambda$$
$$= S_{\alpha}(t).$$

Similarly, we show that

$$t^{\alpha-1}E_{\alpha,\alpha}\left(At^{\alpha}\right) = t^{\alpha-1}\sum_{k=0}^{\infty}\frac{(At^{\alpha})^k}{\Gamma(\alpha+\alpha k)}$$
$$= \frac{t^{\alpha-1}}{2\pi\alpha i}\sum_{k=0}^{\infty}\left\{\int_c e^{\eta^{\frac{1}{\alpha}}}\eta^{-k+\frac{1}{\alpha}-2}\,d\eta\right\}\left(At^{\alpha}\right)^k$$
$$= \frac{t^{\alpha-1}}{2\pi\alpha i}\int_c e^{\eta^{\frac{1}{\alpha}}}\eta^{\frac{1}{\alpha}-2}\left\{\sum_{k=0}^{\infty}\left(At^{\alpha}\eta^{-1}\right)^k\right\}d\eta$$
$$= \frac{1}{2\pi\alpha i}\int_c e^{\eta^{\frac{1}{\alpha}}}t^{-1}\eta^{\frac{1}{\alpha}-1}\left(t^{-\alpha}\eta I - A\right)^{-1}d\eta$$
$$= \frac{1}{2\pi i}\int_c e^{\lambda t}R(\lambda^{\alpha},A)\,d\lambda$$
$$= T_{\alpha}(t)$$

and

$$tE_{\alpha,2}\left(At^{\alpha}\right) = t\sum_{k=0}^{\infty}\frac{(At^{\alpha})^k}{\Gamma(2+\alpha k)}$$
$$= \frac{t}{2\pi\alpha i}\sum_{k=0}^{\infty}\left\{\int_c e^{\eta^{\frac{1}{\alpha}}}\eta^{-k-\frac{1}{\alpha}-1}\,d\eta\right\}\left(At^{\alpha}\right)^k$$
$$= \frac{t}{2\pi\alpha i}\int_c e^{\eta^{\frac{1}{\alpha}}}\eta^{-\frac{1}{\alpha}-1}\left\{\sum_{k=0}^{\infty}\left(At^{\alpha}\eta^{-1}\right)^k\right\}d\eta$$
$$= \frac{1}{2\pi\alpha i}\int_c e^{\eta^{\frac{1}{\alpha}}}t^{-\alpha+1}\eta^{-\frac{1}{\alpha}}\left(t^{-\alpha}\eta I - A\right)^{-1}d\eta$$
$$= \frac{1}{2\pi i}\int_c \lambda^{\alpha-2}e^{\lambda t}R(\lambda^{\alpha},A)\,d\lambda$$
$$= K_{\alpha}(t).$$

□

Remark 2.3 The Mittag-Leffler function $E_{\alpha,1}(x)$ is well known to have a finite number of real zeros in the range $1 < \alpha < 2$. Reference [16] deduces that operator $S_{\alpha}(t)$ is non-positive by Lemma 2.1.

Remark 2.4 It follows from Definition 2.6 and Remark 2.2 that if X is an ordered Banach space and A is a sectorial accretive operator of type (M,θ,α,μ), then the α-resolvent families $\{T_{\alpha}(t)\}_{t\geq 0}$, $\{S_{\alpha}(t)\}_{t\geq 0}$, and $\{K_{\alpha}(t)\}_{t\geq 0}$ are all positive.

Theorem 2.4 ([4]) *Let A be a sectorial operator of type (M,θ,α,μ). Then the following estimates on $\|S_\alpha(t)\|$ hold.*

(i) *Suppose $\mu \geq 0$. Given $\phi \in (0,\pi)$, we have*

$$\|S_\alpha(t)\| \leq \frac{K_1(\theta,\phi)Me^{[K_1(\theta,\phi)(1+\mu t^\alpha)]^{\frac{1}{\alpha}}}[(1+\frac{\sin\phi}{\sin(\phi-\theta)})^{\frac{1}{\alpha}}-1]}{\pi\sin^{1+\frac{1}{\alpha}}\theta}(1+\mu t^\alpha) + \frac{\Gamma(\alpha)M}{\pi(1+\mu t^\alpha)|\cos\frac{\pi-\phi}{\alpha}|^\alpha\sin\theta\sin\phi},$$

for $t>0$, where $K_1(\theta,\phi)=\max\{1,\frac{\sin\theta}{\sin(\theta-\phi)}\}$.

(ii) *Suppose $\mu < 0$. Given $\phi \in (0,\pi)$, we have*

$$\|S_\alpha(t)\| \leq \left(\frac{eM[(1+\sin\phi)^{\frac{1}{\alpha}}-1]}{\pi|\cos\phi|^{1+\frac{1}{\alpha}}} + \frac{\Gamma(\alpha)M}{\pi|\cos\phi||\cos\frac{\pi-\phi}{\alpha}|^\alpha}\right)\frac{1}{1+|\mu|t^\alpha},$$

for $t>0$.

Theorem 2.5 ([4]) *Let A be a sectorial operator of type (M,θ,α,μ). Then the following estimates on $\|T_\alpha(t)\|$, $\|K_\alpha(t)\|$ hold.*

(i) *Suppose $\mu \geq 0$. Given $\phi \in (0,\pi)$, we have*

$$\|T_\alpha(t)\| \leq \frac{Me^{[K_1(\theta,\phi)(1+\mu t^\alpha)]^{\frac{1}{\alpha}}}[(1+\frac{\sin\phi}{\sin(\theta-\phi)})^{\frac{1}{\alpha}}-1]}{\pi\sin\theta}(1+\mu t^\alpha)^{\frac{1}{\alpha}}t^{\alpha-1} + \frac{Mt^{\alpha-1}}{\pi(1+\mu t^\alpha)|\cos\frac{\pi-\phi}{\alpha}|^\alpha\sin\theta\sin\phi},$$

$$\|K_\alpha(t)\| \leq \frac{MK_1(\vartheta,\phi)e^{[K_1(\theta,\phi)(1+\mu t^\alpha)]^{\frac{1}{\alpha}}}[(1+\frac{\sin\phi}{\sin(\theta-\phi)})^{\frac{1}{\alpha}}-1]}{\pi\sin^{\frac{\alpha+2}{\alpha}}\theta}(1+\mu t^\alpha)^{\frac{\alpha-1}{\alpha}}t^{\alpha-1} + \frac{M\alpha\Gamma(\alpha)}{\pi(1+\mu t^\alpha)|\cos\frac{\pi-\phi}{\alpha}|^\alpha\sin\theta\sin\phi},$$

for $t>0$, where $K_1(\theta,\phi)=\max\{1,\frac{\sin\theta}{\sin(\theta-\phi)}\}$.

(ii) *Suppose $\mu < 0$. Given $\phi \in (0,\pi)$, we have*

$$\|T_\alpha(t)\| \leq \left(\frac{eM[(1+\sin\phi)^{\frac{1}{\alpha}}-1]}{\pi|\cos\phi|} + \frac{M}{\pi|\cos\phi||\cos\frac{\pi-\phi}{\alpha}|}\right)\frac{t^{\alpha-1}}{1+|\mu|t^\alpha},$$

$$\|K_\alpha(t)\| \leq \left(\frac{eMt[(1+\sin\phi)^{\frac{1}{\alpha}}-1]}{\pi|\cos\phi|^{1+\frac{2}{\alpha}}} + \frac{\alpha\Gamma(\alpha)M}{\pi|\cos\phi||\cos\frac{\pi-\phi}{\alpha}|}\right)\frac{1}{1+|\mu|t^\alpha},$$

for $t>0$.

3 Main result

Because of the estimation on $\|S_\alpha(t)\|$, $\|K_\alpha(t)\|$, and $\|T_\alpha(t)\|$ in Theorem 2.4 and Theorem 2.5, it is easy to see they are bounded. So we make the following assumptions:

(H_1): There exist positive numbers $\widetilde{M}$ such that for any $t \in J$, we have

$$\sup_{t\in J}\|S_\alpha(t)\| \le \widetilde{M}, \qquad \sup_{t\in J}\|K_\alpha(t)\| \le \widetilde{M}, \qquad \sup_{t\in J}\|T_\alpha(t)\| \le \widetilde{M}.$$

(H_2): The linear operator A is a sectorial accretive operator of type (M,θ,α,μ) and generates compact α-resolvent families $\{T_\alpha(t)\}_{t\ge 0}$, $\{S_\alpha(t)\}_{t\ge 0}$, and $\{K_\alpha(t)\}_{t\ge 0}$.

(H_3): $f,g : J\times X \to P$ is jointly continuous and for any $k>0$ there exist positive functions $\mu_k, \nu_k \in L([0,T],R^+)$ such that

$$\sup_{\|u\|\le k}\|f(t,u)\| \le \mu_k(t), \qquad \sup_{\|u\|\le k}\|g(t,u)\| \le \nu_k(t).$$

(H_4): $u_0 - m(u), u_1 - n(u) \in C(X,P)$ and there exist positive numbers a, b, c, d such that

$$\|m(u)\| \le a\|u\| + b, \qquad \|n(u)\| \le c\|u\| + d, \quad \text{for all } u \in X.$$

Theorem 3.1 *Assume that* (H_1)-(H_4) *hold, if* $\widetilde{M}(a+c)<1$, *then problem* (1.2) *has at least one positive mild solution on* J.

Proof Using Definition 2.2, the mild solution of nonlocal fractional differential evolution equation (1.2) can be expressed as

$$u(t) = S_\alpha(t)\big[u_0 - m(u)\big] + K_\alpha(t)\big[u_1 - n(u)\big] + \int_0^t T_\alpha(t-s)\left[f\big(s,u(s)\big) + \int_0^s q(s-\tau)g\big(\tau,u(\tau)\big)\,d\tau\right] ds.$$

Choose

$$r \ge \frac{\widetilde{M}(T\|\mu_r\|_{L^\infty(J,R^+)} + Tq\|\nu_r\|_{L^\infty(J,R^+)} + \|u_0\| + \|u_1\| + b + d)}{1-\widetilde{M}(a+c)}$$

and consider $Q = \{u\in K : \|u\| \le r\}$. Define the operator $\Gamma : Q \to C(J,X)$ by

$$(\Gamma u)(t) = S_\alpha(t)\big[u_0 - m(u)\big] + K_\alpha(t)\big[u_1 - n(u)\big] + \int_0^t T_\alpha(t-s)\left[f\big(s,u(s)\big) + \int_0^s q(s-\tau)g\big(\tau,u(\tau)\big)\,d\tau\right] ds.$$

Step 1: we prove that $\Gamma Q \subseteq Q$.

For any $u \in Q$, based on assumptions (H_3) and (H_4), for $t \in J$, we have

$$u_0 - m(u) \ge \theta, \qquad u_1 - n(u) \ge \theta, \tag{3.1}$$

$$f\big(t,u(t)\big) \ge \theta, \qquad \int_0^t q(t-s)g\big(s,u(s)\big)\,ds \ge \theta. \tag{3.2}$$

In view of (H_2), we note that A is a sectorial accretive operator of type (M,θ,α,μ) and generates compact and positive α-resolvent families $\{S_\alpha(t)\}_{t\ge 0}$, $\{K_\alpha(t)\}_{t\ge 0}$, and $\{T_\alpha(t)\}_{t\ge 0}$. Then we have

$$S_\alpha(t)\big[u_0 - m(u)\big] \ge \theta, \qquad K_\alpha(t)\big[u_1 - n(u)\big] \ge \theta, \tag{3.3}$$

$$\int_0^t T_\alpha(t-s)\left[f(s,u(s)) + \int_0^s q(a-\tau)g(\tau,u(\tau))\,d\tau\right]ds \geq \theta. \tag{3.4}$$

Consequently, we obtain

$$(\Gamma u) \geq \theta, \quad \text{for } u \in \Omega. \tag{3.5}$$

Also we have

$$\begin{aligned}
\|(\Gamma u)(t)\| &\leq \|S_\alpha(t)\| \cdot \|u_0 - m(u)\| + \|K_\alpha(t)\| \cdot \|u_1 - n(u)\| \\
&\quad + \int_0^t \|T_\alpha(t-s)\| \cdot \left\|f(s,x(s)) + \int_0^s q(s-\tau)g(\tau,u(\tau))\,d\tau\right\| ds \\
&\leq \widetilde{M}\Bigg(\|u_0\| + \|m(u)\| + \|u_1\| + \|n(u)\| + T\|f(t,u(t))\| \\
&\quad + T\int_0^t |q(t-s)| \cdot \|g(s,u(s))\|\,ds\Bigg) \\
&\leq \widetilde{M}\big(\|u_0\| + a\|u\| + b + \|u_1\| + c\|u\| + d \\
&\quad + T\|\mu_r\|_{L^\infty(J,R^+)} + Tq\|\nu_r\|_{L^\infty(J,R^+)}\big) \\
&\leq r.
\end{aligned} \tag{3.6}$$

Combining (3.5) with (3.6), we obtain

$$\Gamma Q \subseteq Q, \quad \text{for all } u \in Q.$$

Step 2: continuity of F.

Let $\{u_n\}$ be a sequence in Q such that $\|u_n - \overline{u}\| \to 0$. Noting that f, g, m, n are continuous, as $n \to \infty$ we have

$$m(u_n) \to m(u), \qquad n(u_n) \to n(u), \tag{3.7}$$

$$f(t,u_n(t)) \to f(t,u(t)), \qquad g(t,u_n(t)) \to g(t,u(t)). \tag{3.8}$$

For all $t \in J$, we get

$$\begin{aligned}
\|(\Gamma u_n)(t) - (\Gamma u)(t)\| &\leq \|S_\alpha(t)\| \cdot \|m(u_n) - m(u)\| + \|K_\alpha(t)\| \cdot \|n(u_n) - n(u)\| \\
&\quad + \int_0^t \|T_\alpha(t-s)\| \cdot \Big[\|f(s,u_n(s)) - f(s,u(s))\| \\
&\quad + \int_0^s |q(s-\tau)| \cdot \|g(\tau,u_n(\tau)) - g(\tau,u(\tau))\|\,d\tau\Big] ds \\
&\leq \widetilde{M}\|m(u_n) - m(u)\| + \widetilde{M}\|n(u_n) - n(u)\| \\
&\quad + \widetilde{M}T\|f(t,u_n(t)) - f(t,u(t))\| \\
&\quad + \widetilde{M}Tq\|g(t,u_n(t)) - g(t,u(t))\|.
\end{aligned}$$

From (3.7) and (3.8), we obtain $\lim_{n\to\infty}(\Gamma u_n)(t) = (\Gamma u)(t)$, that is, the operator Γ is continuous.

Step 3: compactness of Γ.

To this end, we use the Arzela-Ascoli theorem. We prove that $\{(\Gamma u)(t) : u \in Q\}$ is relatively compact in X. First, we prove that $\{(\Gamma u)(t) : u \in Q\}$ is uniformly bounded. We have

$$\begin{aligned}
\|(\Gamma u)(t)\| &\leq \|S_\alpha(t)\| \cdot \|u_0 - m(u)\| + \|K_\alpha(t)\| \cdot \|u_1 - n(u)\| \\
&\quad + \int_0^t \|T_\alpha(t-s)\| \cdot \left[\|f(s,u(s))\| + \int_0^s |q(\tau - s)| \cdot \|g(\tau, u(\tau))\| \, d\tau \right] ds \\
&\leq \widetilde{M}\Big(\|u_0\| + \|m(u)\| + \|u_1\| + \|n(u)\| + T\|f(t,u(t))\| \\
&\quad + T \int_0^t |q(t-s)| \|g(s,u(s))\| \, ds \Big) \\
&\leq \widetilde{M}\big(\|u_0\| + a\|u\| + b + \|u_1\| + c\|u\| + d \\
&\quad + T\|\mu_r\|_{L^\infty(J,R^+)} + Tq\|\nu_r\|_{L^\infty(J,R^+)} \big) \\
&\leq r \\
&\leq \infty.
\end{aligned}$$

Now, let us prove that $\Gamma(Q)$ is equicontinuous. The function $\{(\Gamma u)(t) : u \in Q\}$ are equicontinuous at $t = 0$. For $0 < t_1 < t_2 \leq T$ and $u \in Q$, we have

$$\begin{aligned}
\|(\Gamma u)(t_2) - (\Gamma u)(t_1)\| &\leq \|S_\alpha(t_2) - S_\alpha(t_1)\| \cdot \|u_0 - m(u)\| \\
&\quad + \|K_\alpha(t_2) - K_\alpha(t_1)\| \cdot \|u_1 - n(u)\| \\
&\quad + \int_0^{t_1} \|T_\alpha(t_2 - s) - T_\alpha(t_1 - s)\| \Big[\|f(s,u(s))\| \\
&\quad + \int_0^s |q(\tau - s)| \cdot \|g(\tau, u(\tau))\| \, d\tau \Big] ds \\
&\quad + \int_{t_1}^{t_2} \|T_\alpha(t_2 - s)\| \Big[\|f(s,u(s))\| \\
&\quad + \int_0^s |q(\tau - s)| \cdot \|g(\tau, u(\tau))\| \, d\tau \Big] ds \\
&\leq I_1 + I_2 + I_3,
\end{aligned}$$

where

$$\begin{aligned}
I_1 &= \|S_\alpha(t_2) - S_\alpha(t_1)\| \cdot \|u_0 - m(u)\| + \|K_\alpha(t_2) - K_\alpha(t_1)\| \cdot \|u_1 - n(u)\|, \\
I_2 &= \int_0^{t_1} \|T_\alpha(t_2 - s) - T_\alpha(t_1 - s)\| \cdot \left[\|f(s,u(s))\| + \int_0^s q(\tau - s)\|g(\tau,u(\tau))\| \, d\tau \right] ds, \\
I_3 &= \int_{t_1}^{t_2} \|T_\alpha(t_2 - s)\| \cdot \left[\|f(s,u(s))\| + \int_0^s |q(\tau - s)| \|g(\tau, u(\tau))\| \, d\tau \right] ds.
\end{aligned}$$

The continuity of functions $t \to \|S_\alpha(t)\|$, $t \to \|K_\alpha(t)\|$ for $t \in (0, T]$, allows us to conclude that $\lim_{t_1 \to t_2} I_1 = 0$. Indeed, we have

$$I_2 \leq \int_{t_1}^{t_2} \|T_\alpha(t_2 - s) - T_\alpha(t_1 - s)\| \cdot \left[\|\mu_r\|_{L^\infty(J,R^+)} + q\|\nu_r\|_{L^\infty(J,R^+)} \right] ds.$$

Therefore, the continuity of function $t \to \|T_\alpha(t)\|$ for $t \in (0, T]$ also allows us to conclude that $\lim_{t_1 \to t_2} I_2 = 0$. We have

$$I_3 \leq \int_{t_1}^{t_2} \|T_\alpha(t_2 - s)\| \left(\|\mu_r\|_{L^\infty(J,R^+)} + q\|\nu_r\|_{L^\infty(J,R^+)}\right) ds$$
$$\leq \widetilde{M}\left(\|\mu_r\|_{L^\infty(J,R^+)} + q\|\nu_r\|_{L^\infty(J,R^+)}\right)|t_2 - t_1|.$$

Consequently, $\lim_{t_1 \to t_2} I_3 = 0$.

Thus, for $t \in J$, $\{(\Gamma u)(t) : u \in Q\}$ is a family of equicontinuous function. We have proved that $\{\Gamma(Q)\}$ is relatively compact. Hence by the Arzela-Ascoli theorem, Γ is compact. Schauder's fixed point theorem allows us to conclude that Γ has at least one fixed point on J. As $(\Gamma u)(t) \geq \theta$ when $u \in Q$, problem (1.2) has at least one positive mild solution on J. □

In the following, we give an existence result in the case where (H_4) is not satisfied. We need the following assumptions.

(H_5): $u_0 - m(u), u_1 - n(u) : X \to P$ are continuous and bounded on X.
(H_6): There exist positive numbers l_1, l_2 such that for any $t \in J$, $u, v \in X$ we have

$$\|f(t,u) - f(t,v)\| \leq l_1\|u - v\|, \qquad \|g(t,u) - g(t,v)\| \leq l_2\|u - v\|.$$

Theorem 3.2 *Assume that* (H_1)-(H_3), (H_5)-(H_6) *hold. If* $\widetilde{M}T(l_1 + ql_2) < 1$, *then problem* (1.2) *has at least one positive mild solution on* J.

Proof Choose

$$R \geq \widetilde{M}\left(\|u_0 - m(u)\| + \|u_1 - n(u)\| + T\|\mu_R\|_{L^\infty(J,R^+)} + Tq\|\nu_R\|_{L^\infty(J,R^+)}\right)$$

and consider $\Omega = \{u \in K : \|u\| \leq R\}$. Define operators S, T on Ω by

$$(Su)(t) = S_\alpha(t)\left[u_0 - m(u)\right] + K_\alpha(t)\left[u_1 - n(u)\right],$$
$$(Tu)(t) = \int_0^t T_\alpha(t-s)\left[f(s,u(s)) + \int_0^s q(s-\tau)g(\tau,u(\tau))\,d\tau\right] ds.$$

Firstly, we prove that when $u, v \in \Omega$, we have $Su + Tv \in \Omega$.

Similar to (3.3) and (3.4), for $u, v \in \Omega$, we obtain

$$S_\alpha(t)\left[u_0 - m(u)\right] \geq \theta, \qquad K_\alpha(t)\left[u_1 - n(u)\right] \geq \theta,$$
$$\int_0^t T_\alpha(t-s)\left[f(s,v(s)) + \int_0^s q(s-\tau)g(\tau,v(\tau))\,d\tau\right] ds \geq \theta.$$

Consequently, we have

$$(Su)(t) + (Tv)(t) \geq \theta, \quad \text{for } u, v \in \Omega. \tag{3.9}$$

Following from (H_1), (H_3), and (H_5), we also have

$$\begin{aligned}\left\|(Su)(t)+(Tv)(t)\right\| &\leq \left\|S_\alpha(t)\right\|\cdot\left\|u_0-m(u)\right\|+\left\|K_\alpha(t)\right\|\cdot\left\|u_1-n(u)\right\| \\ &\quad+\int_0^t\left\|T_\alpha(t-s)\right\|\cdot\bigg\|f\big(s,u(s)\big) \\ &\quad+\int_0^s q(s-\tau)g\big(\tau,u(\tau)\big)\,d\tau\bigg\|\,ds \\ &\leq \widetilde{M}\big(\left\|u_0-m(u)\right\|+\left\|u_1-n(u)\right\| \\ &\quad+T\|\mu_R\|_{L^\infty(J,R^+)}+Tq\|\nu_R\|_{L^\infty(J,R^+)}\big) \\ &\leq R.\end{aligned} \tag{3.10}$$

Combining (3.9) with (3.10), we have $Su+Tv\in\Omega$, for $u,v\in\Omega$.

Secondly, we prove that the operator T is a contraction.

For any $u,v\in\Omega$, we get

$$\begin{aligned}\left\|(Tu)(t)-(Tv)(t)\right\| &\leq \int_0^t\left\|T_\alpha(t-s)\right\|\cdot\left\|f\big(s,u(s)\big)-f\big(s,v(s)\big)\right\|ds \\ &\quad+\int_0^t\left\|T_\alpha(t-s)\right\|\cdot\bigg(\int_0^s|q(s-\tau)|\cdot\big\|g\big(\tau,u(\tau)\big) \\ &\quad-g\big(\tau,v(\tau)\big)\big\|\,d\tau\bigg)ds \\ &\leq \widetilde{M}Tl_1\left\|u(t)-v(t)\right\|+\widetilde{M}Tl_2q\left\|u(t)-v(t)\right\| \\ &\leq \widetilde{M}T(l_1+ql_2)\|u-v\|.\end{aligned}$$

Since $\widetilde{M}T(l_1+ql_2)<1$, operator T is a contraction.

Thirdly, we prove that S is continuous.

Let $u_n,u\in\Omega$, $\|u_n(t)-u(t)\|\to 0$ as $n\to\infty$. Noting that m, n are continuous, we have

$$m(u_n)\to m(u),\qquad n(u_n)\to n(u),\quad \text{as } n\to\infty. \tag{3.11}$$

Then

$$\begin{aligned}\left\|(Su_n)(t)-(Su)(t)\right\| &\leq \left\|S_\alpha(t)\right\|\cdot\left\|m(u_n)-m(u)\right\| \\ &\quad+\left\|K_\alpha(t)\right\|\cdot\left\|n(u_n)-n(u)\right\| \\ &\leq \widetilde{M}\big(\left\|m(u_n)-m(u)\right\|+\left\|n(u_n)-n(u)\right\|\big).\end{aligned}$$

Following from (3.11), we have $\lim_{n\to\infty}(Su_n)(t)=(Su)(t)$. That is, operator S is continuous.

Lastly, we prove S is compact.

To this end, we use the Ascoli-Arzela theorem. We prove that $\{(Su)(t):u\in\Omega\}$ is relatively compact for $t\in J$. For $u\in\Omega$, we have

$$\begin{aligned}\left\|(Su)(t)\right\| &\leq \left\|S_\alpha(t)\right\|\cdot\left\|u_0-m(u)\right\| \\ &\quad+\left\|K_\alpha(t)\right\|\cdot\left\|u_1-n(u)\right\|\end{aligned}$$

$$\leq \widetilde{M}\big(\|u_0 - m(u)\| + \|u_1 - n(u)\|\big)$$
$$< \infty.$$

This shows operator S is uniformly bounded.

Now, let us prove that $S(\Omega)$ is equicontinuous. Obviously, the function $(Su)(t)$ is equicontinuous at $t = 0$. For $0 < t_1 < t_2 \leq T$, $u \in \Omega$, we have

$$\|(Su)(t_2) - (Su)(t_1)\| \leq \|S_\alpha(t_2) - S_\alpha(t_1)\| \cdot \|u_0 - m(u)\| + \|K_\alpha(t_2) - K_\alpha(t_1)\| \cdot \|u_1 - n(u)\|.$$

In view of (H_5), $\|u_0 - m(u)\|$, $\|u_1 - n(u)\|$ are bounded, so the continuity of functions $t \to \|S_\alpha(t)\|$, $t \to \|K_\alpha(t)\|$ for $t \in (0, T]$, allows us to conclude that

$$\lim_{t_1 \to t_2} (Su)(t_1) = (Su)(t_2).$$

In short, we have proved that $\{S(\Omega)\}$ is relatively compact for $\{(Su)(t) : u \in \Omega\}$ is a family of equicontinuous function. Hence by the Arzela-Ascoli theorem, S is compact. As all the conditions of Krasnoselskii's fixed point theorem are satisfied, we conclude that Cauchy problem (1.2) has at least one mild solution on J. Given that $Su + Tu \geq \theta$ for $u \in \Omega$, we learn that Cauchy problem (1.2) has at least one positive mild solution on J. □

4 Example

We consider the following fractional differential equation:

$$\begin{cases} D^\alpha u(t,x) = \frac{\partial^2 u(t,x)}{\partial x^2} + \frac{e^t |u(t,x)|}{(24+e^t)(1+|u(t,x)|)} + \int_0^t e^{t-s} \frac{e^s}{\sqrt{48+|u(t,x)|}} ds, \\ u(0,x) - \frac{|u(t,x)|}{8+|u(t,x)|} = 0, \quad \frac{du(t,x)}{dt}\Big|_{t=0} - \frac{|u(t,x)|}{8+|u(t,x)|} = 0, \\ u(t,0) = u(t,\pi) = 0, \quad u'(t,0) = u'(t,\pi) = 0, \end{cases} \tag{4.1}$$

where $t \in J = [0,1]$, $0 \leq x \leq \pi$, $1 < \alpha < 2$, let $X = L^2([0,\pi])$. Then the fractional differential equation (4.1) has at least one positive mild solution on J.

Proof As $X = L^2([0,\pi])$, then the positive cone of X is $P = \{u \in C(J,X) : u(t,x) \geq 0, \text{a.e. } (t,x) \in J \times X\}$. The operator $A : D(A) \subset X \to X$ is given by

$$Ax = x'' \quad \text{with } D(A) := \{x \in X : x' \in X, x(0) = x(\pi) = 0\}.$$

It is well known that A is the infinitesimal generator of an analytic semigroup $\{T(t)\}_{t\geq 0}$ on X. Furthermore, A has discrete spectrum with eigenvalues $-n^2$, $n \in N$, and corresponding normalized eigenfunctions given by $z_n(x) = (\frac{\pi}{2})^{1/2} \sin(nx)$ [17]. In addition, $\{z_n : n \in N\}$ is an orthonormal basis of X and

$$T(t) = \sum_{n=1}^{\infty} e^{-n^2 t} \langle x, z_n \rangle z_n, \quad \text{for } x \in X, t \geq 0.$$

It follows from this representation that $T(t)$ is compact for every $t > 0$ and that $\|T(t)\| \leq e^{-t}$ for every $t \geq 0$ [18].

As indicated in [9], the operator $A = \Delta$ is a sectorial operator of type (M, θ, α, μ) and generates compact α-resolvent families $\{S_\alpha(t)\}_{t\geq 0}$, $\{K_\alpha(t)\}_{t\geq 0}$, and $\{T_\alpha(t)\}_{t\geq 0}$. Since it was proved in [19] that $A = \Delta$ is an m-accretive operator on X with dense domain, assumption (H_2) is satisfied.

In this situation,

$$f(t,u) = \frac{e^t|u|}{(24+e^t)(1+|u|)}, \qquad g(t,u) = \frac{e^t}{\sqrt{48}+|u|},$$

$$m(u) = -\frac{|u|}{8+|u|}, \qquad n(u) = -\frac{|u|}{8+|u|}, \qquad q(t-s) = e^{t-s}.$$

From the estimates on the norms of operators of Theorem 2.4 and Theorem 2.5, we can obtain $\widetilde{M} = 3$ (see [4]). Moreover, for $t \in J$, $u, v \in R$ we have

$$\|f(t,u) - f(t,v)\| = \frac{e^t}{24+e^t}\left\|\frac{u}{1+u} - \frac{v}{1+v}\right\| \leq \frac{e^t}{24+e^t}\|u-v\| \leq \frac{1}{8}\|u-v\|,$$

$$\|g(t,u) - g(t,v)\| = e^t\left\|\frac{1}{\sqrt{48}+u} - \frac{1}{\sqrt{48}+v}\right\| \leq \frac{e^t}{48}\|u-v\| \leq \frac{1}{16}\|u-v\|.$$

So, we have $l_1 = \frac{1}{8}$, $l_2 = \frac{1}{16}$. Meanwhile,

$$\|f(t,u)\| \leq \frac{e^t}{24+e^t} < \frac{1}{8}, \qquad \|g(t,u)\| \leq \frac{e^t}{\sqrt{48}} < \frac{3}{\sqrt{48}} = \frac{\sqrt{3}}{4}.$$

Hence, $\|\mu_1\| = \frac{1}{8}$, $\|\nu_1\| = \frac{\sqrt{3}}{4}$, which means assumptions (H_3) and (H_6) are satisfied. We have

$$\max \int_0^t |q(t-s)|\, ds = \max_{t\in[0,1]} \int_0^t e^{t-s}\, ds = \max_{t\in[0,1]} e^t - 1 \leq 2.$$

For $u \in R$, we have

$$\|u_0 - m(u)\| \leq \left\|\frac{u}{8+u}\right\| \leq \frac{1}{8}, \qquad \|u_1 - n(u)\| \leq \frac{1}{8}.$$

Then assumption (H_5) is satisfied.

Consequently,

$$\widetilde{M}T(l_1 + ql_2) = 3\left(\frac{1}{8} + 2 \times \frac{1}{16}\right) = \frac{3}{4} < 1.$$

It is not difficult to conclude that all the conditions of Theorem 3.2 are satisfied. Hence, the nonlocal fractional differential equation (4.1) has at least one positive mild solution. □

5 Conclusion

In this paper, we discussed the existence of positive mild solution of a kind of fractional differential evolution equation with nonlocal conditions of order $1 < \alpha < 2$. Firstly, we investigated the properties of solution operators by means of the classical Mittag-Leffler function. Secondly, we obtain the existence result by applying Schauder's fixed point theorem and Krasnoselskii's fixed point theorem under some special conditions.

Based on this work, our future work will be devoted to the study of the existence of positive mild solutions of impulsive fractional differential evolution equations of order $1 < \alpha < 2$.

Competing interests

The authors declare that they have no competing interests.

Authors' contributions

The authors declare that the study was realized in collaboration with the same responsibility. All authors read and approved the final manuscript.

Acknowledgements

The work is supported by Hunan Provincial Natural Science Foundation of China Grant No. 14JJ2050.

References

1. Benson, DA: The fractional advection-dispersion equation. PhD thesis, University of Nevada, Reno, NV, USA (1998)
2. Schumer, R, Benson, DA, Meerschaert, MM, Wheatcraft, SW: Eulerian derivation of the fractional advection-dispersion equation. J. Contam. Hydrol. **48**, 69-88 (2001)
3. Henry, BI, Wearne, SL: Existence of Turing instabilities in a two-species fractional reaction-diffusion system. SIAM J. Appl. Math. **62**, 870-887 (2001)
4. Shu, X, Wang, Q: The existence and uniqueness of mild solutions for fractional differential equations with nonlocal conditions of order $1 < \alpha < 2$. Comput. Appl. Math. **64**, 2100-2110 (2012)
5. Mu, J: Monotone iterative technique for fractional evolution equations in Banach spaces. J. Appl. Math. (2011). doi:10.1155/2011/767186
6. Mu, J, Fan, H: Positive mild solutions of periodic boundary value problems for fractional evolution equations. J. Appl. Math. (2012). doi:10.1155/2012/691651
7. Bazhlekova, E: Fractional evolution equations in Banach spaces. PhD thesis, Eindhoven University of Technology (2001)
8. Zhang, X: Positive solutions for a second-order nonlinear impulsive singular integro-differential equation with integral conditions in Banach spaces. J. Math. Res. Appl. **32**(5), 599-614 (2012)
9. Denton, Z, Ng, PW, Vatsala, AS: Quasilinearization method via lower and upper solutions for Riemann-Liouville fractional differential equations. Nonlinear Dyn. Syst. Theory **11**, 239-251 (2011)
10. Ren, Y, Qin, Y, Sakthivel, R: Existence results for fractional order semilinear integro-differential evolution equations with infinite delay. Integral Equ. Oper. Theory **67**, 33-49 (2010)
11. Bai, Z, Lü, H: Positive solutions for boundary value problem of nonlinear fractional differential equation. J. Math. Anal. Appl. **311**, 495-505 (2005)
12. Mainardi, F, Paradisi, P, Gorenflo, R: Probability distributions generated by fractional diffusion equations. In: Kertesz, J, Kondor, I (eds.) Econophysics: An Emerging Science. Kluwer Academic, Dordrecht (2000)
13. Guo, D, Lakshmikantham, V: Nonlinear Problems in Abstract Cones. Academic Press, Boston (1988)
14. Podlubny, I: Fractional Differential Equations. Academic Press, New York (1999)
15. Pazy, A: Semigroups of linear operators and applications to partial differential equations. Appl. Math. Sci. (1983). doi:10.1007/978-1-4612-5561-1
16. Hanneken, JW, Vaught, DM, Narahari Achar, BN: Enumeration of the real zeros of the Mittag-Leffler function $E\alpha(z)$, $1 < \alpha < 2$. In: Advances in Fractional Calculus: Theoretical Developments and Applications, pp. 15-26 (2007)
17. Balachandran, K, Sakthivel, R: Existence of solutions of neutral functional integrodifferential equation in Banach spaces. Proc. Indian Acad. Sci. Math. Sci. **109**, 325-332 (1999)
18. Hernández, E, Sakthivel, R, Aki, ST: Existence results for impulsive evolution differential equations with state-dependent delay. Electron. J. Differ. Equ. **2008**, 28 (2008)
19. Bénilan, P: Equations d'evolution dans un espace de Banach quelconque et appications. Thèse de doctorat d'état, Orasy (1972)

5

Periodic solutions for a seasonally forced SIR model with impact of media coverage

Jian Zu[1] and Lin Wang[2,3*]

*Correspondence:
wanglinmath@hotmail.com
[2]School of Basic Science, Changchun University of Technology, Changchun, 130012, P.R. China
[3]College of Mathematics, Jilin University, Changchun, 130012, P.R. China
Full list of author information is available at the end of the article

Abstract

In this paper, we study periodic solutions for a seasonally forced SIR model with impact of media coverage. Usually, media reports, information processing, and individuals' alerted responses to the information can only arise as the number of infected individuals reaches and exceeds a certain level. The piecewise smooth righthand side is introduced to describe the impact of this kind of media coverage. Using Leray-Schauder degree theory, we establish new results on the existence of at least one positive periodic solution for a seasonally forced SIR model with impact of media coverage. Some numerical simulations are presented to illustrate the effectiveness of such media coverage.

MSC: 34C25; 37J45; 92B05

Keywords: periodic solution; SIR model; media effects; non-smooth righthand sides

1 Introduction

Many infectious diseases, such as measles, chickenpox, mumps, rubella, pertussis and influenza, show seasonal patterns of incidence [1–3]. The cause of seasonal patterns may vary from the periodic contact rates [3, 4], periodic fluctuation in birth and death rates [5–7], and periodic vaccination program [8]. Thus, it is natural to model these diseases by seasonally forced epidemiological models.

The media coverage is an important factor responsible for the transmission of an infectious disease. When a type of contagious disease appears and starts to spread, people's response to the threat of disease is dependent on their perception of risk, which is affected by public and private information disseminated widely by the media. Massive news coverage and fast information flow have played an important role in affecting the outcome of infectious disease outbreak, such as the 2003 severe acute respiratory syndrome (SARS) and the 2009 H1N1 influenza epidemic [9–14].

Recently, Wang and Xiao have used piecewise continuous transmission rate to describe that the media coverage exhibits its effect once the number of infected individuals exceeds a certain critical level [15]. In this paper, we study the following periodic forced SIR model:

$$\begin{cases} \frac{dS(t)}{dt} = \mu - \beta(t)f(I)SI - \mu S, \\ \frac{dI(t)}{dt} = \beta(t)f(I)SI - (\mu + \gamma)I, \\ \frac{dR(t)}{dt} = \gamma I - \mu R, \end{cases} \tag{1.1}$$

in which

- S, I, R are the fractions of the susceptible, infective and recovered population,
- μ and γ denote the birth (death) rate and recovery rate respectively, which are positive constant,
- $\beta(t)$ is the seasonally-dependent transmission rate, which is a positive continuous T-periodic function,
- $f(I)$ is a decreasing piecewise smooth factor which can describe the impact of media coverage on the transmission coefficient, given by

$$f(I) = \begin{cases} 1, & I \le I_c, \\ 1 + \sigma^{-1}(e^{-\alpha(I_c+\sigma)} - 1)(I - I_c), & I_c < I < I_c + \sigma, \\ e^{-\alpha I}, & I \ge I_c + \sigma, \end{cases} \tag{1.2}$$

where α is the factor of influences, σ is a small parameter and I_c is a critical level.

We think that $f(I)$ in (1.2) is a good approximation to the discontinuous factor in [15] provided σ is small enough. Denote the basic reproduction number $\mathcal{R}_0 = \frac{\bar{\beta}}{\gamma+\mu}$ with $\bar{\beta} = \frac{1}{T}\int_0^T \beta(t)\,dt$. When $f(I) \equiv 1$ in (1.1), Katriel [16] got the existence of periodic positive solutions for the periodically forced SIR model by Leray-Schauder degree theory provided $\mathcal{R}_0 > 1$. By Gaines-Mawhin's continuation theorem, Jódar *et al.* [17] obtained that a T-periodic solution exists for a more general system whenever the condition $\min_{t\in\mathbb{R}} \beta(t) > \gamma + \mu$ holds; Bai and Zhou [18], Bai *et al.* [19], and Liu [20] studied the existence of periodic solutions for a periodically forced SIR model with saturated incidence rates. By persistence theory, Zhang and Zhao [21] studied a periodic epidemic model in a patchy environment; Sun *et al.* [22] studied the SEI model with seasonality comprehensively; Rebelo *et al.* [23] extended these results to some delay differential equations and partial differential equations.

When $f(I)$ in (1.1) is a non-smooth function, to the best of our knowledge, there are no results on the existence of periodic solutions. The methods we mentioned above cannot deal with the non-smooth righthand sides directly.

In this paper, we use an integral version of Leray-Schauder degree theory under Katriel's frame to prove the existence of periodic solutions for our SIR model. Some numerical simulations are presented to illustrate the effectiveness of such media coverage. Our main results are as follows.

Theorem 1.1 *Whenever $\mathcal{R}_0 > e^{\alpha}$, there exists at least one T-periodic solution $(S(t), I(t), R(t))$ of (1.1)-(1.2), all of whose components are positive.*

In [24], Liu and Xiao consider the non-periodic coefficient SIR model with

$$f(I) = \begin{cases} e^{-mI}, & I < I_c, \\ e^{-mI_c}, & I > I_c, \end{cases} \tag{1.3}$$

where $m > 0$ is a factor of influence and I_c is a critical level. Using the same method as in Theorem 1.1, we can present the following theorem without a proof.

Theorem 1.2 *Whenever $\mathcal{R}_0 > e^{mI_c}$, there exists at least one T-periodic solution $(S(t), I(t), R(t))$ of (1.1) and (1.3), all of whose components are positive.*

This paper is organized as follows. In Section 2, we study the properties of the homotopy equation from the classical autonomous SIR model to our SIR model. In Section 3, we construct an equivalent integral equation and define a completely continuous operator. In Section 4, we prove the main theorem by Leray-Schauder degree theory. In Section 5, some numerical simulations are presented to illustrate the effectiveness of such media coverage.

2 Homotopy equation and suitable domain

In the rest of this paper, we assume that $f(I)$ has the expression (1.2). Observing system (1.1), we have $\frac{dS}{dt} + \frac{dI}{dt} + \frac{dR}{dt} \equiv 0$. Since $S(t), I(t), R(t)$ are fractions of the population, we have $S(t) + I(t) + R(t) = 1$ for all t. Because R does not appear in the first two equations in (1.1), it is sufficient to consider the existence of periodic solutions of following systems:

$$\begin{cases} \frac{dS(t)}{dt} = \mu - \beta(t)f(I)SI - \mu S, \\ \frac{dI(t)}{dt} = \beta(t)f(I)SI - (\mu + \gamma)I, \end{cases} \tag{2.1}$$

with

$$S(t) > 0, \qquad I(t) > 0, \qquad S(t) + I(t) < 1.$$

In order to prove the existence of periodic solutions of (2.1), we consider the following homotopy system:

$$\begin{cases} \frac{dS(t)}{dt} = \mu - \bar{\beta}SI - \mu S - \lambda(\beta(t)f(I)SI - \bar{\beta}SI), \\ \frac{dI(t)}{dt} = \bar{\beta}SI - (\mu + \gamma)I + \lambda(\beta(t)f(I)SI - \bar{\beta}SI), \end{cases} \tag{2.2}$$

where $\lambda \in [0,1]$. Let

$$D := \{(S,I) \in C[0,T] \times C[0,T] \mid S(t) > 0, I(t) > 0, S(t) + I(t) < 1\}.$$

Lemma 2.1 *$\overline{D}$ is an invariant region with respect to* (2.2). *The disease-free equilibrium* $(S_0, I_0) = (1,0)$ *is the unique periodic solution of* (2.2) *satisfying* $(S,I) \in \partial D$ *for any* $\lambda \in [0,1]$.

Proof First, we will prove that $\overline{D}$ is an invariant region. In fact, it follows from model (2.2) that

$$\frac{dS}{dt}\bigg|_{S=0} = \mu > 0, \qquad \frac{dI}{dt}\bigg|_{I=0} = 0, \qquad \frac{d(S+I)}{dt}\bigg|_{S+I=1} = -\gamma I \le 0.$$

Second, we will prove that the disease-free equilibrium $(S_0, I_0) = (1,0)$ is the unique periodic solution of (2.2) satisfying $(S,I) \in \partial D$. We assume that $(S,I) \in \partial D$ is a solution of (2.2), which means that at least one of the following conditions holds:

(i) There exists $t_0 \in [0,T]$ such that $I(t_0) = 0$.
(ii) There exists $t_0 \in [0,T]$ such that $S(t_0) = 0$.
(iii) There exists $t_0 \in [0,T]$ such that $S(t_0) + I(t_0) = 1$.

We now consider each of these three cases:

In the case of (i), we have $I(t_0) = 0$ and $I'(t_0) = 0$, which implies $I \equiv 0$. Thus, the only possible periodic solution of $S' = \mu(1 - S)$ is $S \equiv 1$.

In the case of (ii), we have $S(t_0) = 0$ and $S'(t_0) = \mu > 0$. Thus, it is easy to obtain that $S(t) < 0$ for $t < t_0$ sufficiently close to t_0, which contradicts the fact that $\overline{D}$ is an invariant region.

In the case of (iii), we get

$$(S + I)'(t_0) = \mu\big(1 - S(t_0) - I(t_0)\big) - \gamma I(t_0) = -\gamma I(t_0) \leq 0.$$

Because the case of $I(t_0) = 0$ has been discussed, we only discuss the case $S(t_0) + I(t_0) = 1$, $(S + I)'(t_0) < 0$, which contradicts the fact that $\overline{D}$ is an invariant region. □

To use the continuity method, we need to choose a bounded open set $\mathcal{U} \subseteq D$ such that there is no solution (S, I) of (2.2) satisfying $(S, I) \in \partial\mathcal{U}$ for any $\lambda \in [0, 1]$. Since $(S_0, I_0) \in \partial D$ is a solution of (2.2), we need to exclude (S_0, I_0) from the boundary of $\mathcal{U}$ we chose. Following the idea of Katriel [16], we take $\mathcal{U}$ to be the open subset of D given by

$$\mathcal{U} = \left\{(S, I) \in D \mid \min_{t\in[0,T]} S(t) < \delta\right\}, \tag{2.3}$$

where $\delta \in (0, 1)$ is to be fixed.

Remark 2.2 For fixed t_0, $\mathcal{U}$ is an open set in $\mathbb{R}^2$. Furthermore, for any $t \in [0, T]$, $\mathcal{U}$ with norm $\|(S, I)\| = \max_{t\in[0,T]}(|S(t)| + |I(t)|)$ is an open set in $C[0, T] \times C[0, T]$.

Lemma 2.3 *Let $\mathcal{R}_0 > e^{\alpha}$. If we choose $\delta \in (\frac{e^{\alpha}}{\mathcal{R}_0}, 1)$, then there is no solution (S, I) of (2.2) with $(S, I) \in \partial\mathcal{U}$ for any $\lambda \in [0, 1]$.*

Proof Suppose $(S, I) \in \partial\mathcal{U}$. Then either $(S, I) \in \partial D$ or $(S, I) \in D$ and

$$\min_{t\in[0,T]} S(t) = \delta. \tag{2.4}$$

In the first case, Lemma 2.1 and the fact that $(S_0, I_0) \notin \partial\mathcal{U}$ imply that (S, I) is not a solution of (2.2).

In the second case, we can infer that $I(t) > 0$ and $S(t) \geq \delta$, $\forall t \in [0, T]$. If (S, I) is a solution of (2.2), we can divide the second equation of (2.2) by I. Integrating over $[0, T]$, we obtain that

$$\frac{1}{T}\int_0^T \bar{\beta} S\,dt - \mu - \gamma + \frac{1}{T}\int_0^T \lambda\big(\beta(t)Sf(I) - \bar{\beta}S\big)\,dt = 0. \tag{2.5}$$

Thus

$$\begin{aligned}
\mu + \gamma &= \frac{1}{T}\int_0^T (1 - \lambda)\bar{\beta}S + \lambda\beta(t)Sf(I)\,dt \\
&\geq \frac{\delta}{T}\int_0^T (1 - \lambda)\bar{\beta} + \lambda\beta(t)f(I)\,dt \\
&\geq \delta e^{-\alpha}\bar{\beta}.
\end{aligned}$$

By the assumption $\delta > \frac{e^{\alpha}}{\mathcal{R}_0}$, we have

$$\mu + \gamma < \delta e^{-\alpha}\bar{\beta},$$

which is a contradiction. □

3 Existence of periodic solutions

3.1 Equivalent integral equation and completely continuous operator

We rewrite (2.2) as

$$\frac{d}{dt}\begin{pmatrix} S \\ I \end{pmatrix} + \begin{pmatrix} \mu & 0 \\ 0 & \mu+\gamma \end{pmatrix}\begin{pmatrix} S \\ I \end{pmatrix} = \begin{pmatrix} F_1(S,I,\lambda) \\ F_2(S,I,\lambda) \end{pmatrix}, \tag{3.1}$$

where $F_1(S,I,\lambda) = \mu - \bar{\beta}SI - \lambda(\beta(t)f(I)SI - \bar{\beta}SI)$, and $F_2(S,I,\lambda) = \bar{\beta}SI + \lambda(\beta(t)f(I)SI - \bar{\beta}SI)$. If $\Phi(t)$ is the fundamental solution matrix of

$$\frac{d}{dt}\begin{pmatrix} S \\ I \end{pmatrix} + \begin{pmatrix} \mu & 0 \\ 0 & \mu+\gamma \end{pmatrix}\begin{pmatrix} S \\ I \end{pmatrix} = 0,$$

satisfying $\Phi(0) = \mathrm{Id}$, we have

$$\Phi(t) = \begin{pmatrix} e^{-\mu t} & 0 \\ 0 & e^{-(\mu+\gamma)t} \end{pmatrix}.$$

Then (3.1) can be transformed into the equivalent integral equation

$$\begin{pmatrix} S \\ I \end{pmatrix}(t) = \Phi(t)\left(\begin{pmatrix} S(0) \\ I(0) \end{pmatrix} + \int_0^t \Phi^{-1}(\tau)\begin{pmatrix} F_1(S,I,\lambda) \\ F_2(S,I,\lambda) \end{pmatrix} d\tau\right). \tag{3.2}$$

If S, I is a T-periodic solution of (3.1), then

$$(I - \Phi(T))\begin{pmatrix} S(0) \\ I(0) \end{pmatrix} = \Phi(T)\int_0^T \Phi^{-1}(\tau)\begin{pmatrix} F_1(S,I,\lambda) \\ F_2(S,I,\lambda) \end{pmatrix} d\tau. \tag{3.3}$$

Since $(I - \Phi(T))$ is invertible, we have

$$\begin{pmatrix} S(0) \\ I(0) \end{pmatrix} = (I - \Phi(T))^{-1}\Phi(T)\int_0^T \Phi^{-1}(\tau)\begin{pmatrix} F_1(S,I,\lambda) \\ F_2(S,I,\lambda) \end{pmatrix} d\tau. \tag{3.4}$$

Substituting (3.4) into (3.2), we have

$$\begin{aligned}\begin{pmatrix} S \\ I \end{pmatrix}(t) &= \Phi(t)(I - \Phi(T))^{-1}\Phi(T)\int_0^T \Phi^{-1}(\tau)\begin{pmatrix} F_1(S,I,\lambda) \\ F_2(S,I,\lambda) \end{pmatrix} d\tau \\ &\quad + \Phi(t)\int_0^t \Phi^{-1}(\tau)\begin{pmatrix} F_1(S,I,\lambda) \\ F_2(S,I,\lambda) \end{pmatrix} d\tau. \end{aligned}\tag{3.5}$$

Define an operator $P_\lambda : C[0,T] \times C[0,T] \to C[0,T] \times C[0,T]$ such that

$$P_\lambda[(S,I)](t) = \Phi(t)(\mathrm{Id} - \Phi(T))^{-1}\Phi(T)\int_0^T \Phi^{-1}(\tau)\begin{pmatrix} F_1(S,I,\lambda) \\ F_2(S,I,\lambda) \end{pmatrix} d\tau$$

$$+ \Phi(t)\int_0^t \Phi^{-1}(\tau)\begin{pmatrix} F_1(S,I,\lambda) \\ F_2(S,I,\lambda) \end{pmatrix} d\tau. \tag{3.6}$$

Lemma 3.1 *P_λ in* (3.6) *is a completely continuous operator.*

Proof Since $f(I)$ is a non-smooth but continuous function, the operator P_λ is continuous with respect to S and I. Since $\beta(t)$, $\Phi(t)$ and $\Phi^{-1}(t)$ are all bounded in $[0,T]$, S and I are bounded on U, $e^{-\alpha} \leq f(I) \leq 1$ on U, it is easy to see that the operator P_λ in (3.6) is uniformly bounded and equicontinuous, which implies P_λ in (3.6) is a completely continuous operator. □

3.2 Main results

We recall that the existence of a periodic solution (S,I) of (3.5) can be assured by Leray-Schauder degree theory [25] if the following conditions hold:

(1) $(\mathrm{Id} - P_\lambda)(S,I) \neq 0$ for all $(S,I) \in \partial U$, $\lambda \in [0,1]$,

(2) $\deg(\mathrm{Id} - P_0, U, 0) \neq 0$.

By Lemma 2.3, there are no solutions (S,I) of (2.2) with $(S,I) \in \partial U$, $\lambda \in [0,1]$. Now we prove that $\deg(\mathrm{Id} - P_0, U, 0) \neq 0$.

Lemma 3.2 *For $\lambda = 0$,* (2.2) *has only one periodic solution in U, which is endemic equilibrium: $(S^*, I^*) = (\frac{\gamma+\mu}{\bar{\beta}}, \mu(\frac{1}{\gamma+\mu} - \frac{1}{\bar{\beta}}))$.*

Proof When $\lambda = 0$, (2.2) is an autonomous system. $\mu(1-S) - \bar{\beta}SI$ and $\bar{\beta}SI - (\mu+\gamma)I$ are both C^1 in D. If we set the Dulac function to be $h = \frac{1}{SI}$ in D, we have

$$\frac{\partial(h(\mu(1-S) - \bar{\beta}SI))}{\partial S} + \frac{\partial(h(\bar{\beta}SI - (\mu+\gamma)I))}{\partial I} = -\frac{\mu}{S^2 I} < 0,$$

so there is no closed orbit in D.

Since $\frac{e^\alpha}{\mathcal{R}_0} < \delta < 1$, γ, μ and $\bar{\beta}$ are positive constant, we have $0 < S^* < \frac{\delta}{e^\alpha} < \delta$ and $0 < I^* < \frac{\mu}{\gamma+\mu} < 1$. For $\lambda = 0$, it is easy to calculate that (S^*, I^*) is the unique constant periodic solution in U, which implies that (2.2) has only one periodic solution in U. □

Lemma 3.3 [26] *Let Ω be a bounded open set in the Banach space X. Assume that a completely continuous field $f = \mathrm{Id} - F : \bar{\Omega} \to X$ has no zero points on $\partial\Omega$, and there are only finite zero points $x_1, x_2, \ldots, x_n$ in Ω. Then we have the index formula*

$$\deg(f, \Omega, 0) = \sum_{i=1}^{n} \mathrm{index}(f, x_i). \tag{3.7}$$

Lemma 3.4 [26] *Let Ω be a bounded open set in the Banach space X. x_0 is the zero point of a completely continuous field $f = \mathrm{Id} - F : \bar{\Omega} \to X$ in Ω. We assume that f is Frechet differentiable at x_0 and 1 is not the eigenvalue of $A = F'(x_0)$. Then x_0 is an isolated zero point* *of f, and*

$$\mathrm{index}(f, x_0) = \mathrm{index}(\mathrm{Id} - A, 0) = (-1)^{\beta}, \tag{3.8}$$

where $\beta = \sum_{\lambda_j > 1 | \lambda_j \in \sigma(A)} \beta_j$ *and* $\beta_j = \dim \bigcup_{k=1}^{\infty} \mathrm{Ker}(\lambda_j I - A)^k$.

Define the operator

$$\begin{aligned} &DP_0\big[(S^*, I^*)\big] \circ (V, W)(t) \\ &\quad = \Phi(t)\big(\mathrm{Id} - \Phi(T)\big)^{-1}\Phi(T)\int_0^T \Phi^{-1}(\tau)\begin{pmatrix} -\bar{\beta}(I^*V + S^*W) \\ \bar{\beta}(I^*V + S^*W) \end{pmatrix} d\tau \\ &\quad + \Phi(t)\int_0^t \Phi^{-1}(\tau)\begin{pmatrix} -\bar{\beta}(I^*V + S^*W) \\ \bar{\beta}(I^*V + S^*W) \end{pmatrix} d\tau. \end{aligned} \tag{3.9}$$

Obviously,

$$\big\|P_0\big[(S^* + V, I^* + W)\big] - P_0\big[(S^*, I^*)\big] - DP_0\big[(S^*, I^*)\big] \circ (V, W)\big\| = o\big[(V, W)\big]. \tag{3.10}$$

Thus, $DP_0[(S^*, I^*)]$ is the Frechet derivative of $P_0[(S, I)]$.

Lemma 3.5 1 *is not the eigenvalue of* $DP_0(S^*, I^*)$.

Proof Let λ be the eigenvalue of operator $DP_0(S^*, I^*)$:

$$DP_0\big[(S^*, I^*)\big] \circ (V, W)(t) = \lambda(V, W). \tag{3.11}$$

Multiplying $\Phi^{-1}(t)$ by both sides of (3.11) and taking the derivative with respect to t, we have

$$\Phi^{-1}(t)\begin{pmatrix} -\bar{\beta}(I^*V + S^*W) \\ \bar{\beta}(I^*V + S^*W) \end{pmatrix} = \frac{d}{dt}\left(\lambda\Phi^{-1}(t)\begin{pmatrix} V \\ W \end{pmatrix}\right), \tag{3.12}$$

which is equal to

$$\lambda\begin{pmatrix} \dot{V} \\ \dot{W} \end{pmatrix} = \begin{pmatrix} -\bar{\beta}I^* - \lambda\mu & -\bar{\beta}S^* \\ \bar{\beta}I^* & \bar{\beta}S^* - \lambda\mu - \lambda\gamma \end{pmatrix}\begin{pmatrix} V \\ W \end{pmatrix}. \tag{3.13}$$

If 1 is an eigenvalue of $DP_0(S^*, I^*)$, we have

$$\frac{d}{dt}\begin{pmatrix} V \\ W \end{pmatrix} = \begin{pmatrix} -\mu\mathcal{R}_0 & -(\gamma + \mu) \\ \mu(\mathcal{R}_0 - 1) & 0 \end{pmatrix}\begin{pmatrix} V \\ W \end{pmatrix}. \tag{3.14}$$

The characteristic polynomial of the above matrix is

$$p(x) = x^2 + \mu\mathcal{R}_0 x + (\gamma + \mu)\mu(\mathcal{R}_0 - 1). \tag{3.15}$$

It is easy to calculate that $p(0) > 0$, and $p(\omega i) \neq 0$ for $\omega \in \mathbb{R}$, which means that the matrix has no imaginary or 0 eigenvalues, so that (3.14) has no periodic solutions except $(V, W) = (0, 0)$, which is a contradiction to (3.11). □

Now we will finish to prove that $\deg(\mathrm{Id} - P_0, \mathcal{U}, 0) \neq 0$. $\mathrm{Id} - P_0$ is a completely continuous field, 1 is not the eigenvalue of $DP_0(S^*, I^*)$. By Lemma 3.4, we have $\mathrm{index}(\mathrm{Id} - P_0, (S^*, I^*)) = (-1)^\beta \neq 0$. Since we have proved that $(\mathrm{Id} - P_0)(S, I) \neq 0$ for all $(S, I) \in \partial\mathcal{U}$, and (S^*, I^*) is the unique zero point in $\mathcal{U}$, by Lemma 3.3 we have

$$\deg(\mathrm{Id} - P_0, \mathcal{U}, 0) = \mathrm{index}\big(\mathrm{Id} - P_0, (S^*, I^*)\big) \neq 0.$$

Finally, by Leray-Schauder degree theory, we obtain

$$\deg(\mathrm{Id} - P_1, \mathcal{U}, 0) = \deg(\mathrm{Id} - P_0, \mathcal{U}, 0) \neq 0.$$

Thus, there exists at least one positive periodic solutions for a seasonally forced SIR model with impact of media coverage.

4 Simulation

In this section, we present some numerical examples to illustrate the effectiveness of such media coverage. Furthermore, we show how various parameters influence the solutions of our SIR model.

With the period $T = 2\pi$ of the forcing representing one year, we take $\gamma = 14\frac{2\pi}{365}$ corresponding to a two-week infectious period. We set $\bar{\beta} = 4\gamma$, $\mu = \frac{0.5}{2\pi}$, $\beta(t) = \bar{\beta}(1 + 0.8\cos(t))$, $I_c = 0.15$ and $\delta = 0.05$. Let $[0, 2\pi]$ be divided into $k = 200$ intervals equally. Given the initial point $(S^{**}, I^{**}) = (\frac{\mu+\gamma}{\bar{\beta}}, \frac{\mu}{\mu+\gamma} - \frac{u}{\bar{\beta}})$, which is the endemic equilibrium of SIR model without periodic transmission rate and impact of media coverage. The periodic solutions of system (1.1) can be solved by the Newton iteration method.

In Figure 1, we make ten steps of Newton iteration to get the approximate infective population and susceptible population of system (1.1) with different α. Obviously, the infective

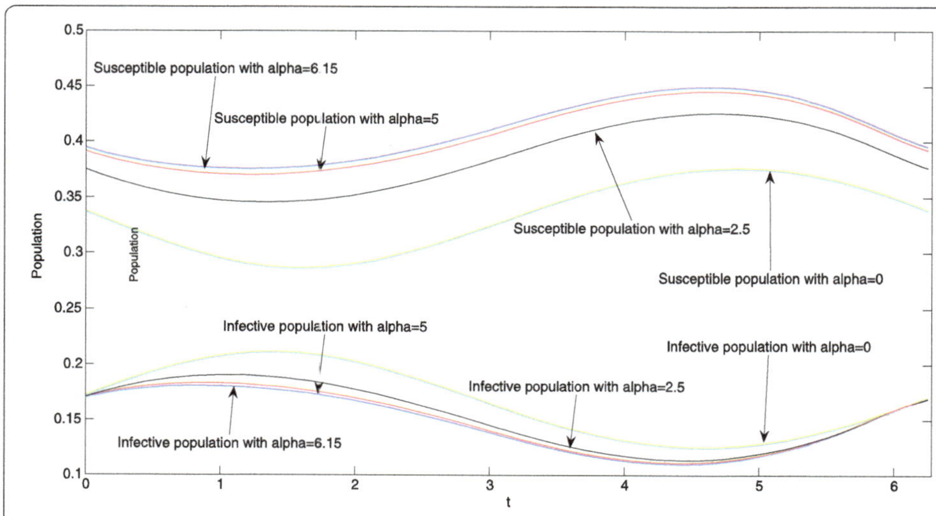

Figure 1 Infective population and susceptible population with $\alpha = 0$, $\alpha = 2.5$, $\alpha = 5$ and $\alpha = 6.15$.

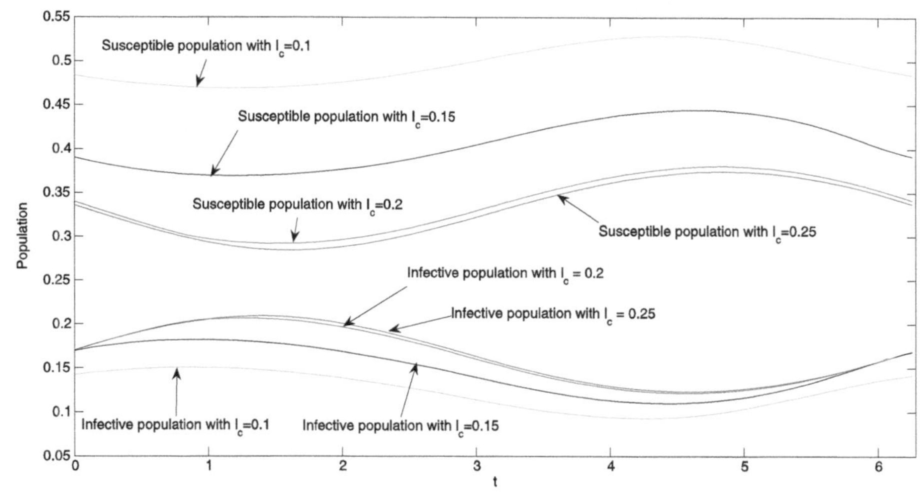

Figure 2 Infective population and susceptible population with $I_c = 0.1$, $I_c = 0.15$, $I_c = 0.2$ and $I_c = 0.25$.

population of system (1.1) with impact of media coverage ($\alpha > 0$) is lower than the infective population of system (1.1) without impact of media coverage ($\alpha = 0$). The effectiveness of impact of media coverage grows as α grows. The solutions in both cases are locally stable and the error is about 10^{-6}.

In Figure 2, we fix $\alpha = 5$ and make ten steps of Newton iteration to get the approximate infective population and susceptible population of system (1.1) with different I_c. Obviously, the infective population of system (1.1) becomes smaller as I_c decreases. The solutions in both cases are locally stable and the error is about 10^{-6}.

In a word, it is effective to reduce the infective population by media coverage.

5 Conclusion

In this paper, we study the existence of positive periodic solutions for a seasonally forced SIR model with impact of media coverage. This paper can be divided into two parts. In the first part, we construct a homotopy equation from an autonomous system to our SIR model. Using Leray-Schauder degree theory, we establish a new result on the existence of at least one positive periodic solution for our SIR model. In the final part, some numerical simulations are presented to illustrate the effect of media coverage.

Competing interests
The authors declare that they have no competing interests.

Authors' contributions
All authors contributed equally to the writing of this paper. All authors read and approved the final manuscript.

Author details
[1]Center for Mathematics and Interdisciplinary Sciences, and School of Mathematics and Statistics, Northeast Normal University, Changchun, 130024, P.R. China. [2]School of Basic Science, Changchun University of Technology, Changchun, 130012, P.R. China. [3]College of Mathematics, Jilin University, Changchun, 130012, P.R. China.

Acknowledgements
The authors would like to express their sincere gratitude to Professor Yong Li for his enthusiastic guidance and constant encouragement. Jian Zu was supported by NSFC (Grant No. 11401089).

References

1. Altizer, S, et al.: Seasonality and the dynamics of infectious diseases. Ecol. Lett. **9**, 467-484 (2006)
2. Grassly, NC, Fraser, C: Seasonal infectious disease epidemiology. Proc. - Royal Soc., Biol. Sci. **273**, 2541-2550 (2006)
3. Hethcote, HW, Levin, SA: Periodicity in epidemiological models. In: Applied Mathematical Ecology (Trieste, 1986). Biomathematics, vol. 18, pp. 193-2`1. Springer, Berlin (1989)
4. Grossman, Z, Gumowski, I, Dietz, K: The incidence of infectious diseases under the influence of seasonal fluctuations - analytical approach. In: Nonlinear Systems and Applications (Proc. Internat. Conf., Univ. Texas, Arlington, Tex., 1976), pp. 525-546. Academic Press, New York (1977)
5. Lou, YJ, Zhao, XQ: Threshold dynamics in a time-delayed periodic SIS epidemic model. Discrete Contin. Dyn. Syst., Ser. B **12**(1), 169-186 (2009)
6. Lou, YJ, Zhao, XQ: A climate-based malaria transmission model with structured vector population. SIAM J. Appl. Math. **70**(6), 2023-2044 (2010)
7. Ma, JL, Ma, ZE: Epidemic threshold conditions for seasonally forced SEIR models. Math. Biosci. Eng. **3**(1), 161-172 (2006)
8. Earn, DJ, Rohani, P, Bolker, BM, Grenfell, BT: A simple model for complex dynamical transitions in epidemics. Science **287**(5453), 667-670 (2000)
9. Liu, RS, Wu, JH, Zhu, HP: Media/psychological impact on multiple outbreaks of emerging infectious diseases. Comput. Math. Methods Med. **8**(3), 153-164 (2007)
10. Cui, JA, Sun, YH, Zhu, HP: The impact of media on the control of infectious diseases. J. Dyn. Differ. Equ. **20**(1), 31-53 (2008)
11. Cui, JA, Tao, X, Zhu, HP: An SIS infection model incorporating media coverage. Rocky Mt. J. Math. **38**(5), 1323-1334 (2008)
12. Sun, CJ, Yang, W, Arino, J, Khan, K: Effect of media-induced social distancing on disease transmission in a two patch setting. Math. Biosci. **230**(2), 87-95 (2011)
13. Li, YF, Cui, JA: The effect of constant and pulse vaccination on SIS epidemic models incorporating media coverage. Commun. Nonlinear Sci. Numer. Simul. **14**(5), 2353-2365 (2009)
14. Cai, LM, Li, XZ: Analysis of a SEIV epidemic model with a nonlinear incidence rate. Appl. Math. Model. **33**(7), 2919-2926 (2009)
15. Wang, AL, Xiao, YN: A Filippov system describing media effects on the spread of infectious diseases. Nonlinear Anal. Hybrid Syst. **11**, 84-97 (2014)
16. Katriel, G: Existence of periodic solutions for the periodically forced SIR model. Nelīnīĭnī Koliv. **16**(3), 359-366 (2013)
17. Jódar, L, Villanueva, RJ, Arenas, A: Modeling the spread of seasonal epidemiological diseases: theory and applications. Math. Comput. Model. **48**(3-4), 548-557 (2008)
18. Bai, ZG, Zhou, YC: Existence of two periodic solutions for a non-autonomous SIR epidemic model. Appl. Math. Model. **35**(1), 382-391 (2011)
19. Bai, ZG, Zhou, YC, Zhang, TL: Existence of multiple periodic solutions for an SIR model with seasonality. Nonlinear Anal. **74**(11), 3548-3555 (2011)
20. Liu, ZJ: Dynamics of positive solutions to SIR and SEIR epidemic models with saturated incidence rates. Nonlinear Anal., Real World Appl. **14**(3), 1286-1299 (2013)
21. Zhang, F, Zhao, XQ: A periodic epidemic model in a patchy environment. J. Math. Anal. Appl. **325**(1), 496-516 (2007)
22. Sun, GQ, Bai, ZG, Zhang, ZK, Zhou, T, Jin, Z: Positive periodic solutions of an epidemic model with seasonality. Sci. World J. **2013**, Article ID 470646 (2013)
23. Rebelo, C, Margheri, A, Bacaër, N: Persistence in some periodic epidemic models with infection age or constant periods of infection. Discrete Contin. Dyn. Syst., Ser. B **19**(4), 1155-1170 (2014)
24. Liu, YY, Xiao, YN: An epidemic model with saturated media/psychological impact. Appl. Math. Mech. **34**(4), 399-407 (2013)
25. Zeidler, E: Nonlinear Functional Analysis and Its Applications. I. Fixed-Point Theorems. Springer, New York (1986); Translated from the German by Peter R. Wadsack
26. Chang, KC: Methods in Nonlinear Analysis. Springer Monographs in Mathematics. Springer, Berlin (2005)

Some identities of higher order Barnes-type q-Bernoulli polynomials and higher order Barnes-type q-Euler polynomials

Lee-Chae Jang[1*], Sang-Ki Choi[2] and Hyuck In Kwon[3]

*Correspondence:
lcjang@konkuk.ac.kr
[1]Graduate School of Education, Konkuk University, Seoul, 143-701, Korea
Full list of author information is available at the end of the article

Abstract

In this paper, we consider higher order Barnes-type q-Bernoulli polynomials and numbers and investigate some identities of them. Furthermore, we discuss some identities of higher order Barnes-type q-Euler polynomials and numbers.

MSC: 11B68; 11S40

Keywords: p-adic invariant integral; Bernoulli polynomials; Euler polynomials; higher order Barnes-type q-Bernoulli polynomials and numbers

1 Introduction

Let p be a given odd prime number. Throughout this paper, we assume that $\mathbb{Z}_p$, $\mathbb{Q}_p$ and $\mathbb{C}_p$ will, respectively, denote the rings of p-adic integers, the fields of p-adic numbers and the completion of algebraic closure of $\mathbb{Q}-p$. The p-adic norm $|p|_p = \frac{1}{p}$. Let $\mathrm{UD}(\mathbb{Z}_p)$ be the space of uniformly differentiable functions on $\mathbb{Z}_p$. For $f \in \mathrm{UD}(\mathbb{Z}_p)$, the bosonic p-adic integral on $\mathbb{Z}_p$ is defined as

$$I_0(f) = \int_{\mathbb{Z}_p} f(x)\,d\mu_0(x) = \lim_{N\to\infty} \frac{1}{p^N} \sum_{x=0}^{p^N-1} f(x) \quad \text{(see [1–12])}. \tag{1}$$

It is well known that an integral equation of the bosonic p-adic integral I_0 on $\mathbb{Z}_p$,

$$I_0(f_1) - I_0(f) = f'(0), \tag{2}$$

where $f_1(x) = f(x+1)$. Higher order Bernoulli polynomials are defined by Kim to be

$$\left(\frac{t}{e^t-1}\right)^r e^{xt} = \sum_{n=0}^{\infty} B_n^{(r)}(x)\frac{t^n}{n!} \quad \text{(see [5, 13–16])}. \tag{3}$$

When $x = 0$, $B_n^{(r)} = B_n^{(r)}(0)$ is called higher order Bernoulli numbers. Higher order Barnes-type Bernoulli polynomials are defined by Kim to be

$$\prod_{i=1}^{r}\left(\frac{t}{e^{a_i t}-1}\right)^r e^{xt} = \sum_{n=0}^{\infty} B_n^{(r)}(x|a_1,\ldots,a_r)\frac{t^n}{n!} \quad \text{(see [11–15, 17–21])}. \tag{4}$$

When $x = 0$, $B_n^{(r)}(a_1, \ldots, a_r) = B_n^{(r)}(0|a_1, \ldots, a_r)$ is called higher order Barnes-type Bernoulli numbers.

In this paper we consider higher order Barnes-type q-Bernoulli polynomials and numbers and investigate some identities of them. We also discuss some identities of higher order Barnes-type q-Euler polynomials and numbers.

2 Higher order Barnes-type q-Bernoulli polynomials and numbers

In this section, we assume that $q \in \mathbb{C}_p$ with $|1-q|_p < p^{-\frac{1}{p-1}}$. By (2), if we take $f(x) = q^y e^{(x+y)t}$, then we get

$$\int_{\mathbb{Z}_p} q^y e^{(x+y)t} \, d\mu_0(y) = \frac{t + \log q}{qe^t - 1} e^{xt}, \tag{5}$$

where $f_1(x) = f(x+1)$. q-Bernoulli polynomials are defined by Kim to be

$$\frac{t + \log q}{qe^t - 1} e^{xt} = \sum_{n=0}^{\infty} B_{n,q}(x) \frac{t^n}{n!} \quad \text{(see [13–15, 17, 19–21])}. \tag{6}$$

When $x = 0$, $B_{n,q} = B_{n,q}(0)$ is called q-Bernoulli numbers.

Higher order q-Bernoulli polynomials are defined as

$$\left(\frac{t + \log q}{qe^t - 1} \right)^r e^{xt} = \sum_{n=0}^{\infty} B_{n,q}^{(r)}(x) \frac{t^n}{n!}. \tag{7}$$

When $x = 0$, $B_{n,q}^{(r)} = B_{n,q}^{(r)}(0)$ is called higher order q-Bernoulli numbers.

We define higher order Barnes-type q-Bernoulli polynomials as follows:

$$\frac{(t + \log q)^r}{(q^{a_1} e^{a_1 t} - 1) \cdots (q^{a_r} e^{a_r t} - 1)} e^{xt} = \sum_{n=0}^{\infty} B_{n,q}(x|a_1, \ldots, a_r) \frac{t^n}{n!}. \tag{8}$$

When $x = 0$, $B_{n,q}(a_1, \ldots, a_r) = B_{n,q}(0|a_1, \ldots, a_r)$ is called higher order Barnes-type q-Bernoulli numbers. By (5), we get

$$\begin{aligned} &\int_{\mathbb{Z}_p} q^{a_1 x_1 + \cdots + a_r t_r} e^{(a_1 x_1 + \cdots + a_r x_r + x)t} \, d\mu_0(x_1) \cdots d\mu_0(x_r) \\ &= \left(\prod_{i=1}^{r} a_i \right) \frac{(t - \log q)^r}{(q^{a_1} e^{a_1 t} - 1) \cdots (q^{a_r} e^{a_r t} - 1)} e^{xt}. \end{aligned} \tag{9}$$

By (9) and (8), we get

$$\begin{aligned} &\sum_{n=0}^{\infty} B_{n,q}(x|a_1, \ldots, a_r) \frac{t^n}{n!} \\ &= \frac{(t + \log q)^r}{(q^{a_1} e^{a_1 t} - 1) \cdots (q^{a_r} e^{a_r t} - 1)} e^{xt} \\ &= \left(\prod_{i=1}^{r} a_i \right)^{-1} \int_{\mathbb{Z}_p} \cdots \int_{\mathbb{Z}_p} q^{a_1 x_1 + \cdots + a_r x_r} e^{(a_1 x_1 + \cdots + a_r x_r + x)t} \, d\mu_0(x_1) \cdots d\mu_0(x_r) \end{aligned}$$

$$= \sum_{n=0}^{\infty}\left(\left(\prod_{i=1}^{r} a_i\right)^{-1}\int_{\mathbb{Z}_p}\cdots\int_{\mathbb{Z}_p} q^{a_1x_1+\cdots+a_rx_r+x} \times (a_1x_1+\cdots+a_rx_r+x)^n\, d\mu_0(x_1)\cdots d\mu_0(x_r)\right)\frac{t^n}{n!}. \tag{10}$$

From (10), we obtain the following theorem.

Theorem 2.1 *Let $n \in \mathbb{N}\cup\{0\}$. Then we have*

$$B_{n,q}(x|a_1,\ldots,a_r) = \left(\prod_{i=1}^{r} a_i\right)^{-1}\int_{\mathbb{Z}_p}\cdots\int_{\mathbb{Z}_p} q^{a_1x_1+\cdots+a_rx_r+x} \times (a_1x_1+\cdots+a_rx_r+x)^n\, d\mu_0(x_1)\cdots d\mu_0(x_r). \tag{11}$$

From (1), we have

$$\begin{aligned}\int_{\mathbb{Z}_p} f(x)\,d\mu_0(x) &= \lim_{N\to\infty}\frac{1}{p^N}\sum_{x=0}^{dp^N-1} f(x)\\ &= \frac{1}{d}\lim_{N\to\infty}\frac{1}{p^N}\sum_{a=0}^{d-1}\sum_{x=0}^{p^N-1} f(a+dx)\\ &= \frac{1}{d}\sum_{a=0}^{d-1}\int_{\mathbb{Z}_p} f(a+dx)\,d\mu_0(x).\end{aligned} \tag{12}$$

By (12), we have

$$\begin{aligned}&\left(\prod_{i=1}^{r} a_i\right)^{-1}\int_{\mathbb{Z}_p}\cdots\int_{\mathbb{Z}_p} q^{a_1x_1+\cdots+a_rx_r} e^{(a_1dx_1+\cdots+a_rdx_r+x)t}\, d\mu_0(x_1)\cdots d\mu_0(x_r)\\ &= \frac{1}{d^r}\sum_{l_1,\ldots,l_r=0}^{d-1} q^{a_1l_1\cdots+a_rl_r}\left(\prod_{i=1}^{r} a_i\right)^{-1}\int_{\mathbb{Z}_p}\cdots\int_{\mathbb{Z}_p} q^{a_1\,dx_1+\cdots+a_r\,dx_r}\\ &\quad\times e^{(a_1l_1\cdots+a_rl_r+x+a_1\,dx_1+\cdots+a_r\,dx_r)t}\, d\mu_0(x_1)\cdots d\mu_0(x_r)\\ &= \frac{1}{d^r}\sum_{l_1,\ldots,l_r=0}^{d-1} q^{a_1l_1\cdots+a_rl_r}\left(\prod_{i=1}^{r} a_i\right)^{-1}\int_{\mathbb{Z}_p}\cdots\int_{\mathbb{Z}_p} q^{a_1\,dx_1+\cdots+a_r\,dx_r}\\ &\quad\times e^{(\frac{a_1l_1+\cdots+a_rl_r+x}{d}+a_1x_1+\cdots+a_rx_r)\,dt}\, d\mu_0(x_1)\cdots d\mu_0(x_r)\\ &= \sum_{l_1,\ldots,l_r=0}^{d-1} q^{a_1l_1\cdots+a_rl_r}\sum_{n=0}^{\infty}\frac{d^n}{d^r}\left(\prod_{i=1}^{r} a_i\right)^{-1}\int_{\mathbb{Z}_p}\cdots\int_{\mathbb{Z}_p} q^{a_1x_1+\cdots+a_rx_r}\\ &\quad\times\left(\frac{a_1l_1+\cdots+a_rl_r+x}{d}+a_1x_1+\cdots+a_rx_r\right)^n d\mu_0(x_1)\cdots d\mu_0(x_r)\frac{t^n}{n!}\\ &= \sum_{n=0}^{\infty} d^{n-r}\sum_{l_1,\ldots,l_r=0}^{d-1} q^{a_1x_1+\cdots+a_rx_r} B_{n,q}\left(\frac{a_1l_1+\cdots+a_rl_r+x}{d}\middle|a_1,\ldots,a_r\right)\frac{t^n}{n!}.\end{aligned} \tag{13}$$

By (8), (9), (11) and (13), we obtain the following theorem.

Theorem 2.2 *Let $n \in \mathbb{N} \cup \{0\}$. Then we have*

$$B_{n,q}(x|a_1,\ldots,a_r) = d^{n-r} \sum_{l_1,\ldots,l_r=0}^{d-1} q^{a_1x_1+\cdots+a_rx_r} B_{n,q}\left(\frac{l_1x_1+\cdots+l_rx_r+x}{d}\middle|a_1,\ldots,a_r\right). \tag{14}$$

It is well known that an integral equation of the bosonic p-adic integral I_0 on $\mathbb{Z}_p$ satisfies the following integral equation:

$$I_0(f_n) - I_0(f) = \sum_{i=1}^{n-1} f'(i). \tag{15}$$

If we take $f(x_i) = q^{a_ix_i}e^{a_ix_it}$ for $i = 1,\ldots,r$, then we have

$$\int_{\mathbb{Z}_p} q^{a_ix_i}e^{a_ix_it}\,d\mu_0(x_i) = \frac{a_i(t+\log q)}{q^{a_in}e^{a_int}-1}\sum_{l=0}^{n-1} q^{a_il}e^{a_ilt}. \tag{16}$$

By (16), we get

$$\begin{aligned}
&\left(\prod_{i=1}^{r} a_i\right)^{-1} \int_{\mathbb{Z}_p}\cdots\int_{\mathbb{Z}_p} q^{a_1x_1+\cdots+a_rx_r} e^{(a_1x_1+\cdots+a_rx_r)t}\,d\mu_0(x_1)\cdots d\mu_0(x_r)\\
&= \frac{(t+\log q)^r}{(q^{a_1n}e^{a_1nt}-1)\cdots(q^{a_rn}e^{a_rnt}-1)} \sum_{l_1,\ldots,l_r=0}^{n-1} q^{a_1l_1+\cdots+a_rl_r} e^{(a_1l_1+\cdots+a_rl_r)t}\\
&= \left(\sum_{k=0}^{\infty} B_{k,q}(na_1,\ldots,na_r)\frac{t^k}{k!}\right) \sum_{l_1,\ldots,l_r=0}^{n-1} q^{a_1l_1+\cdots+a_rl_r} \sum_{j=0}^{\infty}(a_1l_1+\cdots+a_rl_r)^j\frac{t^j}{j!}\\
&= \sum_{l_1,\ldots,l_r=0}^{n-1}\sum_{k=0}^{\infty}\sum_{j=0}^{\infty} q^{a_1l_1+\cdots+a_rl_r}(a_1l_1+\cdots+a_rl_r)^j B_{k,q}(na_1,\ldots,na_r)\frac{t^{k+j}}{k!j!}\\
&= \sum_{m=0}^{\infty}\sum_{l_1,\ldots,l_r=0}^{n-1}\sum_{j=0}^{m}\binom{m}{j} q^{a_1l_1+\cdots+a_rl_r}(a_1l_1+\cdots+a_rl_r)^j\\
&\quad\times B_{m-j,q}(na_1,\ldots,na_r)\frac{t^m}{m!}.
\end{aligned} \tag{17}$$

Thus, by (11) and (17), we obtain the following theorem.

Theorem 2.3 *Let $n \in \mathbb{N} \cup \{0\}$. Then we have*

$$B_{n,q}(a_1,\ldots,a_r) = \sum_{l_1,\ldots,l_r=0}^{n-1}\sum_{j=0}^{m}\binom{m}{j} q^{a_1l_1+\cdots+a_rl_r}(a_1l_1+\cdots+a_rl_r)^j B_{m-j,q}(na_1,\ldots,na_r). \tag{18}$$

By (16), we get

$$\left(\prod_{i=1}^{r} a_i\right)^{-1} \int_{\mathbb{Z}_p} \cdots \int_{\mathbb{Z}_p} q^{a_1x_1+\cdots+a_rx_r} e^{(a_1x_1+\cdots+a_rx_r)t} \, d\mu_0(x_1)\cdots d\mu_0(x_r)$$
$$= \frac{(t+\log q)^r}{(q^{a_1n}e^{a_1nt}-1)\cdots(q^{a_rn}e^{a_rnt}-1)} \sum_{l_1,\ldots,l_r=0}^{n-1} q^{a_1l_1+\cdots+a_rl_r} e^{(a_1l_1+\cdots+a_rl_r)t}$$
$$= \sum_{l_1,\ldots,l_r=0}^{n-1} \frac{(t+\log q)^r}{(q^{a_1n}e^{a_1nt}-1)\cdots(q^{a_rn}e^{a_rnt}-1)} q^{a_1l_1+\cdots+a_rl_r} e^{(a_1l_1+\cdots+a_rl_r)t}$$
$$= \sum_{l_1,\ldots,l_r=0}^{n-1} q^{a_1l_1+\cdots+a_rl_r} \frac{(t+\log q)^r}{(q^{a_1n}e^{a_1nt}-1)\cdots(q^{a_rn}e^{a_rnt}-1)} e^{\frac{a_1l_1+\cdots+a_rl_r}{n}nt}$$
$$= \sum_{l_1,\ldots,l_r=0}^{n-1} q^{a_1l_1+\cdots+a_rl_r} \sum_{m=0}^{\infty} B_{m,q^n}\left(\frac{a_1l_1+\cdots+a_rl_r}{n}\middle| a_1,\ldots,a_r\right)\frac{n^m t^m}{m!}$$
$$= \sum_{m=0}^{\infty} n^m \sum_{l_1,\ldots,l_r=0}^{n-1} q^{a_1l_1+\cdots+a_rl_r} B_{m,q^n}\left(\frac{a_1l_1+\cdots+a_rl_r}{n}\middle| a_1,\ldots,a_r\right)\frac{t^m}{m!}. \tag{19}$$

Thus, by (11) and (19), we obtain the following theorem.

Theorem 2.4 *Let $n \in \mathbb{N} \cup \{0\}$. Then we have*

$$B_{m,q}(a_1,\ldots,a_r) = n^m \sum_{l_1,\ldots,l_r=0}^{n-1} q^{a_1l_1+\cdots+a_rl_r} B_{m,q^n}\left(\frac{a_1l_1+\cdots+a_rl_r}{n}\middle| a_1,\ldots,a_r\right). \tag{20}$$

3 Higher order Barnes-type q-Euler polynomials

Higher Euler polynomials are defined as

$$\left(\frac{2}{e^t+1}\right)^r e^{xt} = \sum_{n=0}^{\infty} E_n(x)\frac{t^n}{n!} \quad \text{(see [17–19, 22–24])}. \tag{21}$$

When $x = 0$, $E_n = E_n(0)$ is called higher Euler numbers. For $f \in \mathrm{UD}(\mathbb{Z}_p)$, the fermionic p-adic integral on $\mathbb{Z}_p$ is defined by Kim to be

$$I_{-1}(f) = \int_{\mathbb{Z}_p} f(x)\,d\mu_{-1}(x) = \lim_{N\to\infty} \sum_{x=0}^{p^N-1} f(x)(-1)^x \quad \text{(see [4])}. \tag{22}$$

It is well known that an integral equation of the fermionic p-adic integral on $\mathbb{Z}_p$ is

$$I_{-1}(f_1) + I_{-1}(f) = 2f(0), \tag{23}$$

where $f_1(x) = f(x+1)$.

Let $a_1,\ldots,a_r \in \mathbb{C}_p \setminus \{0\}$. Higher order Barnes-type Euler polynomials are defined as

$$\frac{2^r}{(e^{a_1t}+1)\cdots(e^{a_rt}+1)} e^{xt} = \sum_{n=0}^{\infty} E_n(x|a_1,\ldots,a_r)\frac{t^n}{n!} \quad \text{(see [18, 19, 23])}. \tag{24}$$

When $x = 0$, $E_n(a_1, \ldots, a_r) = E_n(0|a_1, \ldots, a_r)$ is called higher order Barnes-type Euler numbers. We define higher order Barnes-type q-Euler polynomials as follows:

$$\frac{2^r}{(q^{a_1}e^{a_1 t} + 1)\cdots(q^{a_r}e^{a_r t} + 1)}e^{xt} = \sum_{n=0}^{\infty} E_{n,q}(x|a_1, \ldots, a_r)\frac{t^n}{n!}. \tag{25}$$

When $x = 0$, $E_{n,q}(a_1, \ldots, a_r) = E_{n,q}(0|a_1, \ldots, a_r)$ is called higher order Barnes-type q-Euler numbers.

By (23), if we take $f(x_i) = q^{a_i x_i}e^{a_i x_i t}$ for $i = 1, \ldots, r$, then we have

$$\int_{\mathbb{Z}_p} q^{a_i x_i}e^{a_i x_i t}\, d\mu_{-1}(x_i) = \frac{2}{q^{a_i x_i}e^{a_i x_i t} + 1}. \tag{26}$$

By (26), we get

$$\int_{\mathbb{Z}_p} q^{a_1 x_1 + \cdots + a_r t_r} e^{(a_1 x_1 + \cdots + a_r x_r + x)t}\, d\mu_0(x_1)\cdots d\mu_0(x_r)$$
$$= \frac{2^r}{(q^{a_1}e^{a_1 t} + 1)\cdots(q^{a_r}e^{a_r t} + 1)}e^{xt}. \tag{27}$$

By (24) and (27), we get

$$\sum_{n=0}^{\infty} E_{n,q}(x|a_1, \ldots, a_r)\frac{t^n}{n!}$$
$$= \frac{2^r}{(q^{a_1}e^{a_1 t} + 1)\cdots(q^{a_r}e^{a_r t} + 1)}e^{xt}$$
$$= \int_{\mathbb{Z}_p}\cdots\int_{\mathbb{Z}_p} q^{a_1 x_1 + \cdots + a_r x_r} e^{(a_1 x_1 + \cdots + a_r x_r + x)t}\, d\mu_{-1}(x_1)\cdots d\mu_{-1}(x_r)$$
$$= \sum_{n=0}^{\infty}\int_{\mathbb{Z}_p}\cdots\int_{\mathbb{Z}_p} q^{a_1 x_1 + \cdots + a_r x_r}(a_1 x_1 + \cdots + a_r x_r + x)^n\, d\mu_{-1}(x_1)\cdots d\mu_{-1}(x_r)\frac{t^n}{n!}. \tag{28}$$

From (28), we obtain the following theorem.

Theorem 3.1 *Let $n \in \mathbb{N} \cup \{0\}$. Then we have*

$$E_{n,q}(x|a_1, \ldots, a_r)$$
$$= \int_{\mathbb{Z}_p}\cdots\int_{\mathbb{Z}_p} q^{a_1 x_1 + \cdots + a_r x_r}(a_1 x_1 + \cdots + a_r x_r + x)^n\, d\mu_{-1}(x_1)\cdots d\mu_{-1}(x_r). \tag{29}$$

From (22), we have

$$\int_{\mathbb{Z}_p} f(x)\, d\mu_{-1}(x) = \lim_{N\to\infty}\sum_{x=0}^{dp^N-1} f(x)(-1)^x$$
$$= \frac{1}{d}\lim_{N\to\infty}\sum_{a=0}^{d-1}\sum_{x=0}^{p^N-1}(-1)^{a+x}f(a + dx)$$
$$= \frac{1}{d}\sum_{a=0}^{d-1}(-1)^a\int_{\mathbb{Z}_p} f(a + dx)\, d\mu_{-1}(x). \tag{30}$$

By (30), if we take $f(x_i) = q^{a_i x_i} e^{a_i x_i t}$ for $i = 1, \ldots, r$, then we have

$$\begin{aligned}\int_{\mathbb{Z}_p} q^{a_i x_i} e^{a_i x_i t}\, d\mu_{-1}(x) &= \sum_{a=0}^{d-1} (-1)^a \int_{\mathbb{Z}_p} q^{a_i(a+dx-i)} e^{a_i(a+dx_i)t}\, d\mu_{-1}(x_i) \\ &= \sum_{a=0}^{d-1} (-1)^a q^{a_i a} e^{a_i a t} \int_{\mathbb{Z}_p} q^{a_i d x_i} e^{a_i d x_i t}\, d\mu_{-1}(x_i). \end{aligned} \tag{31}$$

By (31), we get

$$\begin{aligned}&\int_{\mathbb{Z}_p} \cdots \int_{\mathbb{Z}_p} q^{a_1 x_1 + \cdots + a_r x_r} e^{(a_1 x_1 + \cdots + a_r x_r + x)t}\, d\mu_{-1}(x_1) \cdots d\mu_{-1}(x_r) \\ &= \sum_{l_1, \ldots, l_r = 0}^{d-1} (-1)^{l_1 + \cdots + l_r} q^{a_1 l_1 + \cdots + a_r l_r} \\ &\quad \times \int_{\mathbb{Z}_p} \cdots \int_{\mathbb{Z}_p} q^{a_1\, dx_1 + \cdots + a_r\, dx_r} e^{(\frac{a_1 l_1 + \cdots + a_1 l_r + x}{d} + a_1 x_1 + \cdots + a_r x_r)\, dt}\, d\mu_{-1}(x_1) \cdots d\mu_{-1}(x_r) \\ &= \sum_{n=0}^{\infty} d^n \sum_{l_1, \ldots, l_r = 0}^{d-1} (-1)^{l_1 + \cdots + l_r} q^{a_1 l_1 + \cdots + a_r l_r} \int_{\mathbb{Z}_p} \cdots \int_{\mathbb{Z}_p} q^{a_1\, dx_1 + \cdots + a_r\, dx_r} \\ &\quad \times \left(\frac{a_1 l_1 + \cdots + a_1 l_r + x}{d} + a_1 x_1 + \cdots + a_r x_r \right)^n d\mu_{-1}(x_1) \cdots d\mu_{-1}(x_r) \frac{t^n}{n!} \\ &= \sum_{n=0}^{\infty} d^n \sum_{l_1, \ldots, l_r = 0}^{d-1} (-1)^{l_1 + \cdots + l_r} q^{a_1 l_1 + \cdots + a_r l_r} \\ &\quad \times E_{n,q^d}\left(\frac{a_1 l_1 + \cdots + a_r l_r + x}{d} \middle| a_1, \ldots, a_r \right) \frac{t^n}{n!}. \end{aligned} \tag{32}$$

By (27) and (32), we obtain the following theorem.

Theorem 3.2 *Let $n \in \mathbb{N} \cup \{0\}$. Then we have*

$$\begin{aligned}&E_{n,q}(x | a_1, \ldots, a_r) \\ &= d^n \sum_{l_1, \ldots, l_r = 0}^{d-1} (-1)^{l_1 + \cdots + l_r} q^{a_1 l_1 + \cdots + a_r l_r} E_{n,q^d}\left(\frac{a_1 l_1 + \cdots + a_r l_r + x}{d} \middle| a_1, \ldots, a_r \right).\end{aligned}$$

Competing interests
The authors declare that they have no competing interests.

Authors' contributions
All authors contributed equally to this work. All authors read and approved the final manuscript.

Author details
[1]Graduate School of Education, Konkuk University, Seoul, 143-701, Korea. [2]Department of Mathematics Education, Konkuk University, Seoul, 143-701, Korea. [3]Department of Mathematics, Kwangwoon University, Seoul, Korea.

References
1. Kim, T: Barnes' type multiple degenerate Bernoulli and Euler polynomials. Appl. Math. Comput. **258**, 556-564 (2015)
2. Kim, T, Kim, DS, Bayad, A, Rim, S-H: Identities on the Bernoulli and the Euler numbers and polynomials. Ars Comb. **107**, 455-463 (2012)

3. Kim, T, Dolgy, DV, Kim, DS, Rim, S-H: A note on the identities of special polynomials. Ars Comb. **113A**, 97-106 (2014)
4. Kim, T: On the analogs of Euler numbers and polynomials associated with p-adic q-integral on $\mathbb{Z}_p$ at $q = -1$. J. Math. Anal. Appl. **331**(2), 779-792 (2007)
5. Kim, DS, Kim, T: A note on poly-Bernoulli and higher-order poly-Bernoulli polynomials. Russ. J. Math. Phys. **22**(1), 26-33 (2015)
6. Lim, D, Do, Y: Some identities of Barnes-type special polynomials. Adv. Differ. Equ. **2015**, 42 (2015)
7. Luo, Q-M, Qi, F: Relationships between generalized Bernoulli numbers and polynomials and generalized Euler numbers and polynomials. Adv. Stud. Contemp. Math. **7**(1), 11-18 (2003)
8. Ozden, H: p-Adic distribution of the unification of the Bernoulli, Euler and Genocchi polynomials. Appl. Math. Comput. **218**(3), 970-973 (2011)
9. Park, J-W, Rim, S-H: On the modified q-Bernoulli polynomials with weight. Proc. Jangjeon Math. Soc. **17**(2), 231-236 (2014)
10. Ryoo, CS, Kwon, HI, Yoon, J, Jang, YS: Fourier series of the periodic Bernoulli and Euler functions. Abstr. Appl. Anal. **2014**, Article ID 856491 (2014)
11. Ding, D, Yang, J: Some identities related to the Apostol-Euler and Apostol-Bernoulli polynomials. Proc. Jangjeon Math. Soc. **17**(1), 115-123 (2014)
12. Jang, LC: A family of Barnes-type multiple twisted q-Euler numbers and polynomials related to fermionic p-adic invariant integrals on $\mathbb{Z}_p$. J. Comput. Anal. Appl. **13**(2), 376-387 (2011)
13. Andrews, GE, Askey, R, Roy, R: Special Functions. Encyclopedia of Mathematics and Its Applications, vol. 71. Cambridge University Press, Cambridge (1999). ISBN:0-521-62321-9; 0-521-78988-5
14. Bayad, A, Kim, T: Results on values of Barnes polynomials. Rocky Mt. J. Math. **43**(6), 1857-1869 (2013)
15. Bayad, A, Kim, T, Kim, W, Lee, SH: Arithmetic properties of q-Barnes polynomials. J. Comput. Anal. Appl. **15**(1), 111-117 (2013)
16. Kang, D, Jeong, J-J, Lee, BJ, Rim, S-H, Choi, SH: Some identities of higher order Genocchi polynomials arising from higher order Genocchi basis. J. Comput. Anal. Appl. **17**(1), 141-146 (2014)
17. Chen, C-P, Srivastava, HM: Some inequalities and monotonicity properties associated with the gamma and psi functions and the Barnes G-function. Integral Transforms Spec. Funct. **22**(1), 1-15 (2011)
18. Kim, T: p-Adic q-integrals associated with the Changhee-Barnes' q-Bernoulli polynomials. Integral Transforms Spec. Funct. **15**(5), 415-420 (2004)
19. Kim, T: Barnes-type multiple q-zeta functions and q-Euler polynomials. J. Phys. A **43**(25), 255201 (2010)
20. Kim, T: On the multiple q-Genocchi and Euler numbers. Russ. J. Math. Phys. **15**(4), 482-486 (2008)
21. Kim, T, Rim, S-H: On Changhee-Barnes' q-Euler numbers and polynomials. Adv. Stud. Contemp. Math. **9**(2), 81-86 (2004)
22. Can, M, Cenkci, M, Kurt, V, Simsek, Y: Twisted Dedekind type sums associated with Barnes' type multiple Frobenius-Euler l-functions. Adv. Stud. Contemp. Math. **18**(2), 135-160 (2009)
23. Jang, L, Kim, T, Kim, Y-H, Hwang, K-W: Note on the q-extension of Barnes' type multiple Euler polynomials. J. Inequal. Appl. **2009**, Article ID 13532 (2009)
24. Kim, T: On Euler-Barnes multiple zeta functions. Russ. J. Math. Phys. **10**(3), 261-267 (2003)

7

Existence of positive solutions for a system of Caputo fractional difference equations depending on parameters

Shugui Kang*, Huiqin Chen and Jianmin Guo

*Correspondence:
dtkangshugui@126.com
School of Mathematics and Computer Sciences, Shanxi Datong University, Datong, Shanxi 037009, P.R. China

Abstract

We consider the existence of at least two positive solutions for a system of Caputo fractional difference equations $\Delta_C^{\nu_j} y_j(t) = -\lambda_j f_j(y_1(t+\nu_1-1), \dots, y_n(t+\nu_n-1))$, subject to boundary conditions $y_j(\nu_j - 3) = \Delta y_j(\nu_j + b) = \Delta^2 y_j(\nu_j - 3) = 0$, where $2 < \nu_j \leq 3$, $j = 1, \dots, n$. We use the Krasnosel'skiĭ fixed point theorem to obtain the sufficient conditions of the existence of two positive solutions for this boundary value problem of Caputo fractional difference equations depending on parameters.

MSC: 26A33; 39A05; 39A12

Keywords: Caputo fractional difference; boundary value problem; fixed point theory; positive solution

1 Introduction

In this paper we consider a system of Caputo fractional difference boundary value problem (FBVP) of the form:

$$\Delta_C^{\nu_j} y_j(t) = -\lambda_j f_j\big(y_1(t+\nu_1-1), \dots, y_n(t+\nu_n-1)\big), \tag{1.1}$$

$$y_j(\nu_j - 3) = \Delta y_j(\nu_j + b) = \Delta^2 y_j(\nu_j - 3) = 0, \tag{1.2}$$

where $t \in [0, b+1]_{\mathbb{N}_0} := \{0, 1, \dots, b+1\}$, $b > 3$, $\lambda_j > 0$, $2 < \nu_j \leq 3$, $f_j : [0, +\infty) \times \cdots \times [0, +\infty) \to [0, +\infty)$ are continuous functions for each j ($j = 1, 2, \dots, n$). $\Delta_C^{\nu} y(t)$ is the standard Caputo difference.

Fractional difference equations have been of great interest recently. It is caused by intensive development of the theory of discrete fractional calculus itself, see [1–19] and the references therein. Abdeljawad [1] defined left and right Caputo fractional sums and differences, studied some of their properties. Holm [2] introduced the fractional sum and difference operators. He developed and presented a complete and precise theory for composing fractional sums and differences. Atici and Sengül [3] provided some analysis of discrete fractional variational problems, their paper also provided some initial attempts at using the discrete fractional calculus to model biological processes. Abdeljawad and Baleanu [4] defined the right fractional sum and difference operators and obtained many of their properties. Then by using those properties they obtained a by-part formula analogous to that in the usual fractional calculus. In [5] the authors studied the stability of discrete nonautonomous systems within the frame of the Caputo fractional difference by using the Lyapunov direct method. They discussed the conditions for uniform stability, uniform asymptotic stability, and uniform global stability. Mohammadi and Rezapour [6] discussed the existence and uniqueness of solutions for some nonlinear fractional differential equations via some boundary value problems by using fixed point results on ordered complete gauge spaces. Recently, Wu and Baleanu introduced some applications of the Caputo fractional difference to discrete chaotic maps in [7, 8].

In particular, the authors [9–19] developed some of the basic theory of fractional difference both IVPs and BVPs with delta derivative on the time scale $\mathbb{Z}$. In [10], we obtained some results on the existence of one or more positive solutions for the Caputo fractional boundary value problems by means of cone theoretic fixed point theorems. Thus, the fractional difference equation has recently attracted increasing attention from a growing number of researchers. However, systems of discrete fractional boundary value problems are limited (see [14–19]). Among them, Atici and Eloe [14] studied a linear system of fractional nabla difference equation with constant coefficients of the form

$$\nabla_0^{\nu} y(t) = Ay(t) + f(t), \quad t = 1, 2, \ldots,$$

where $0 < \nu < 1$, A is an $n \times n$ matrix with constant entries, and f are n-vector valued functions. The operator ∇_0^{ν} is a Riemann-Liouville fractional difference. They constructed the fundamental matrix for the homogeneous system and the causal Green's function for the nonhomogeneous system.

In [15], the authors investigated the existence of solutions for a k-dimensional system of fractional finite difference equations:

$$\begin{aligned}
&\Delta^{\nu_1} y_1(t) + f_1\big(y_1(t + \nu_1 - 1), y_2(t + \nu_2 - 1), \ldots, y_k(t + \nu_k - 1)\big) = 0,\\
&\Delta^{\nu_2} y_2(t) + f_2\big(y_1(t + \nu_1 - 1), y_2(t + \nu_2 - 1), \ldots, y_k(t + \nu_k - 1)\big) = 0,\\
&\ldots,\\
&\Delta^{\nu_k} y_k(t) + f_k\big(y_1(t + \nu_1 - 1), y_2(t + \nu_2 - 1), \ldots, y_k(t + \nu_k - 1)\big) = 0,\\
&y_1(\nu_1 - 2) = \Delta y_1(\nu_1 + b) = 0,\\
&y_2(\nu_2 - 2) = \Delta y_2(\nu_2 + b) = 0,\\
&\ldots,\\
&y_k(\nu_k - 2) = \Delta y_k(\nu_k + b) = 0,
\end{aligned}$$

where $b \in \mathbb{N}_0, 1 < \nu_i \leq 2, f_i : \mathbb{R}^k \to \mathbb{R}$ are continuous functions for $k = 1, 2, \ldots$. They investigated the existence of solutions for this k-dimensional system of fractional finite difference equations by using the Krasnosel'skiĭ fixed point theorem.

In [16], Goodrich studied the following pair of discrete fractional boundary value problems:

$$\begin{aligned}
&-\Delta^{\nu_1} y_1(t) = \lambda_1 a_1(t + \nu_1 - 1) f_1\big(y_1(t + \nu_1 - 1), y_2(t + \nu_n - 1)\big),\\
&-\Delta^{\nu_2} y_2(t) = \lambda_2 a_2(t + \nu_1 - 1) f_2\big(y_1(t + \nu_1 - 1), y_2(t + \nu_n - 1)\big),
\end{aligned}$$

$$y_1(\nu_1-2)=\psi_1(y_1),\qquad y_2(\nu_2-2)=\psi_2(y_2),$$
$$y_1(\nu_1+b)=\phi_1(y_1),\qquad y_2(\nu_2+b)=\phi_2(y_2),$$

where $t\in[0,b]_{\mathbb{N}_0}:=\{0,1,\dots,b\}$, $\lambda_1,\lambda_2>0$, $\nu_1,\nu_2\in(1,2]$. Goodrich obtained the existence of at least one positive solution to this problem by means of the Krasnosel'skiĭ theorem for cons.

In [18], the authors considered the existence of at least one positive solution to the discrete fractional system:

$$\begin{cases}-\Delta^{\nu_1}y_1(t)=\lambda_1 f_1(t+\nu_1-1,y_1(t+\nu_1-1),y_2(t+\nu_2-1)), & t\in[1,b+1],\\ -\Delta^{\nu_2}y_2(t)=\lambda_2 f_2(t+\nu_2-1,y_1(t+\nu_1-1),y_2(t+\nu_2-1)), & t\in[1,b+1],\\ y_1(\nu_1-2)=y_1(\nu_1+b+1)=0,\\ y_2(\nu_2-2)=y_2(\nu_2+b+1)=0,\end{cases}$$

where $\nu_1,\nu_2\in(1,2]$.

Following this trend, in [19], we discussed the boundary value problems of fractional difference system of the form

$$-\Delta^{\nu_j}y_j(t)=\lambda_j f_j\big(y_1(t+\nu_1-1),\dots,y_n(t+\nu_n-1)\big),$$
$$y_j(\nu_j-2)=\psi_j(y_j),\qquad y_j(\nu_j+b)=\phi_j(y_j),$$

where $t\in[0,b]_{\mathbb{N}_0}:=\{0,1,\dots,b\}$, $\lambda_j>0$, $1<\nu_j\leq 2$, $f_j:[0,+\infty)\times\cdots\times[0,+\infty)\to[0,+\infty)$ are continuous functions. For each j we have that $\psi_j,\phi_j:\mathbb{R}^{b+3}\to\mathbb{R}$ ($j=1,2,\dots,n$) are given functions. We obtained the sufficient conditions for the existence of two positive solutions to the boundary value problem of a fractional difference system. In this paper we open our studies in this field. We establish some conditions on parameters λ_i which are able to guarantee that FBVP (1.1)-(1.2) has at least two positive solutions and one positive solution, respectively, based on the Krasnosel'skiĭ theorem.

This paper is organized as follows. In Section 2, we provide basic definitions and demonstrate some lemmas in order to prove our main results. In Section 3, we establish some results for the existence of at least two positive solutions to FBVP (1.1)-(1.2), and we conclude with an example explicating our main result.

2 Preliminaries

In this section, we present some basic definitions in the discrete fractional calculus and establish some lemmas.

Definition 2.1 [1] We define

$$t^{(\nu)}=\frac{\Gamma(t+1)}{\Gamma(t+1-\nu)}$$

for any t and ν for which the right-hand side is defined. We also appeal to the convention that if $t+1-\nu$ is a pole of the gamma function and $t+1$ is not a pole, then $t^{(\nu)}=0$.

Definition 2.2 [1] The νth fractional sum of a function f is defined by

$$\Delta^{-\nu}f(t)=\frac{1}{\Gamma(\nu)}\sum_{s=a}^{t-\nu}(t-s-1)^{\underline{\nu-1}}f(s)$$

for $\nu > 0$ and $t \in \{a+\nu, a+\nu+1, \ldots\} = \mathbb{N}_{a+\nu}$. We also define the νth Caputo fractional difference for $\nu > 0$ by

$$\Delta_C^{\nu} f(t) = \Delta^{-(n-\nu)} \Delta^n f(t) = \frac{1}{\Gamma(n-\nu)} \sum_{s=a}^{t-(n-\nu)} (t-s-1)^{\underline{n-\nu-1}} \Delta^n f(s),$$

where $n-1 < \nu \leq n$.

Lemma 2.3 [1, 2] *Assume that $\nu > 0$ and f is defined on domains $\mathbb{N}_a$, then*

$$\Delta_{a+(n-\nu)}^{-\nu} \Delta_C^{\nu} f(t) = f(t) - \sum_{k=0}^{n-1} c_k (t-a)^{\underline{k}},$$

where $c_i \in \mathbb{R}$, $i = 0, 1, \ldots, n-1$; $n-1 < \nu \leq n$.

Lemma 2.4 [2] *Let $f : \mathbb{N}_{a+\nu} \times \mathbb{N}_a \to \mathbb{R}$ be given. Then*

$$\Delta\left(\sum_{s=a}^{t-\nu} f(t,s)\right) = \sum_{s=a}^{t-\nu} \Delta_t f(t,s) + f(t+1, t+1-\nu) \quad \text{for } t \in \mathbb{N}_{a+\nu}.$$

In order to get our main results, we now state an important lemma. This lemma gives a representation for the solution of (1.1)-(1.2), provided that the solution exists.

Lemma 2.5 [10] *Let $2 < \nu \leq 3$ and $g : [\nu-2, \nu-1, \ldots, \nu+b]_{\mathbb{N}_{\nu-2}} \to \mathbb{R}$ be given. Then the solution of the FBVP*

$$\Delta_C^{\nu} y(t) = -g(t+\nu-1), \tag{2.1}$$

$$y(\nu-3) = \Delta y(\nu+b) = \Delta^2 y(\nu-3) = 0 \tag{2.2}$$

is given by

$$y(t) = \sum_{s=0}^{b+1} G(t,s) g(s+\nu-1),$$

where the Green's function $G : [\nu-2, \nu-1, \ldots, \nu+b]_{\mathbb{N}_{\nu-2}} \times [0, b+1]_{\mathbb{N}_0} \to \mathbb{R}$ is defined by

$$G(t,s) = \frac{1}{\Gamma(\nu)} \begin{cases} (\nu-1)(t-\nu+3)(\nu+b-s-1)^{\underline{\nu-2}} \\ \quad -(t-s-1)^{\underline{\nu-1}}, & 0 \leq s < t-\nu+1 \leq b+1, \\ (\nu-1)(t-\nu+3)(\nu+b-s-1)^{\underline{\nu-2}}, & 0 \leq t-\nu+1 \leq s \leq b+1. \end{cases}$$

Remark Notice that $G(\nu-3, s) = 0$, $G(t, b+2) = 0$. G could be extended to $[\nu-3, \nu+b]_{\mathbb{N}_{\nu-3}} \times [0, b+2]_{\mathbb{N}_0}$, so we only discuss on $(t,s) \in [\nu-2, \nu+b]_{\mathbb{N}_{\nu-2}} \times [0, b+1]_{\mathbb{N}_0}$.

Lemma 2.6 [10] *The Green's function G satisfies the following conditions*:

(i) $G(t,s) > 0$, $(t,s) \in [\nu-2, \nu+b]_{\mathbb{N}_{\nu-2}} \times [0, b+1]_{\mathbb{N}_0}$.

(ii) $\max_{t \in [\nu-2, \nu+b]_{\mathbb{N}_{\nu-2}}} G(t,s) = G(\nu+b, s)$, $s \in [0, b+1]_{\mathbb{N}_0}$.

(iii) $\min_{\frac{\nu+b}{4} \leq t \leq \frac{3(\nu+b)}{4}} G(t,s) \geq \frac{1}{4} \max_{t \in [\nu-2, \nu+b]_{\mathbb{N}_{\nu-2}}} G(t,s) = \frac{1}{4} G(\nu+b, s)$, $s \in [0, b+1]_{\mathbb{N}_0}$.

The proofs of Lemma 2.5 and Lemma 2.6 can be found in [10], so we omit their proofs. Let

$$\mathcal{B}_j := \left\{y_j : [\nu_j - 3, \nu_j + b]_{\mathbb{N}_{\nu_j-3}} \to \mathbb{R}, y_j(\nu_j - 3) = \Delta y_j(\nu_j + b) = \Delta^2 y_j(\nu_j - 3) = 0\right\}$$

be equipped with the usual maximum norm $\|\cdot\|$, it is easy to verify that $\mathcal{B}_j$ is the Banach space. Then we put $\mathcal{K} := \mathcal{B}_1 \times \mathcal{B}_2 \times \cdots \times \mathcal{B}_n$. By equipping $\mathcal{K}$ with the norm

$$\left\|(y_1, \dots, y_n)\right\| = \|y_1\| + \cdots + \|y_n\|,$$

it follows that $(\mathcal{K}, \|\cdot\|)$ is a Banach space.

Now consider the operator $T : \mathcal{K} \to \mathcal{K}$ defined by

$$T(y_1, \dots, y_n)(t_1, \dots, t_n) = \left(T_1(y_1, \dots, y_n)(t_1), \dots, T_n(y_1, \dots, y_n)(t_n)\right), \tag{2.3}$$

where we define $T_j : \mathcal{K} \to \mathcal{B}_j$ by

$$T_j(y_1, \dots, y_n)(t_j) = \lambda_j \sum_{s=0}^{b+1} G_j(t_j, s) f_j\left(y_1(s + \nu_1 - 1), \dots, y_n(s + \nu_n - 1)\right). \tag{2.4}$$

Let $I := [\frac{\nu_1+b}{4}, \frac{3(\nu_1+b)}{4}] \times \cdots \times [\frac{\nu_n+b}{4}, \frac{3(\nu_n+b)}{4}]$. In the sequel, we shall also make use of the cone

$$\Lambda := \left\{(y_1, \dots, y_n) \in \mathcal{K} : y_1, \dots, y_n \geq 0, \right.$$
$$\left. \min_{(t_1,\dots,t_n)\in I} \left[y_1(t_1) + \cdots + y_n(t_n)\right] \geq \frac{1}{4}\left\|(y_1, \dots, y_n)\right\|\right\}.$$

Lemma 2.7 *Let T be the operator defined as in* (2.3). *Then $T : \Lambda \to \Lambda$.*

Proof We show first that for $(y_1, \dots, y_n) \in \mathcal{K}$, by the definitions T_j $(j = 1, 2, \dots, n)$, it is clear that

$$T_j(y_1, \dots, y_n)(t_j) \geq 0, \quad j = 1, 2, \dots, n.$$

On the other hand, we show that

$$\min_{(t_1,\dots,t_n)\in I} \left[T_1(y_1, \dots, y_n)(t_1) + \cdots + T_n(y_1, \dots, y_n)(t_n)\right] \geq \frac{1}{4}\left\|T(y_1, \dots, y_n)\right\|$$

for $(y_1, \dots, y_n) \in \mathcal{K}$. In fact, by Lemma 2.6(iii), we have

$$\min_{t_j\in[\frac{\nu_j+b}{4}, \frac{3(\nu_j+b)}{4}]} T_j(y_1, \dots, y_n)(t_j)$$
$$\geq \min_{t_j\in[\frac{b+\nu_j}{4}, \frac{3(\nu_j+b)}{4}]} \lambda_j \sum_{s=0}^{b+1} G_j(t_j, s) f_j\left(y_1(s + \nu_1 - 1), \dots, y_n(s + \nu_n - 1)\right)$$

$$\geq \lambda_j \sum_{s=0}^{b+1} \frac{1}{4} G_j(v_j + b, s) f_j\big(y_1(s + v_1 - 1), \ldots, y_n(s + v_n - 1)\big)$$

$$\geq \frac{1}{4} \max_{t_j \in [v_j - 2, v_j + b]_{\mathbb{N}_{v_j - 2}}} \lambda_j \sum_{s=0}^{b+1} G_j(t_j, s) f_j\big(y_1(s + v_1 - 1), \ldots, y(s + v_n - 1)\big)$$

$$= \frac{1}{4} \big\| T_j(y_1, \ldots, y_n) \big\|$$

for $(y_1, \ldots, y_n) \in \mathcal{K}$ and $j = 1, 2, \ldots, n$. Then we obtain

$$\begin{aligned}
&\min_{(t_1,\ldots,t_n)\in I} \big[T_1(y_1, \ldots, y_n)(t_1) + \cdots + T_n(y_1, \ldots, y_n)(t_n)\big] \\
&\quad \geq \min_{(t_1,\ldots,t_n)\in I} T_1(y_1, \ldots, y_n)(t_1) + \cdots + \min_{(t_1,\ldots,t_n)\in I} T_n(y_1, \ldots, y_n)(t_n) \\
&\quad \geq \frac{1}{4} \big\| T_1(y_1, \ldots, y_n) \big\| + \cdots + \frac{1}{4} \big\| T_n(y_1, \ldots, y_n) \big\| \\
&\quad \geq \frac{1}{4} \big\{ \big\| T_1(y_1, \ldots, y_n) \big\| + \cdots + \big\| T_n(y_1, \ldots, y_n) \big\| \big\} \\
&\quad = \frac{1}{4} \big\| T(y_1, \ldots, y_n) \big\|
\end{aligned}$$

for $(y_1, \ldots, y_n) \in \mathcal{K}$. So, we conclude that $T : \Lambda \to \Lambda$. This completes the proof. □

Theorem 2.8 *Let* $f_j : [0, +\infty) \times \cdots \times [0, +\infty) \to [0, +\infty)$ *be given for* $j = 1, \ldots, n$. *If* $(y_1, \ldots, y_n) \in \mathcal{K}$ *is a fixed point of* T, *then* $(y_1, \ldots, y_n) \in \mathcal{K}$ *is a solution of FBVP* (1.1)-(1.2).

Proof Suppose that the operator T has a fixed point, say $(y_1, \ldots, y_n) \in \mathcal{K}$. Let $(t_1, \ldots, t_n) \in \mathbb{N}_{v_1 - 2} \times \cdots \times \mathbb{N}_{v_n - 2}$, then we have

$$y_j(t_j) = T_j(y_1, \ldots, y_n)(t_j), \quad j = 1, 2, \ldots, n,$$

where T_j is defined as in (2.4). It is easy to check that

$$T_j(y_1, \ldots, y_n)(v_j - 3) = 0$$

and

$$\begin{aligned}
&\Delta T_j(y_1, \ldots, y_n)(v_j + b) \\
&\quad = T_j(y_1, \ldots, y_n)(v_j + b - 1) - T_j(y_1, \ldots, y_n)(v_j + b) \\
&\quad = \lambda_j \sum_{s=0}^{b+1} G_j(v_j + b + 1, s) f_j\big(y_1(s + v_1 - 1), \ldots, y_n(s + v_n - 1)\big) \\
&\qquad - \lambda_j \sum_{s=0}^{b+1} G_j(v_j + b, s) f_j\big(y_1(s + v_1 - 1), \ldots, y_n(s + v_n - 1)\big) \\
&\quad = \lambda_j \sum_{s=0}^{b+1} \big[G_j(v_j + b + 1, s) - G_j(v_j + b, s)\big] f_j\big(y_1(s + v_1 - 1), \ldots, y_n(s + v_n - 1)\big) \\
&\quad = \frac{\lambda_j}{\Gamma(v)} \sum_{s=0}^{b+1} \big[(v_j - 1)(v_j + b + 1 - v_j + 3)(v_j + b - s - 1)^{\underline{v_j - 2}} - (v_j + b + 1 - s - 1)^{\underline{v_j - 1}}
\end{aligned}$$

$$
\begin{aligned}
&- (\nu_j - 1)(\nu_j + b - \nu_j + 3)(\nu_j + b - s - 1)^{\underline{\nu_j - 2}} + (\nu_j + b - s - 1)^{\underline{\nu_j - 1}}\big] \\
&\times f_j\big(y_1(s + \nu_1 - 1), \dots, y_n(s + \nu_n - 1)\big) \\
&= \frac{\lambda_j}{\Gamma(\nu)} \sum_{s=0}^{b+1} \bigg\{ (\nu_j - 1)(\nu_j + b - s - 1)^{\underline{\nu_j - 2}} \\
&- \left[\frac{(\nu_j + b - s)\Gamma(\nu_j + b - s)}{(b - s + 1)\Gamma(b - s + 1)} - \frac{\Gamma(\nu_j + b - s)}{\Gamma(b - s + 1)} \right] \bigg\} \\
&\times f_j\big(y_1(s + \nu_1 - 1), \dots, y_n(s + \nu_n - 1)\big) \\
&= \frac{\lambda_j}{\Gamma(\nu)} \sum_{s=0}^{b+1} \left[(\nu_j - 1)(\nu_j + b - s - 1)^{\underline{\nu_j - 2}} - \frac{(\nu_j - 1)\Gamma(\nu_j + b - s)}{(b - s + 1)\Gamma(b - s + 1)} \right] \\
&\times f_j\big(y_1(s + \nu_1 - 1), \dots, y_n(s + \nu_n - 1)\big) \\
&= 0.
\end{aligned}
$$

Finally, when $0 \le t_j - \nu_j + 1 \le s \le b + 1$,

$$
G_j(t_j, s) = \frac{1}{\Gamma(\nu_j)} (\nu_j - 1)(t_j - \nu_j + 3)(\nu_j + b - s - 1)^{\underline{\nu_j - 2}},
$$

then

$$
\Delta_{t_j}^2 G_j(t_j, s) = 0.
$$

Therefore, we can get

$$
\Delta^2 T_j(y_1, \dots, y_n)(\nu_j - 3) = 0.
$$

So the boundary conditions are satisfied, which completes the proof. □

Finally, to accomplish proof of our main results, we state cones theory. In particular, we require the following well-known fixed point theorem for cons in [20].

Theorem 2.9 [20] *Let $\mathcal{B}$ a Banach space and let $\mathcal{K} \subseteq \mathcal{B}$ be a cone. Assume that Ω_1 and Ω_2 are bounded open sets contained in $\mathcal{B}$ such that $0 \in \Omega_1$ and $\overline{\Omega_1} \subseteq \Omega_2$. Assume further that $T : \mathcal{K} \cap (\overline{\Omega}_2 \backslash \Omega_1) \to \mathcal{K}$ is a completely continuous operator. If either*

(i) *$\|Ty\| \le \|y\|$ for $y \in \mathcal{K} \cap \partial\Omega_1$ and $\|Ty\| \ge \|y\|$ for $y \in \mathcal{K} \cap \partial\Omega_2$, or*
(ii) *$\|Ty\| \ge \|y\|$ for $y \in \mathcal{K} \cap \partial\Omega_1$ and $\|Ty\| \le \|y\|$ for $y \in \mathcal{K} \cap \partial\Omega_2$,*

then the operator T has at least one fixed point in $\mathcal{K} \cap (\overline{\Omega}_2 \backslash \Omega_1)$.

3 Main results

In this section, we state and prove the existence of at least two positive solutions regarding FBVP (1.1)-(1.2). Then we conclude this section with examples to illustrate our main results. First, denote

$$
\alpha_j = \sum_{s=0}^{b+1} G_j(\nu_j + b, s), \qquad \beta_j = \sum_{s=[\frac{\nu_j+b}{4} - \nu_j + 1]}^{[\frac{3(\nu_j+b)}{4} - \nu_j + 1]} G_j\left(\left[\frac{b - \nu_j}{2} \right] + \nu_j, s \right),
$$

$$\Omega_\xi = \{(y_1,\ldots,y_n) \in \Lambda : \|(y_1,\ldots,y_n)\| < \xi\},$$
$$\partial\Omega_\xi = \{(y_1,\ldots,y_n) \in \Lambda : \|(y_1,\ldots,y_n)\| = \xi\}.$$

For convenience, we now present the conditions that we presume in the sequel.

(H_1) $\lim_{(y_1+\cdots+y_n)\to 0^+} \frac{f_j(y_1,\ldots,y_n)}{y_1+\cdots+y_n} = \infty,\ t_j \in [\nu_j - 2, \nu_j + b]_{\mathbb{N}_{\nu_j-2}}, j = 1,2,\ldots,n,$

(H_2) $\lim_{(y_1+\cdots+y_n)\to \infty} \frac{f_j(y_1,\ldots,y_n)}{y_1+\cdots+y_n} = \infty,\ t_j \in [\nu_j - 2, \nu_j + b]_{\mathbb{N}_{\nu_j-2}}, j = 1,2,\ldots,n,$

(H_3) $\lim_{(y_1+\cdots+y_n)\to 0^+} \frac{f_j(y_1,\ldots,y_n)}{y_1+\cdots+y_n} = 0,\ t_j \in [\nu_j - 2, \nu_j + b]_{\mathbb{N}_{\nu_j-2}}, j = 1,2,\ldots,n,$

(H_4) $\lim_{(y_1+\cdots+y_n)\to \infty} \frac{f_j(y_1,\ldots,y_n)}{y_1+\cdots+y_n} = 0,\ t_j \in [\nu_j - 2, \nu_j + b]_{\mathbb{N}_{\nu_j-2}}, j = 1,2,\ldots,n,$

(H_5) $\lim_{(y_1+\cdots+y_n)\to 0^+} \frac{f_j(y_1,\ldots,y_n)}{y_1+\cdots+y_n} = l_j,\ t_j \in [\nu_j - 2, \nu_j + b]_{\mathbb{N}_{\nu_j-2}}, j = 1,2,\ldots,n,$

(H_6) $\lim_{(y_1+\cdots+y_n)\to \infty} \frac{f_j(y_1,\ldots,y_n)}{y_1+\cdots+y_n} = L_j,\ t_j \in [\nu_j - 2, \nu_j + b]_{\mathbb{N}_{\nu_j-2}}, j = 1,2,\ldots,n,$

where $0 < l_j, L_j < +\infty$.

Theorem 3.1 *Suppose that there exist two different positive numbers r_1 and r_2 ($r_1 < r_2$) such that*

$$\max_{0\le y_1+\cdots+y_n\le r_1} f_j(y_1,\ldots,y_n) \le \frac{r_1}{n\lambda_j\alpha_j},$$
$$\min_{\frac{1}{4}r_2\le y_1+\cdots+y_n\le r_2} f_j(y_1,\ldots,y_n) \ge \frac{r_2}{n\lambda_j\beta_j}.$$

Then the operator T has a fixed point $(\overline{y}_1,\ldots,\overline{y}_n) \in \Lambda$ such that

$$r_1 \le \|(\overline{y}_1,\ldots,\overline{y}_n)\| \le r_2.$$

Proof For any $(y_1,\ldots,y_n) \in \Omega_{r_1}$ and $\|(y_1,\ldots,y_n)\| = r_1$, we have

$$\begin{aligned}
&\|T_j(y_1,\ldots,y_n)\| \\
&= \max_{t_j\in[\nu_j-2,\nu_j+b]_{\mathbb{N}_{\nu_j-2}}} \left|\lambda_j \sum_{s=0}^{b+1} G_j(t_j,s) f_j\big(y_1(s+\nu_1-1),\ldots,y_n(s+\nu_n-1)\big)\right| \\
&\le \lambda_j \sum_{s=0}^{b+1} \max_{t_j\in[\nu_j-2,\nu_j+b]_{\mathbb{N}_{\nu_j-2}}} G_j(t_j,s) f_j\big(y_1(s+\nu_1-1),\ldots,y_n(s+\nu_n-1)\big) \\
&\le \lambda_j \sum_{s=0}^{b+1} G_j(\nu_j+b,s)\frac{r_1}{n\lambda_j\alpha_j} \\
&= \frac{r_1}{n} \\
&= \frac{1}{n}\|(y_1,\ldots,y_n)\|,
\end{aligned}$$

$j = 1,\ldots,n$. That is,

$$\begin{aligned}
&\|T(y_1,\ldots,y_n)(t_1,\ldots,t_n)\| \\
&= \|(T_1(y_1,\ldots,y_n)(t_1),\ldots,T_n(y_1,\ldots,y_n)(t_n))\|
\end{aligned}$$

$$= \left\| T_1(y_1,\ldots,y_n) \right\| + \cdots + \left\| T_n(y_1,\ldots,y_n) \right\|$$
$$\leq \frac{1}{n}\left\| (y_1,\ldots,y_n) \right\| + \cdots + \frac{1}{n}\left\| (y_1,\ldots,y_n) \right\|$$
$$= \left\| (y_1,\ldots,y_n) \right\|$$

for $(y_1,\ldots,y_n) \in \partial\Omega_{r_1}$.

On the other hand, for any $(y_1,\ldots,y_n) \in \Omega_{r_2}$ and $\frac{\nu_j+b}{4} \leq t_j \leq \frac{3(\nu_j+b)}{4}$, note that $[\frac{b-\nu_j}{2}] + \nu_j \in [\frac{\nu_j+b}{4}, \frac{3(\nu_j+b)}{4}]$, we have

$$\left(T_j(y_1,\ldots,y_n)\right)\left(\left[\frac{b-\nu_j}{2}\right] + \nu_j\right)$$
$$= \lambda_j \sum_{s=0}^{b+1} G_j\left(\left[\frac{b-\nu_j}{2}\right] + \nu_j, s\right) f_j\left(y_1(s+\nu_1-1),\ldots,y_n(s+\nu_n-1)\right)$$
$$\geq \lambda_j \sum_{s=[\frac{\nu_j+b}{4}-\nu_j+1]}^{[\frac{3(\nu_j+b)}{4}-\nu_j+1]} G_j\left(\left[\frac{b-\nu_j}{2}\right] + \nu_j, s\right) f_j\left(y_1(s+\nu_1-1),\ldots,y_n(s+\nu_n-1)\right)$$
$$\geq \lambda_j \sum_{s=[\frac{\nu_j+b}{4}-\nu_j+1]}^{[\frac{3(\nu_j+b)}{4}-\nu_j+1]} G_j\left(\left[\frac{b-\nu_j}{2}\right] + \nu_j, s\right) \frac{r_2}{n\lambda_j\beta_j}$$
$$= \frac{r_2}{n}.$$

Then

$$\left\| T_j(y_1,\ldots,y_n) \right\| \geq \frac{r_2}{n} = \frac{1}{n}\left\| (y_1,\ldots,y_n) \right\|, \quad j = 1,\ldots,n.$$

That is,

$$\left\| T(y_1,\ldots,y_n)(t_1,\ldots,t_n) \right\|$$
$$= \left\| \left(T_1(y_1,\ldots,y_n)(t_1),\ldots,T_n(y_1,\ldots,y_n)(t_n)\right) \right\|$$
$$= \left\| T_1(y_1,\ldots,y_n) \right\| + \cdots + \left\| T_n(y_1,\ldots,y_n) \right\|$$
$$\geq \frac{1}{n}\left\| (y_1,\ldots,y_n) \right\| + \cdots + \frac{1}{n}\left\| (y_1,\ldots,y_n) \right\|$$
$$= \left\| (y_1,\ldots,y_n) \right\|$$

for $(y_1,\ldots,y_n) \in \partial\Omega_{r_2}$.

By the use of Theorem 2.9, there exists $(\overline{y}_1,\ldots,\overline{y}_n) \in \Lambda$ such that $T(\overline{y}_1,\ldots,\overline{y}_n) = (\overline{y}_1,\ldots,\overline{y}_n)$, the proof is complete. □

Theorem 3.2 *Suppose that conditions* (H_1) *and* (H_2) *hold. Then, for every* $\lambda_j \in (0,\lambda_j^*)$, *FBVP* (1.1)-(1.2) *has at least two positive solutions, where*

$$\lambda_j^* = \frac{1}{n\alpha_j} \sup_{r>0} \frac{r}{\max_{0\leq y_1+\cdots+y_n\leq r} f_j(y_1,\ldots,y_n)}.$$

Proof Define the function

$$\varphi_j(r) = \frac{r}{n\alpha_j \max_{0\le y_1+\cdots+y_n\le r} f_j(y_1,\ldots,y_n)}, \quad j = 1,\ldots,n.$$

It is easy to know that $\varphi_j : (0,+\infty) \to (0,+\infty)$ is a continuous function. From (H_1), we see that $\lim_{r\to 0}\frac{r}{f_j(r)} = 0$, that is, $\lim_{r\to 0}\frac{r}{n\alpha_j f_j(r)} = 0$, and

$$0 < \varphi_j(r) = \frac{r}{n\alpha_j \max_{0\le y_1-\cdots+y_n\le r} f_j(y_1,\ldots,y_n)} \le \frac{r}{n\alpha_j f_j(r)},$$

so $\lim_{r\to 0}\varphi_j(r) = 0$.

From (H_2), we see further that $\lim_{r\to\infty}\varphi_j(r) = 0$. Then there exists $r_0 > 0$ such that $\varphi_j(r_0) = \max_{r>0}\varphi_j(r) = \lambda_j^*$, $j = 1,\ldots,n$. For any $\lambda_j \in (0,\lambda_j^*)$, by the intermediate value theorem, there exist two points $d_1 \in (0,r_0)$, $d_2 \in (r_0,\infty)$ such that $\varphi_j(d_1) = \varphi_j(d_2) = \lambda_j$. Thus, we have

$$f_j(y_1,\ldots,y_n) \le \frac{d_1}{n\lambda_j\alpha_j}, \quad y_1+\cdots+y_n \in [0,d_1],$$

$$f_j(y_1,\ldots,y_n) \le \frac{d_2}{n\lambda_j\alpha_j}, \quad y_1+\cdots+y_n \in [0,d_2].$$

On the other hand, since (H_1) and (H_2) hold, there exist $e_1 \in (0,d_1)$, $e_2 \in (d_2,\infty)$ such that

$$\frac{f_j(y_1,\ldots,y_n)}{y_1+\cdots+y_n} \ge \frac{4}{n\lambda_j\beta_j}, \quad y_1+\cdots+y_n \in (0,e_1] \cup \left[\frac{1}{4}e_2,\infty\right).$$

Thus

$$f_j(y_1,\ldots,y_n) \ge \frac{e_1}{n\lambda_j\beta_j}, \quad y_1+\cdots+y_n \in \left[\frac{1}{4}e_1,e_1\right],$$

$$f_j(y_1,\ldots,y_n) \ge \frac{e_2}{n\lambda_j\beta_j}, \quad y_1+\cdots+y_n \in \left[\frac{1}{4}e_2,e_2\right].$$

Application of Theorem 3.1 and Theorem 2.8 leads to two distinct positive solutions of FBVP (1.1)-(1.2) which satisfy

$$e_1 \le \left\|(\overline{y}_1,\ldots,\overline{y}_n)\right\| \le d_1, \qquad d_2 \le \left\|(\overline{y}'_1,\ldots,\overline{y}'_n)\right\| \le e_2.$$

The proof is complete. □

By the proof of Theorem 3.2, we obtain the following.

Corollary 3.3 *If one of conditions* (H_1) *and* (H_2) *is satisfied, then for every* $0 < \lambda_j < \lambda_j^*$, *FBVP* (1.1)-(1.2) *has at least one positive solution.*

Theorem 3.4 *Suppose that* (H_3), (H_4) *hold. Then, for any* $\lambda_j \ge \lambda_j^{**}$, *FBVP* (1.1)-(1.2) *has at least two positive solutions, where*

$$\lambda_j^{**} = \frac{1}{n\beta_j}\inf_{r>0}\frac{r}{\min_{\frac{1}{4}r\le y_1+\cdots+y_n\le r} f_j(y_1,\ldots,y_n)}.$$

Proof Define the function

$$\psi_j(r) = \frac{r}{n\beta_j \min_{r/4 \le y_1+\cdots+y_n \le r} f_j(y_1,\ldots,y_n)}, \quad j = 1,\ldots,n.$$

We know that $\psi_j : (0,+\infty) \to (0,+\infty)$ is a continuous function. For $\lambda_j > \lambda_j^{**}$, there exists $0 < e_3 < +\infty$ such that

$$f_j(y_1,\ldots,y_n) \ge \frac{e_3}{n\lambda_j\beta_j}, \quad y_1+\cdots+y_n \in \left[\frac{1}{4}e_3, e_3\right].$$

By condition (H_3), there exists $0 < d_3 < e_3$ such that

$$f_j(y_1,\ldots,y_n) \le \frac{d_3}{n\lambda_j\alpha_j}, \quad y_1+\cdots+y_n \in [0,d_3].$$

From condition (H_4), there exists $e_3 < d_0 < +\infty$ such that

$$\frac{f_j(y_1,\ldots,y_n)}{y_1+\cdots+y_n} \le \frac{1}{n\lambda_j\alpha_j}, \quad y_1+\cdots+y_n \in [d_0,+\infty).$$

Let $M_j = \max_{0 \le y_1+\cdots+y_n \le d_0} f_j(y_1,\ldots,y_n)$. Choose $d_4 > d_0$ such that $d_4 \ge \lambda_j M_j \alpha_j$. Then

$$f_j(y_1,\ldots,y_n) \le \frac{d_4}{n\lambda_j\alpha_j}, \quad y_1+\cdots+y_n \in [0,d_4].$$

By Theorem 3.1 and Theorem 2.8, the proof is complete. □

From the proof of Theorem 3.4, we get the following.

Corollary 3.5 *Suppose that one of conditions* (H_3) *and* (H_4) *holds. Then, for every* $\lambda_j > \lambda_j^{**}$, *FBVP* (1.1)-(1.2) *has at least one positive solution.*

Theorem 3.6 *Suppose that one of the following cases is satisfied*:
(1) (H_1), (H_6) *hold, and* $0 < \lambda_j < \frac{1}{n\alpha_j L_j}$;
(2) (H_2), (H_5) *hold, and* $0 < \lambda_j < \frac{1}{n\alpha_j l_j}$.
Then FBVP (1.1)-(1.2) *has at least one positive solution.*

Proof (1) From (H_6) (namely $\lim_{(y_1+\cdots+y_n)\to\infty} \frac{f_j(y_1,\ldots,y_n)}{y_1+\cdots+y_n} = L_j$, $t_j \in [\nu_j-2, \nu_j+b]_{\mathbb{N}_{\nu_j-2}}$, $j = 1,2,\ldots,n$), for any $\epsilon_j > 0$, there exists a number $R_0 > 0$, for $y_1+\cdots+y_n \in (R_0,+\infty)$, we have

$$f_j(y_1,\ldots,y_n) < (L_j+\epsilon_j)(y_1+\cdots+y_n).$$

Let $M_j = \max_{0 \le y_1+\cdots+y_n \le R_0} f_j(y_1,\ldots,y_n)$. Choose $R > \max\{R_0, \frac{M_j}{L_j+\epsilon_j}\}$, then

$$f_j(y_1,\ldots,y_n) < (L_j+\epsilon_j)(y_1+\cdots+y_n)$$

for $y_1+\cdots+y_n \in [0,R]$. As arbitrarily of ϵ_j, so $f_j(y_1,\ldots,y_n) \le L_j R$. Note that $0 < \lambda_j < \frac{1}{n\alpha_j L_j}$, then

$$\lambda_j^* = \frac{1}{n\alpha_j} \sup_{R>0} \frac{R}{\max_{0 \le y_1+\cdots+y_n \le R} f(y)} > \frac{1}{n\alpha_j}\frac{R}{LR} = \frac{1}{n\alpha_j L_j} > \lambda_j > 0.$$

Namely $0 < \lambda_j < \lambda_j^*$. By means of Corollary 3.3, FBVP (1.1)-(1.2) has at least one positive solution.

The proof is similar to (2) and hence omitted. □

Similarly, we have the following.

Theorem 3.7 *Suppose that one of the following cases is satisfied:*

(1) (H_3), (H_6) *hold, and* $\frac{4}{nL_j\beta_j} < \lambda_j < +\infty$;

(2) (H_4), (H_5) *hold, and* $\frac{4}{nl_j\beta_j} < \lambda_j < +\infty$.

Then FBVP (1.1)-(1.2) *has at least one positive solution.*

Theorem 3.8 *Suppose that conditions* (H_5) *and* (H_6) *hold. If* λ_j *satisfies* $\frac{4}{n\beta_j L_j} < \lambda_j < \frac{1}{n\alpha_j l_j}$ *or* $\frac{4}{n\beta_j l_j} < \lambda_j < \frac{1}{n\alpha_j L_j}$, *then FBVP* (1.1)-(1.2) *has at least one positive solution.*

Proof Suppose that $\frac{4}{n\beta_j L_j} < \lambda_j < \frac{1}{n\alpha_j l_j}$ holds. Choose $\epsilon_j > 0$ such that $\frac{4}{n\beta_j(L_j-\epsilon_j)} \le \lambda_j \le \frac{1}{n\alpha_j(l_j+\epsilon_j)}$. With condition (H_5), there exists $\tau > 0$ such that $f_j(y_1,\dots,y_n) \le (l_j+\epsilon_j)(y_1+\cdots+y_n)$ for $y_1+\cdots+y_n \in (0,\tau)$. Thus, for $y_1+\cdots+y_n \in \partial\Omega_\tau$,

$$
\begin{aligned}
&\left\|T_j(y_1,\dots,y_n)\right\| \\
&\quad = \max_{t_j\in[\nu_j-2,\nu_j+b]_{\mathbb{N}_{\nu_j-2}}} \left|\lambda_j \sum_{s=0}^{b+1} G_j(t_j,s) f_j\big(y_1(s+\nu_1-1),\dots,y_n(s+\nu_n-1)\big)\right| \\
&\quad \le \lambda_j(l_j+\epsilon_j)(y_1+\cdots+y_n)\sum_{s=0}^{b+1} \max_{t_j\in[\nu_j-2,\nu_j+b]_{\mathbb{N}_{\nu_j-2}}} G_j(t_j,s) \\
&\quad = \lambda_j(l_j+\epsilon_j)(y_1+\cdots+y_n)\sum_{s=0}^{b+1} G_j(\nu_j+b,s) \\
&\quad \le \lambda_j\alpha_j \frac{\tau}{n\lambda_j\alpha_j} \\
&\quad = \frac{\tau}{n} \\
&\quad = \frac{1}{n}\left\|(y_1,\dots,y_n)\right\|,
\end{aligned}
$$

$j = 1,\dots,n$. That is,

$$
\begin{aligned}
&\left\|T(y_1,\dots,y_n)(t_1,\dots,t_n)\right\| \\
&\quad = \left\|\big(T_1(y_1,\dots,y_n)(t_1),\dots,T_n(y_1,\dots,y_n)(t_n)\big)\right\| \\
&\quad = \left\|T_1(y_1,\dots,y_n)\right\| + \cdots + \left\|T_n(y_1,\dots,y_n)\right\| \\
&\quad \le \frac{1}{n}\left\|(y_1,\dots,y_n)\right\| + \cdots + \frac{1}{n}\left\|(y_1,\dots,y_n)\right\| \\
&\quad = \left\|(y_1,\dots,y_n)\right\|
\end{aligned}
$$

for $(y_1,\dots,y_n) \in \partial\Omega_\tau$.

By condition (H_6), there exists $R_1 > 0$ such that $f_j(y_1,\dots,y_n) \ge (L_j-\epsilon_j)(y_1+\cdots+y_n)$ for $y_1+\cdots+y_n \ge \frac{1}{4}R_1$.

Let $R_2 = \max\{2\tau, R_1\}$, for $(y_1, \dots, y_n) \in \partial\Omega_{R_2}$, we get

$$\begin{aligned}
&\left(T_j(y_1, \dots, y_n)\right)\left(\left[\frac{b - \nu_j}{2}\right] + \nu_j\right)\\
&= \lambda_j \sum_{s=0}^{b+1} G_j\left(\left[\frac{b - \nu_j}{2}\right] + \nu_j, s\right) f_j(y_1(s + \nu_1 - 1), \dots, y_n(s + \nu_n - 1))\\
&\geq \lambda_j(L_j - \epsilon_j) \sum_{s=[\frac{\nu_j+b}{4} - \nu_j + 1]}^{[\frac{3(\nu_j+b)}{4} - \nu_j + 1]} G_j\left(\left[\frac{b - \nu_j}{2}\right] + \nu_j, s\right)(y_1 + \dots + y_n)\\
&\geq \lambda_j \frac{1}{4}(L_j - \epsilon_j)R_2 \sum_{s=[\frac{\nu_j+b}{4} - \nu_j + 1]}^{[\frac{3(\nu_j+b)}{4} - \nu_j + 1]} G_j\left(\left[\frac{b - \nu_j}{2}\right] + \nu_j, s\right)\\
&\geq \frac{R_2}{n}.
\end{aligned}$$

Then

$$\left\| T_j(y_1, \dots, y_n) \right\| \geq \frac{R_2}{n} = \frac{1}{n}\left\|(y_1, \dots, y_n)\right\|, \quad j = 1, \dots, n.$$

That is,

$$\begin{aligned}
&\left\| T(y_1, \dots, y_n)(t_1, \dots, t_n) \right\|\\
&= \left\| \left(T_1(y_1, \dots, y_n)(t_1), \dots, T_n(y_1, \dots, y_n)(t_n)\right) \right\|\\
&= \left\| T_1(y_1, \dots, y_n) \right\| + \dots + \left\| T_n(y_1, \dots, y_n) \right\|\\
&\geq \frac{1}{n}\left\|(y_1, \dots, y_n)\right\| + \dots + \frac{1}{n}\left\|(y_1, \dots, y_n)\right\|\\
&= \left\|(y_1, \dots, y_n)\right\|
\end{aligned}$$

for $(y_1, \dots, y_n) \in \partial\Omega_{R_2}$

By using Theorem 2.9, we obtain the conclusion.

A similar proof holds when $\frac{4}{n\beta_j l_j} < \lambda_j < \frac{1}{n\alpha_j L_j}$. The proof is complete. □

We now present an example illustrating the sorts of boundary conditions that can be treated by Theorem 3.2.

Example 3.1 Consider the following boundary value problems:

$$\begin{cases}
\Delta^{\frac{17}{8}} y_1(t) = -\lambda_1 f_1(y_1(t + \frac{9}{8}), y_2(t + \frac{17}{16})),\\
\Delta^{\frac{33}{16}} y_2(t) = -\lambda_2 f_2(y_1(t + \frac{9}{8}), y_2(t + \frac{17}{16})),\\
y_1(-\frac{7}{8}) = \Delta y_1(\frac{169}{8}) = \Delta^2 y_1(-\frac{7}{8}) = 0,\\
y_2(-\frac{15}{16}) = \Delta y_2(\frac{337}{16}) = \Delta^2 y_2(-\frac{15}{16}) = 0,
\end{cases} \tag{3.1}$$

where $b = 19$, $\nu_1 = \frac{17}{8}$, $\nu_2 = \frac{33}{16}$, we take

$$f_1(y_1, y_2) = \frac{1}{64\sqrt{2}}(y_1 + y_2)^{\frac{1}{2}} + (y_1 + y_2)^2, \qquad f_2(y_1, y_2) = \frac{1}{25}(y_1 + y_2)^{\frac{1}{2}} + \frac{1}{64}(y_1 + y_2)^{\frac{3}{2}},$$

$f_1, f_2 : [0,+\infty) \times [0,+\infty) \to [0,+\infty)$, and y_1 is defined on the time scale $\{-\frac{7}{8}, \frac{1}{8}, \ldots, \frac{169}{8}\}$, y_2 is defined on the time scale $\{-\frac{15}{16}, \frac{1}{16}, \ldots, \frac{337}{16}\}$. f_1 and f_2 satisfy conditions of Theorem 3.2. A computation shows that $\lambda_1^* \approx 0.01456$, $\lambda_2^* \approx 0.032845$, then, for every $\lambda_j \in (0, \lambda_j^*)$ $(j = 1, 2)$, problem (3.1) has at least two positive solutions.

Competing interests
The authors declare that they have no competing interests.

Authors' contributions
SK conceived of the study and participated in its design. HC drafted the manuscript and participated in its design and coordination. JG participated in the sequence correction. All authors read and approved the final manuscript.

Acknowledgements
The authors are very grateful to the reviewers for their valuable suggestions and useful comments, which led to an improvement of this paper. Project was supported by the National Natural Science Foundation of China (Grant No. 11271235) and Shanxi Datong University Institute (2009-Y-15, 2010-B-01, 2013K5) and the Development Foundation of Higher Education Department of Shanxi Province (20101109, 20111020).

References
1. Abdeljawad, T: On Riemann and Caputo fractional differences. Appl. Comput. Math. **62**, 1602-1611 (2011)
2. Holm, M: Sum and difference compositions in discrete fractional calculus. CUBO **13**(3), 153-184 (2011)
3. Atici, FM, Sengül, S: Modeling with fractional difference equations. J. Math. Anal. Appl. **369**, 1-9 (2010)
4. Abdeljawad, T, Baleanu, D: Fractional differences and integration by parts. J. Comput. Anal. Appl. **13**(3), 574-582 (2011)
5. Jarad, F, Abdeljawad, T, Baleanu, D, Biçen, K: On the stability of some discrete fractional nonautonomous systems. Abstr. Appl. Anal. **2012**, Article ID 476581 (2012)
6. Mohammadi, H, Rezapour, S: Existence results for nonlinear fractional differential equations on ordered gauge spaces. J. Adv. Math. Stud. **6**(2), 154-158 (2013)
7. Wu, GC, Baleanu, D: Discrete fractional logistic map and its chaos. Nonlinear Dyn. **75**, 283-287 (2014)
8. Wu, GC, Baleanu, D: Chaos synchronization of the discrete fractional logistic map. Signal Process. **102**, 96-99 (2014)
9. Chen, HQ, Cui, YQ, Zhao, XL: Multiple solutions to fractional difference boundary value problems. Abstr. Appl. Anal. **2014**, Article ID 879380 (2014)
10. Chen, HQ, Jin, Z, Kang, SG: Existence of positive solutions for Caputo fractional difference equation. Adv. Differ. Equ. **2015**, 44 (2015)
11. Goodrich, CS: On a fractional boundary value problem with fractional boundary conditions. Appl. Math. Lett. **25**, 1101-1105 (2012)
12. Goodrich, CS: On discrete sequential fractional boundary value problems. J. Math. Anal. Appl. **385**, 111-124 (2012)
13. Goodrich, CS: On a discrete fractional three-point boundary value problem. J. Differ. Equ. Appl. **18**, 397-415 (2012)
14. Atici, FM, Eloe, PW: Linear systems of fractional nabla difference equations. Rocky Mt. J. Math. **41**, 353-370 (2011)
15. Baleanu, D, Rezapour, S, Salehi, S: A *k*-dimensional system of fractional finite difference equations. Abstr. Appl. Anal. **2014**, Article ID 312578 (2014)
16. Goodrich, CS: Existence of a positive solution to a system of discrete fractional boundary value problems. Appl. Comput. Math. **217**, 4740-4753 (2011)
17. Goodrich, CS: Nonlocal systems of BVPs with asymptotically superlinear boundary conditions. Comment. Math. Univ. Carol. **53**, 79-97 (2012)
18. Dahal, R, Duncan, D, Goodrich, CS: Systems of semipositone discrete fractional boundary value problems. J. Differ. Equ. Appl. (2013). doi:10.1080/10236198.2013.856073
19. Kang, SG, Zhao, XH, Chen, HQ: Positive solutions for boundary value problems of fractional difference equations depending on parameters. Adv. Differ. Equ. **2013**, 376 (2013)
20. Agarwal, R, Meehan, M, O'Regan, D: Fixed Point Theory and Applications. Cambridge University Press, Cambridge (2001)

8

Some unified formulas and representations for the Apostol-type polynomials

Da-Qian Lu[1] and Qiu-Ming Luo[2*]

*Correspondence: luomath2007@163.com
[2]Department of Mathematics, Chongqing Normal University, Chongqing Higher Education Mega Center, Huxi Campus, Chongqing, 401331, People's Republic of China
Full list of author information is available at the end of the article

Abstract

Recently, a family of the Apostol-type polynomials was introduced by Luo and Srivastava (Appl. Math. Comput. 217:5702-5728 (2011)). In this paper, we further investigate the Apostol-type polynomials and obtain their unified multiplication formula and explicit representations in terms of the Gaussian hypergeometric function and the generalized Hurwitz zeta function. We also show some special cases, which include the corresponding results of Luo, Garg, Srivastava, Ozden, and Özarslan *etc.*

MSC: Primary 11B68; secondary 11M35; 11B73; 33C05

Keywords: (generalized) Apostol-type polynomials; Apostol-Euler polynomials; Apostol-Bernoulli polynomials; Genocchi polynomials of high order; multiplication formula; Gaussian hypergeometric function; (generalized) Hurwitz zeta function

1 Introduction, definitions, and motivation

Throughout this paper, we always make use of the following notations: $\mathbb{N} = \{1,2,3,\ldots\}$ denotes the set of natural numbers, $\mathbb{N}_0 = \{0,1,2,3,\ldots\}$ denotes the set of nonnegative integers, $\mathbb{Z}_0^- = \{0,-1,-2,-3,\ldots\}$ denotes the set of nonpositive integers, $\mathbb{Z}$ denotes the set of integers, $\mathbb{R}$ denotes the set of real numbers, and $\mathbb{C}$ denotes the set of complex numbers.

The symbol $(a)_k$ denotes the shifted factorial (or the Pochhammer symbol), defined, $a \in \mathbb{C}$, by

$$(a)_k = \frac{\Gamma(a+k)}{\Gamma(a)} = \begin{cases} 1, & k = 0, \\ a(a+1)\cdots(a+k-1), & k \in \mathbb{N}. \end{cases} \tag{1.1}$$

The symbol $\{n\}_k$ denotes the falling factorial, defined, $a \in \mathbb{C}$, by

$$\{a\}_k = \begin{cases} 1, & k = 0, \\ a(a-1)\cdots(a-k+1) = \frac{\Gamma(a+1)}{\Gamma(a-k+1)}, & k \in \mathbb{N}, \end{cases} \tag{1.2}$$

where $\Gamma(x)$ is the usual gamma function.

The classical Bernoulli polynomials $B_n(x)$, Euler polynomials $E_n(x)$, and Genocchi polynomials $G_n(x)$, together with their familiar generalizations $B_n^{(\alpha)}(x)$, $E_n^{(\alpha)}(x)$, and $G_n^{(\alpha)}(x)$ of order α, are usually defined by means of the following generating functions (see, for details, [1], pp.532-533 and [2]):

$$\left(\frac{z}{e^z-1}\right)^\alpha e^{xz} = \sum_{n=0}^{\infty} B_n^{(\alpha)}(x)\frac{z^n}{n!} \quad \left(|z| < 2\pi\right), \tag{1.3}$$

$$\left(\frac{2}{e^z+1}\right)^\alpha e^{xz} = \sum_{n=0}^{\infty} E_n^{(\alpha)}(x)\frac{z^n}{n!} \quad \left(|z| < \pi\right) \tag{1.4}$$

and

$$\left(\frac{2z}{e^z+1}\right)^\alpha e^{xz} = \sum_{n=0}^{\infty} G_n^{(\alpha)}(x)\frac{z^n}{n!} \quad \left(|z| < \pi\right). \tag{1.5}$$

Thus, the Bernoulli polynomials $B_n(x)$, Euler polynomials $E_n(x)$, and Genocchi polynomials $G_n(x)$ are given, respectively, by

$$B_n(x) := B_n^{(1)}(x), \qquad E_n(x) := E_n^{(1)}(x) \quad \text{and} \quad G_n(x) := G_n^{(1)}(x) \quad (n \in \mathbb{N}_0). \tag{1.6}$$

The Bernoulli numbers B_n, Euler numbers E_n, and Genocchi numbers G_n are, respectively,

$$B_n := B_n(0) = B_n^{(1)}(0), \qquad E_n := E_n(0) = E_n^{(1)}(0) \quad \text{and} \quad G_n := G_n(0) = G_n^{(1)}(0). \tag{1.7}$$

Some interesting analogs of the classical Bernoulli polynomials and numbers were first investigated by Apostol (see [3], p.165, Eq. (3.1)) and (more recently) by Srivastava (see [4], pp.83-84). We begin by recalling here Apostol's definitions as follows.

Definition 1.1 (Apostol [3]; see also Srivastava [4]) The Apostol-Bernoulli polynomials $\mathcal{B}_n(x;\lambda)$ $(\lambda \in \mathbb{C})$ are defined by means of the following generating function:

$$\frac{ze^{xz}}{\lambda e^z - 1} = \sum_{n=0}^{\infty} \mathcal{B}_n(x;\lambda)\frac{z^n}{n!} \quad \left(|z| < 2\pi \text{ when } \lambda = 1;\, |z| < |\log\lambda| \text{ when } \lambda \neq 1\right) \tag{1.8}$$

with, of course,

$$B_n(x) = \mathcal{B}_n(x;1) \quad \text{and} \quad \mathcal{B}_n(\lambda) := \mathcal{B}_n(0;\lambda), \tag{1.9}$$

where $\mathcal{B}_n(\lambda)$ denotes the so-called Apostol-Bernoulli numbers.

Recently, Luo and Srivastava [5] further extended the Apostol-Bernoulli polynomials as the so-called Apostol-Bernoulli polynomials of order α.

Definition 1.2 (Luo and Srivastava [5]) The Apostol-Bernoulli polynomials $\mathcal{B}_n^{(\alpha)}(x;\lambda)$ $(\lambda \in \mathbb{C})$ of order α $(\alpha \in \mathbb{N})$ are defined by means of the following generating function:

$$\left(\frac{z}{\lambda e^z-1}\right)^\alpha \cdot e^{xz} = \sum_{n=0}^{\infty} \mathcal{B}_n^{(\alpha)}(x;\lambda)\frac{z^n}{n!}$$

$$\left(|z| < 2\pi \text{ when } \lambda = 1;\, |z| < |\log\lambda| \text{ when } \lambda \neq 1\right) \tag{1.10}$$

with, of course,

$$B_n^{(\alpha)}(x) = \mathcal{B}_n^{(\alpha)}(x;1) \quad \text{and} \quad \mathcal{B}_n^{(\alpha)}(\lambda) := \mathcal{B}_n^{(\alpha)}(0;\lambda), \tag{1.11}$$

where $\mathcal{B}_n^{(\alpha)}(\lambda)$ denotes the so-called Apostol-Bernoulli numbers of order α.

On the other hand, Luo [6] gave an analogous extension of the generalized Euler polynomials as the so-called Apostol-Euler polynomials of order α.

Definition 1.3 (Luo [6]) The Apostol-Euler polynomials $\mathcal{E}_n^{(\alpha)}(x;\lambda)$ of order α $(\alpha, \lambda \in \mathbb{C})$ are defined by means of the following generating function:

$$\left(\frac{2}{\lambda e^z + 1}\right)^\alpha \cdot e^{xz} = \sum_{n=0}^{\infty} \mathcal{E}_n^{(\alpha)}(x;\lambda)\frac{z^n}{n!} \quad \left(|z| < \left|\log(-\lambda)\right|\right) \tag{1.12}$$

with, of course,

$$E_n^{(\alpha)}(x) = \mathcal{E}_n^{(\alpha)}(x;1) \quad \text{and} \quad \mathcal{E}_n^{(\alpha)}(\lambda) := \mathcal{E}_n^{(\alpha)}(0;\lambda), \tag{1.13}$$

where $\mathcal{E}_n^{(\alpha)}(\lambda)$ denotes the so-called Apostol-Euler numbers of order α.

On the subject of the Genocchi polynomials $G_n(x)$ and their various extensions, a remarkably large number of investigations have appeared in the literature (see, for example, [7–11]). Moreover, Luo (see [12]) introduced and investigated the Apostol-Genocchi polynomials of (real or complex) order α, which are defined as follows.

Definition 1.4 The Apostol-Genocchi polynomials $\mathcal{G}_n^{(\alpha)}(x;\lambda)$ $(\lambda \in \mathbb{C})$ of order α $(\alpha \in \mathbb{N})$ are defined by means of the following generating function:

$$\left(\frac{2z}{\lambda e^z + 1}\right)^\alpha \cdot e^{xz} = \sum_{n=0}^{\infty} \mathcal{G}_n^{(\alpha)}(x;\lambda)\frac{z^n}{n!} \quad \left(|z| < \left|\log(-\lambda)\right|\right) \tag{1.14}$$

with, of course,

$$\begin{aligned} &G_n^{(\alpha)}(x) = \mathcal{G}_n^{(\alpha)}(x;1), \qquad \mathcal{G}_n^{(\alpha)}(\lambda) := \mathcal{G}_n^{(\alpha)}(0;\lambda), \\ &\mathcal{G}_n(x;\lambda) := \mathcal{G}_n^{(1)}(x;\lambda) \quad \text{and} \quad \mathcal{G}_n(\lambda) := \mathcal{G}_n^{(1)}(\lambda), \end{aligned} \tag{1.15}$$

where $\mathcal{G}_n(\lambda)$, $\mathcal{G}_n^{(\alpha)}(\lambda)$, and $\mathcal{G}_n(x;\lambda)$ denote the so-called Apostol-Genocchi numbers, the Apostol-Genocchi numbers of order α, and the Apostol-Genocchi polynomials, respectively.

Ozden *et al.* [13] investigated the following unification (and generalization) of the generating functions of the three families of Apostol-type polynomials:

$$\frac{2^{1-\kappa} z^\kappa}{\beta^b e^z - a^b} e^{xz} = \sum_{n=0}^{\infty} \mathcal{Y}_{n,\beta}(x;\kappa,a,b)\frac{z^n}{n!}$$

$$\left(|z| < 2\pi \text{ when } \beta = a; |z| < \left|b\log(\beta/a)\right| \text{ when } \beta \neq a; \kappa, \beta \in \mathbb{C}; a, b \in \mathbb{C}\setminus\{0\}\right). \tag{1.16}$$

In [14] Özarslan further gave an extension of the above definition (1.16) as follows:

$$\left(\frac{2^{1-\kappa}z^{\kappa}}{\beta^b e^z - a^b}\right)^{\alpha} e^{xz} = \sum_{n=0}^{\infty} \mathcal{Y}_{n,\beta}^{(\alpha)}(x;\kappa,a,b)\frac{z^n}{n!}$$

$$\big(\alpha \in \mathbb{N}; |z| < 2\pi \text{ when } \beta = a; |z| < \left|b\log(\beta/a)\right| \text{ when } \beta \neq a; \kappa,\beta \in \mathbb{C}; a,b \in \mathbb{C}\setminus\{0\}\big) \tag{1.17}$$

and gave some identities for $\mathcal{Y}_{n,\beta}^{(\alpha)}(x;\kappa,a,b)$.

Recently, Luo and Srivastava [15] further extended the Apostol-type polynomials as follows.

Definition 1.5 (Luo and Srivastava [15]) The generalized Apostol-type polynomials $\mathcal{F}_n^{(\alpha)}(x;\lambda;\mu;\nu)$ of order α $(\alpha,\lambda,\mu;\nu \in \mathbb{C})$ are defined by means of the following generating function:

$$\left(\frac{2^{\mu}z^{\nu}}{\lambda e^z + 1}\right)^{\alpha} e^{xz} = \sum_{n=0}^{\infty} \mathcal{F}_n^{(\alpha)}(x;\lambda;\mu;\nu)\frac{z^n}{n!} \quad \big(|z| < \left|\log(-\lambda)\right|\big). \tag{1.18}$$

By comparing Definition 1.5 with Definitions 1.2, 1.3 and 1.4, we readily find that

$$\mathcal{B}_n^{(\alpha)}(x;\lambda) = (-1)^{\alpha}\mathcal{F}_n^{(\alpha)}(x;-\lambda;0;1) \quad (\alpha \in \mathbb{N}), \tag{1.19}$$

$$\mathcal{E}_n^{(\alpha)}(x;\lambda) = \mathcal{F}_n^{(\alpha)}(x;\lambda;1;0) \quad (\alpha \in \mathbb{C}) \tag{1.20}$$

and

$$\mathcal{G}_n^{(\alpha)}(x;\lambda) = \mathcal{F}_n^{(\alpha)}(x;\lambda;1;1) \quad (\alpha \in \mathbb{N}). \tag{1.21}$$

Furthermore, if we compare the generating functions (1.16), (1.17) and (1.18), we readily see that

$$\mathcal{Y}_{n,\beta}(x;\kappa,a,b) = -\frac{1}{a^b}\mathcal{F}_n^{(1)}\left(x;-\left(\frac{\beta}{a}\right)^b;1-\kappa;\kappa\right), \tag{1.22}$$

$$\mathcal{Y}_{n,\beta}^{(\alpha)}(x;\kappa,a,b) = (-1)^{\alpha}\frac{1}{a^{b\alpha}}\mathcal{F}_n^{(\alpha)}\left(x;-\left(\frac{\beta}{a}\right)^b;1-\kappa;\kappa\right). \tag{1.23}$$

More investigations of this subject can be found in [5, 6, 12–22].

The aim of this paper is to give the multiplication formula for the Apostol-type polynomials $\mathcal{F}_n^{(\alpha)}(x;\lambda;\mu;\nu)$ and obtain an explicit representation of $\mathcal{F}_n^{(\alpha)}(x;\lambda;\mu;\nu)$ in terms of the Gauss hypergeometric function ${}_2F_1(a,b;c;z)$. We study some relations between the family of Apostol-type polynomials $\mathcal{F}_n^{(\alpha)}(x;\lambda;\mu;\nu)$ and the family of Hurwitz zeta functions $\Phi_{\mu}(z,s,a)$. Some special cases also are shown.

2 Multiplication formula for the Apostol-type polynomials

In this section we give a unified multiplication formula for the Apostol-type polynomials $\mathcal{F}_n^{(\alpha)}(x;\lambda;\mu;\nu)$. We will see that some well-known results are the corresponding special cases of our result.

First we need the following lemmas.

Lemma 2.1 (Multinomial identity [23], p.28, Theorem B) *If $x_1, x_2, \ldots, x_m$ are commuting elements of a ring ($\Longleftrightarrow$ $x_i x_j = x_j x_i$, $1 \le i < j \le m$), then we have for all integers $n \ge 0$:*

$$(x_1 + x_2 + \cdots + x_m)^n = \sum_{\substack{a_1, a_2, \ldots, a_m \ge 0 \\ a_1 + a_2 + \cdots + a_m = n}} \binom{n}{a_1, a_2, \ldots, a_m} x_1^{a_1} x_2^{a_2} \cdots x_m^{a_m}, \tag{2.1}$$

the last summation takes place over all positive or zero integers $a_i \ge 0$ such that $a_1 + a_2 + \cdots + a_m = n$, where

$$\binom{n}{a_1, a_2, \ldots, a_m} := \frac{n!}{a_1! a_2! \cdots a_m!},$$

are called multinomial coefficients defined by [23], *p.28, Definition B.*

Lemma 2.2 (Generalized multinomial identity [23], p.41, Eq. (12m)) *If $x_1, x_2, \ldots, x_m$ are commuting elements of a ring ($\Longleftrightarrow$ $x_i x_j = x_j x_i$, $1 \le i < j \le m$), then we have for all real or complex variable α:*

$$(1 + x_1 + x_2 + \cdots + x_m)^\alpha = \sum_{v_1, v_2, \ldots, v_m \ge 0} \binom{\alpha}{v_1, v_2, \ldots, v_m} x_1^{v_1} x_2^{v_2} \cdots x_m^{v_m}, \tag{2.2}$$

the last summation takes place over all positive or zero integers $v_i \ge 0$, where

$$\binom{\alpha}{v_1, v_2, \ldots, v_m} := \frac{\{\alpha\}_{v_1 + v_2 + \cdots + v_m}}{v_1! v_2! \cdots v_m!} = \frac{\alpha(\alpha - 1)(\alpha - 2) \cdots (\alpha - v_1 - v_2 - \cdots - v_m + 1)}{v_1! v_2! \cdots v_m!}$$

are called generalized multinomial coefficients defined by [23], *p.27, Eq.* (10 C'').

Theorem 2.3 (Multiplication formula) *For $\mu, v, r \in \mathbb{N}$ and $v \le 1$, $n, l \in \mathbb{N}_0$, $\alpha, \lambda \in \mathbb{C}$, we have*

$$\mathcal{F}_n^{(\alpha)}(rx; \lambda; \mu; v) = r^{n - v\alpha} \sum_{v_1, v_2, \ldots, v_{r-1} \ge 0} \binom{\alpha}{v_1, v_2, \ldots, v_{r-1}} \times (-\lambda)^m \mathcal{F}_n^{(\alpha)}\left(x + \frac{m}{r}; \lambda^r; \mu; v\right), \quad r \text{ odd}, \tag{2.3}$$

$$\mathcal{F}_n^{(l)}(rx; \lambda; \mu; v) = \frac{(-1)^l 2^{\mu l} r^{n - vl}}{(n+1)_{(1-v)l}} \sum_{\substack{0 \le v_1, v_2, \ldots, v_{r-1} \le l \\ v_1 + v_2 + \cdots + v_{r-1} = l}} \binom{l}{v_1, v_2, \ldots, v_{r-1}} \times (-\lambda)^m \mathcal{B}_{n + (1-v)l}^{(l)}\left(x + \frac{m}{r}; \lambda^r\right), \quad r \text{ even}, \tag{2.4}$$

where $m = v_1 + 2v_2 + \cdots + (r-1)v_{r-1}$.

Proof It is not difficult to show that

$$\frac{1}{\lambda e^z + 1} = -\frac{1 - \lambda e^z + \lambda^2 e^{2z} + \cdots + (-\lambda)^{r-1} e^{(r-1)z}}{(-\lambda)^r e^{rz} - 1}. \tag{2.5}$$

When r is odd, by (1.18) and (2.5) we get

$$\begin{aligned}
&\sum_{n=0}^{\infty} \mathcal{F}_n^{(\alpha)}(rx;\lambda;\mu;\nu)\frac{z^n}{n!} \\
&= \frac{1}{r^{\nu\alpha}}\left(\frac{2^{\mu}(rz)^{\nu}}{\lambda^r e^{rz}+1}\right)^{\alpha}\left(\frac{\lambda^r e^{rz}+1}{\lambda e^z+1}\right)^{\alpha} e^{rxz} \\
&= \frac{1}{r^{\nu\alpha}}\left(\frac{2^{\mu}(rz)^{\nu}}{\lambda^r e^{rz}+1}\right)^{\alpha}\left(\sum_{k=0}^{r-1}(-\lambda e^z)^k\right)^{\alpha} e^{rxz} \\
&= \frac{1}{r^{\nu\alpha}} \sum_{\nu_1,\nu_2,\ldots,\nu_{r-1}\geq 0}\binom{\alpha}{\nu_1,\nu_2,\ldots,\nu_{r-1}}(-\lambda)^m\left(\frac{2^{\mu}(rz)^{\nu}}{\lambda^r e^{rz}+1}\right)^{\alpha} e^{(x+\frac{m}{r})rz} \\
&= \sum_{n=0}^{\infty}\left[r^{n-\nu\alpha}\sum_{\nu_1,\nu_2,\ldots,\nu_{r-1}\geq 0}\binom{\alpha}{\nu_1,\nu_2,\ldots,\nu_{r-1}}(-\lambda)^m\mathcal{F}_n^{(\alpha)}\left(x+\frac{m}{r};\lambda^r;\mu;\nu\right)\right]\frac{z^n}{n!}. \qquad (2.6)
\end{aligned}$$

Comparing the coefficients of $\frac{z^n}{n!}$ on both sides of (2.6), we obtain the assertion (2.3) of Theorem 2.3.

When r is even, we can similarly prove the assertion (2.4) of Theorem 2.3. The proof is complete. □

It follows that we can deduce the well-known formulas from Theorem 2.3.

Letting $\lambda \longmapsto -\lambda$, taking $\mu = 0$ and $\nu = 1$ in (2.3) and (2.4) and noting (1.19), we can obtain the following main result of Luo (see [24], p.380, Theorem 2.1).

Corollary 2.4 *For $r, \alpha \in \mathbb{N}$, $n \in \mathbb{N}_0$, $\lambda \in \mathbb{C}$, the following multiplication formula for the Apostol-Bernoulli polynomials of higher order holds true:*

$$\mathcal{B}_n^{(\alpha)}(rx;\lambda) = r^{n-\alpha}\sum_{\nu_1,\nu_2,\ldots,\nu_{r-1}\geq 0}\binom{\alpha}{\nu_1,\nu_2,\ldots,\nu_{r-1}}\lambda^m\mathcal{B}_n^{(\alpha)}\left(x+\frac{m}{r};\lambda^r\right), \qquad (2.7)$$

where $m = \nu_1 + 2\nu_2 + \cdots + (r-1)\nu_{r-1}$.

Taking $\mu = 1$ and $\nu = 0$ in (2.3) and (2.4), and noting (1.20), we can obtain the following main result of Luo (see [24], p.385, Theorem 3.1).

Corollary 2.5 *For $r \in \mathbb{N}$, $n, l \in \mathbb{N}_0$, $\alpha, \lambda \in \mathbb{C}$, the following multiplication formula for the Apostol-Euler polynomials of higher order holds true:*

$$\mathcal{E}_n^{(\alpha)}(rx;\lambda) = r^n\sum_{\nu_1,\nu_2,\ldots,\nu_{r-1}\geq 0}\binom{\alpha}{\nu_1,\nu_2,\ldots,\nu_{r-1}}(-\lambda)^m\mathcal{E}_n^{(\alpha)}\left(x+\frac{m}{r};\lambda^r\right), \quad r \text{ odd}, \qquad (2.8)$$

$$\begin{aligned}
\mathcal{E}_n^{(l)}(rx;\lambda) = \frac{(-2)^l r^n}{(n+1)_l}&\sum_{\substack{0\leq\nu_1,\nu_2,\ldots,\nu_{r-1}\leq l\\ \nu_1+\nu_2+\cdots+\nu_{r-1}=l}}\binom{l}{\nu_1,\nu_2,\ldots,\nu_{r-1}} \\
&\times(-\lambda)^m\mathcal{B}_{n+l}^{(l)}\left(x+\frac{m}{r};\lambda^r\right), \quad r \text{ even}, \qquad (2.9)
\end{aligned}$$

where $m = \nu_1 + 2\nu_2 + \cdots + (r-1)\nu_{r-1}$.

Taking $\mu = \nu = 1$ in (2.3) and (2.4), and noting (1.21), we can obtain the following main result (see [14], p.2462, Corollary 4.6).

Corollary 2.6 *For $\alpha, r \in \mathbb{N}$, $n, l \in \mathbb{N}_0$, $\lambda \in \mathbb{C}$, the following multiplication formula for the Apostol-Genocchi polynomials of higher order holds true:*

$$\mathcal{G}_n^{(\alpha)}(rx;\lambda) = r^{n-\alpha} \sum_{\nu_1,\nu_2,\ldots,\nu_{r-1}\geq 0} \binom{\alpha}{\nu_1,\nu_2,\ldots,\nu_{r-1}} (-\lambda)^m \mathcal{G}_n^{(\alpha)}\left(x+\frac{m}{r};\lambda^r\right), \quad r \text{ odd}, \tag{2.10}$$

$$\mathcal{G}_n^{(l)}(rx;\lambda) = (-2)^l r^{n-l} \sum_{\substack{0\leq \nu_1,\nu_2,\ldots,\nu_{r-1}\leq l\\ \nu_1+\nu_2+\cdots+\nu_{r-1}=l}} \binom{l}{\nu_1,\nu_2,\ldots,\nu_{r-1}} \times (-\lambda)^m \mathcal{B}_n^{(l)}\left(x+\frac{m}{r};\lambda^r\right), \quad r \text{ even}, \tag{2.11}$$

where $m = \nu_1 + 2\nu_2 + \cdots + (r-1)\nu_{r-1}$.

Taking $\lambda = -(\frac{\beta}{a})^b$, $\mu = 1-\kappa$, $\nu = \kappa$ in (2.3), and noting (1.23), we can obtain the following multiplication formulas for the polynomials $\mathcal{Y}_{n,\beta}^{(\alpha)}(x;\kappa,a,b)$ and $\mathcal{Y}_{n,\beta}(x;\kappa,a,b)$ defined by (1.16) and (1.17), respectively.

Corollary 2.7 *For $\kappa, \mu, \nu, m, n, l, r \in \mathbb{N}_0$, $\alpha, \lambda \in \mathbb{C}$, we have*

$$\begin{aligned}\mathcal{Y}_{n,\beta}^{(\alpha)}(rx;\kappa,a,b) &= r^{n-\kappa\alpha} \sum_{\nu_1,\nu_2,\ldots,\nu_{r-1}\geq 0} \binom{\alpha}{\nu_1,\nu_2,\ldots,\nu_{r-1}} \left(\frac{\beta}{a}\right)^{bm} a^{(r-1)b\alpha} \mathcal{Y}_{n,\beta}^{(\alpha)}\left(x+\frac{m}{r};\kappa;a;br\right) &(2.12)\\ &= r^{n-\kappa\alpha} \sum_{\nu_1,\nu_2,\ldots,\nu_{r-1}\geq 0} \binom{\alpha}{\nu_1,\nu_2,\ldots,\nu_{r-1}} \left(\frac{\beta}{a}\right)^{bm} a^{(r-1)b\alpha} \mathcal{Y}_{n,\beta^r}^{(\alpha)}\left(x+\frac{m}{r};\kappa;a^r;b\right), &(2.13)\end{aligned}$$

where $m = \nu_1 + 2\nu_2 + \cdots + (r-1)\nu_{r-1}$.

Setting $\alpha = l = 1$ in (2.12) and (2.13), respectively, we have (see [13], p.2786, Theorem 8) the following.

Corollary 2.8 *For $\kappa, \mu, \nu, n, r \in \mathbb{N}_0$, $\lambda \in \mathbb{C}$, we have*

$$\begin{aligned}\mathcal{Y}_{n,\beta}(rx;\kappa,a,b) &= r^{n-\kappa} \sum_{j=0}^{r-1} \left(\frac{\beta}{a}\right)^{bj} a^{(r-1)b} \mathcal{Y}_{n,\beta}\left(x+\frac{j}{r};\kappa;a;br\right) &(2.14)\\ &= r^{n-\kappa} \sum_{j=0}^{r-1} \left(\frac{\beta}{a}\right)^{bj} a^{(r-1)b} \mathcal{Y}_{n,\beta^r}\left(x+\frac{j}{r};\kappa;a^r;b\right). &(2.15)\end{aligned}$$

Remark 2.9 In [14], p.2460, Theorem 4.3, one of the main result of Özarslan is not right, the correct form should be (2.12) and (2.13) of Corollary 2.7.

Remark 2.10 In fact, setting $\lambda = -(\frac{\beta}{a})^b$, $\mu = 1-\kappa$, $\nu = \kappa$ in (2.3) and noting (1.23), we deduce the multiplication formulas which are right only when r is odd. In the same way as the proof of [24], p.380, Theorem 2.1, we can obtain the multiplication formulas (2.12) and (2.13) of Corollary 2.7.

3 A unified representation in conjunction with the Gauss hypergeometric function

In this section we obtain a unified representation of the Apostol-type polynomials $\mathcal{F}_n^{(l)}(x;\lambda;\mu;\nu)$ with the Gaussian hypergeometric functions.

Theorem 3.1 *For $\mu,\nu,n,l\in\mathbb{N}_0$, $\lambda\in\mathbb{C}$, we have*

$$\begin{aligned}\mathcal{F}_n^{(l)}(x;\lambda;\mu;\nu)\\ &= 2^{\mu l}(\nu l)!\binom{n}{\nu l}\sum_{k=0}^{n-\nu l}\binom{l+k-1}{k}\binom{n-\nu l}{k}\frac{(-\lambda)^k}{(\lambda+1)^{l+k}}\\ &\times\sum_{m=0}^{k}(-1)^m\binom{k}{m}m^k(x+m)^{n-\nu l-k}{}_2F_1\left(-n+\nu l+k,k;k+1;\frac{m}{m+x}\right),\end{aligned}\tag{3.1}$$

where $F(a,b;c;z)$ denotes Gaussian hypergeometric functions defined by (see [2], p.44, Eq. (4))

$$F(a,b;c;z):=\sum_{n=0}^{\infty}\frac{(a)_n(b)_n}{(c)_n}\frac{z^n}{n!},\quad |z|<1.\tag{3.2}$$

Proof Letting $\alpha=l\in\mathbb{N}$ in (1.18), we have

$$\sum_{n=0}^{\infty}\mathcal{F}_n^{(l)}(x;\lambda;\mu;\nu)\frac{z^n}{n!}=\left(\frac{2^{\mu}z^{\nu}}{\lambda e^z+1}\right)^l e^{xz}.\tag{3.3}$$

Differentiating both sides of (3.3) with respect to the variable z yields

$$\begin{aligned}\mathcal{F}_n^{(l)}(x;\lambda;\mu;\nu)&=D_z^n\left[\left(\frac{2^{\mu}z^{\nu}}{\lambda e^z+1}\right)^l e^{xz}\right]_{z=0}\\ &=2^{\mu l}\sum_{s=0}^{n}\binom{n}{s}x^{n-s}D_z^s\left[z^{\nu l}(\lambda e^z+1)^{-l}\right]_{z=0}\\ &=2^{\mu l}\sum_{s=\nu l}^{n}\binom{n}{s}x^{n-s}(\nu l)!\binom{s}{\nu l}D_z^{s-\nu l}\left[(\lambda e^z+1)^{-l}\right]_{z=0}\\ &=2^{\mu l}\sum_{s=\nu l}^{n}\binom{n}{s}x^{n-s}(\nu l)!\binom{s}{\nu l}D_z^{s-\nu l}\left[(\lambda+1+\lambda(e^z-1))^{-l}\right]_{z=0},\end{aligned}$$

where $D_z=\frac{d}{dz}$ is the differential operator.

Applying the generalized binomial theorem

$$(a+b)^{-\alpha}=\sum_{l=0}^{\infty}\binom{\alpha+l-1}{l}a^{-\alpha-l}(-b)^l\quad\left(\alpha\in\mathbb{C},\left|\frac{b}{a}\right|<1\right)$$

and the generating function of the Stirling numbers of the second kind $S(n,k)$ (see, for details, [23], p.206, Theorem A),

$$\frac{(e^z-1)^k}{k!}=\sum_{n=0}^{\infty}S(n,k)\frac{z^n}{n!},$$

we find that

$$\begin{aligned}
&\mathcal{F}_n^{(l)}(x;\lambda;\mu;\nu)\\
&=2^{\mu l}\sum_{s=\nu l}^{n}\binom{n}{s}x^{n-s}(\nu l)!\binom{s}{\nu l}\sum_{k=0}^{\infty}\binom{l+k-1}{k}(\lambda+1)^{-l-k}(-\lambda)^k D_z^{s-\nu l}\left[(e^z-1)^k\right]_{z=0}\\
&=2^{\mu l}\sum_{s=\nu l}^{n}\binom{n}{s}x^{n-s}(\nu l)!\binom{s}{\nu l}\sum_{k=0}^{s-\nu l}\binom{l+k-1}{k}(-\lambda)^k(\lambda+1)^{-l-k}k!S(s-\nu l,k).
\end{aligned}$$

Noting (see [2], p.58, Eq. (20))

$$S(n,k)=\frac{1}{k!}\sum_{j=0}^{k}(-1)^{k-j}\binom{k}{j}j^n$$

and the well-known combinatorial identity

$$\binom{n}{k}\binom{k}{s}=\binom{n}{s}\binom{n-s}{n-k},$$

we readily obtain

$$\begin{aligned}
&\mathcal{F}_n^{(l)}(x;\lambda;\mu;\nu)\\
&=2^{\mu l}\sum_{s=\nu l}^{n}\binom{n}{s}x^{n-s}(\nu l)!\binom{s}{\nu l}\sum_{k=0}^{s-\nu l}\binom{l+k-1}{k}\\
&\quad\times(-\lambda)^k(\lambda+1)^{-l-k}\sum_{m=0}^{k}(-1)^{k-m}\binom{k}{m}m^{s-\nu l}\\
&=2^{\mu l}(\nu l)!\binom{n}{\nu l}\sum_{k=0}^{n-\nu l}\sum_{s=k+\nu l}^{n}\binom{n-\nu l}{n-s}\binom{l+k-1}{k}\frac{(-\lambda)^k x^{n-s}}{(\lambda+1)^{l+k}}\sum_{m=0}^{k}(-1)^{k-m}\binom{k}{m}m^{s-\nu l}\\
&=2^{\mu l}(\nu l)!\binom{n}{\nu l}\sum_{k=0}^{n-\nu l}\sum_{s=0}^{n-k-\nu l}\binom{n-\nu l}{n-s-\nu l-k}\binom{l+k-1}{k}\\
&\quad\times\frac{(-\lambda)^k x^{n-s-k-\nu l}}{(\lambda+1)^{l+k}}\sum_{m=0}^{k}(-1)^{k-m}\binom{k}{m}m^{s+k}\\
&=2^{\mu l}(\nu l)!\binom{n}{\nu l}\sum_{k=0}^{n-\nu l}\binom{l+k-1}{k}\frac{(-\lambda)^k x^{n-k-\nu l}}{(\lambda+1)^{l+k}}\sum_{m=0}^{k}(-1)^{k-m}\binom{k}{m}m^k\\
&\quad\times\sum_{s=0}^{n-k-\nu l}\binom{n-\nu l}{n-s-\nu l-k}\left(\frac{m}{x}\right)^s.
\end{aligned}$$

Noting that (in view of $\binom{n}{k} = 0$ when $k > n$ or $k < 0$)

$$\sum_{k=0}^{n} \binom{n}{k} = \sum_{k=0}^{\infty} \binom{n}{k},$$

and combining the definition of the Gaussian hypergeometric function

$${}_2F_1(a, b; c; z) := \sum_{n=0}^{\infty} \frac{(a)_n (b)_n}{(c)_n} \frac{z^n}{n!},$$

we obtain

$$\begin{aligned} \mathcal{F}_n^{(l)}(x; \lambda; \mu; \nu) = 2^{\mu l} (\nu l)! \binom{n}{\nu l} \sum_{k=0}^{n-\nu l} \binom{l+k-1}{k} \binom{n-\nu l}{k} \\ \times \frac{(-\lambda)^k x^{n-k-\nu l}}{(\lambda+1)^{l+k}} \sum_{m=0}^{k} (-1)^m \binom{k}{m} m^k \\ \times {}_2F_1\left(-n+\nu l+k, 1; k+1; -\frac{m}{x}\right). \end{aligned} \tag{3.4}$$

Applying the Pfaff-Kummer hypergeometric transformation [25], p.559, Eq. (15.3.4),

$${}_2F_1(a, b; c; z) = (1-z)^{-a} {}_2F_1\left(a, c-b; c; \frac{z}{z-1}\right) \quad \left(c \notin \mathbb{Z}_0^- : |\arg(1-z)| \leq \pi - \epsilon \ (0 < \epsilon < \pi)\right),$$

to (3.4), we arrive at the desired equation, (3.1). This completes our proof. □

Below we show some special cases of (3.1).

Letting $\lambda \longmapsto -\lambda$, taking $\mu = 0$ and $\nu = 1$ in (3.1) and noting (1.19), we easily obtain the following explicit formula for the Apostol-Bernoulli polynomials:

$$\begin{aligned} \mathcal{B}_n^{(l)}(x; \lambda) = l! \binom{n}{l} \sum_{k=0}^{n-l} \binom{l+k-1}{k} \binom{n-l}{k} \frac{\lambda^k}{(\lambda-1)^k} \\ \times \sum_{m=0}^{k} (-1)^m \binom{k}{m} m^k (x+m)^{n-l-k} {}_2F_1\left(-n+l+k, k; k+1; \frac{m}{m+x}\right), \end{aligned} \tag{3.5}$$

with $n, l \in \mathbb{N}_0$, $\lambda \in \mathbb{C} \setminus \{1\}$, which is just the main result of Luo and Srivastava (see [5], p.294, Theorem 1).

Taking $\mu = 1$ and $\nu = 0$ in (3.1) and noting (1.20), we can obtain the following explicit formula for the Apostol-Euler polynomials:

$$\begin{aligned} \mathcal{E}_n^{(l)}(x; \lambda) = 2^l \sum_{k=0}^{n} \binom{l+k-1}{k} \binom{n}{k} \frac{(-\lambda)^k}{(\lambda+1)^{l+k}} \\ \times \sum_{m=0}^{k} (-1)^m \binom{k}{m} m^k (x+m)^{n-k} {}_2F_1\left(-n+k, k; k+1; \frac{m}{m+x}\right), \end{aligned} \tag{3.6}$$

with $n, l \in \mathbb{N}_0$, $\lambda \in \mathbb{C} \setminus \{-1\}$, which is just the main result of Luo (see [6], p.920, Theorem 1).

Taking $\mu = \nu = 1$ in (3.1), and noting (1.21), we can obtain the following explicit representation of the generalized Apostol-Genocchi polynomials:

$$\mathcal{G}_n^{(l)}(x;\lambda) = 2^l l! \binom{n}{l} \sum_{k=0}^{n-l} \binom{l+k-1}{k} \binom{n-l}{k} \frac{(-\lambda)^k}{(\lambda+1)^{l+k}} \times \sum_{m=0}^{k} (-1)^m \binom{k}{m} m^k (x+m)^{n-l-k} {}_2F_1\left(-n+l+k, k; k+1; \frac{m}{m+x}\right), \tag{3.7}$$

with $n, l \in \mathbb{N}_0$, $\lambda \in \mathbb{C} \setminus \{-1\}$, which is just one of the results of Luo and Srivastava (see [15], p.5708, Theorem 1).

Taking $\lambda = -(\frac{\beta}{a})^b$, $\mu = 1-\kappa$, $\nu = \kappa$ in (3.1), and noting (1.23), we deduce the following well-known formula:

$$\mathcal{Y}_{n,\beta}^{(l)}(x;\kappa,a,b) = 2^{l(1-\kappa)}(l\kappa)! \binom{l+k-1}{k} \binom{n}{l\kappa} \sum_{k=0}^{n-l\kappa} \binom{n-l\kappa}{k} \frac{\beta^{bk}}{(\beta^b - a^b)^{k+1}} \times \sum_{m=0}^{k} (-1)^m \binom{k}{m} m^k (x+m)^{n-k-l\kappa} \times {}_2F_1\left(-n+l\kappa+k, k; k+1; \frac{m}{m+x}\right), \tag{3.8}$$

with $n, l, \kappa \in \mathbb{N}_0$, $\beta \in \mathbb{C}$, $a, b \in \mathbb{C} \setminus \{0\}$, $\beta \neq a$, which is just main result of Özarslan (see [14], p.2454, Theorem 2.1).

Further setting $l = 1$ in (3.8) we deduce the following formula for $\mathcal{Y}_{n,\beta}(x;\kappa,a,b)$:

$$\mathcal{Y}_{n,\beta}(x;\kappa,a,b) = 2^{1-\kappa}\kappa! \binom{n}{\kappa} \sum_{k=0}^{n-\kappa} \binom{n-\kappa}{k} \frac{\beta^{bk}}{(\beta^b - a^b)^{k+1}} \times \sum_{m=0}^{k} (-1)^m \binom{k}{m} m^k (x+m)^{n-k-\kappa} \times {}_2F_1\left(-n+\kappa+k, k; k+1; \frac{m}{m+x}\right). \tag{3.9}$$

4 Some explicit relationships between the generalized Apostol-type polynomials and generalized Hurwitz-Lerch zeta function

A general Hurwitz-Lerch zeta function $\Phi(z,s,a)$ defined by (*cf.*, *e.g.*, [2], p.121, *et seq.*)

$$\Phi(z,s,a) := \sum_{n=0}^{\infty} \frac{z^n}{(n+a)^s} \quad \left(a \in \mathbb{C} \setminus \mathbb{Z}_0^-; s \in \mathbb{C} \text{ when } |z| < 1; \Re(s) > 1 \text{ when } |z| = 1\right) \tag{4.1}$$

contains, as *special* cases, not only the Hurwitz (or generalized) zeta function $\zeta(s,a)$ defined by (*cf.* [26], p.249 and [4], p.88)

$$\zeta(s,a) := \Phi(1,s,a) = \sum_{n=0}^{\infty} \frac{1}{(n+a)^s} \quad \left(\Re(s) > 1; a \in \notin \mathbb{Z}_0^-\right) \tag{4.2}$$

and the Riemann zeta function $\zeta(s)$,

$$\zeta(s) := \Phi(1,s,1) = \zeta(s,1) = \frac{1}{2^s-1}\zeta\left(s,\frac{1}{2}\right) \quad \left(\Re(s)>1; a \notin \mathbb{Z}_0^-\right), \tag{4.3}$$

and the Lerch zeta function:

$$\ell_s(\xi) := \sum_{n=1}^{\infty} \frac{e^{2n\pi i\xi}}{n^s} = e^{2\pi i\xi}\,\Phi\left(e^{2\pi i\xi},s,1\right) \quad \left(\xi \in \mathbb{R}; \Re(s)>1\right), \tag{4.4}$$

but also such other functions as the polylogarithm function:

$$\mathrm{Li}_s(z) := \sum_{n=1}^{\infty} \frac{z^n}{n^s} = z\Phi(z,s,1)$$
$$\left(s \in \mathbb{C} \text{ when } |z|<1; \Re(s)>1 \text{ when } |z|=1\right) \tag{4.5}$$

and the Lipschitz-Lerch zeta function (*cf.* [2], p.122, Eq. 2.5(11)):

$$\phi(\xi,a,s) := \sum_{n=0}^{\infty} \frac{e^{2n\pi i\xi}}{(n+a)^s} = \Phi\left(e^{2\pi i\xi},s,a\right) =: L(\xi,s,a)$$
$$\left(a \in \mathbb{C}\setminus\mathbb{Z}_0^-; \Re(s)>0 \text{ when } \xi \in \mathbb{R}\setminus\mathbb{Z}; \Re(s)>1 \text{ when } \xi \in \mathbb{Z}\right), \tag{4.6}$$

which was first studied by Rudolf Lipschitz (1832-1903) and Matyáš Lerch (1860-1922) in connection with Dirichlet's famous theorem on primes in arithmetic progressions.

A family of the Hurwitz-Lerch zeta functions $\Phi_{\mu,\nu}^{(\rho,\sigma)}(z,s,a)$ defined by (see *e.g.* [27], p.727, Eq. (8))

$$\Phi_{\mu,\nu}^{(\rho,\sigma)}(z,s,a) := \sum_{n=0}^{\infty} \frac{(\mu)_{\rho n}}{(\nu)_{\sigma n}} \frac{z^n}{(n+a)^s}$$
$$\left(\mu \in \mathbb{C}; a,\nu \in \mathbb{C}\setminus\mathbb{Z}_0^-; \rho,\sigma \in \mathbb{R}^+; \rho<\sigma \text{ when } s,z \in \mathbb{C};\right.$$
$$\left.\rho=\sigma \text{ and } s \in \mathbb{C} \text{ when } |z|<1; \rho=\sigma \text{ and } \Re(s-\mu+\nu)>1 \text{ when } |z|=1\right), \tag{4.7}$$

contains, as *special* cases, not only the Hurwitz-Lerch zeta function

$$\Phi_{\nu,\nu}^{(\sigma,\sigma)}(z,s,a) = \Phi_{\mu,\nu}^{(0,0)}(z,s,a) = \Phi(z,s,a) = \sum_{n=0}^{\infty} \frac{z^n}{(n+a)^s} \tag{4.8}$$

and the Lipschitz-Lerch zeta function $\phi(\xi,a,s) := \Phi(e^{2\pi i\xi},s,a)$, but also the following generalized Hurwitz-Lerch zeta functions introduced and studied earlier by Goyal and Laddha [28], p.100, Eq. (1.5):

$$\Phi_{\mu,1}^{(1,1)}(z,s,a) = \Phi_\mu(z,s,a) := \sum_{n=0}^{\infty} \frac{(\mu)_n}{n!} \frac{z^n}{(n+a)^s}, \tag{4.9}$$

which, are called the *Goyal-Laddha-Hurwitz-Lerch zeta functions.*

Below we give an explicit relationship between the Apostol-type polynomials $\mathcal{F}_n^{(\alpha)}(x;\lambda;\mu;\nu)$ and the Hurwitz-Lerch zeta function $\Phi_\mu(z,s,a)$.

Theorem 4.1 *For $n,\nu,\alpha \in \mathbb{N}_0$; $-1 < \lambda \leqq 1$; $n \geqq \nu\alpha$; $x \in \mathbb{C} \setminus \mathbb{Z}_0^-$, $\mu \in \mathbb{C}$, the relationship*

$$\mathcal{F}_n^{(\alpha)}(x;\lambda;\mu;\nu) = 2^{\mu\alpha}(\nu\alpha)!\binom{n}{\nu\alpha}\Phi_\alpha(-\lambda,\nu\alpha-n,x) \tag{4.10}$$

holds true.

Proof Applying the generalized binomial theorem

$$(1+w)^{-\alpha} = \sum_{r=0}^{\infty}\binom{\alpha+r-1}{r}(-w)^r \quad (|w|<1)$$

in (1.18), we have

$$\begin{aligned}
\sum_{n=0}^{\infty}\mathcal{F}_n^{(\alpha)}(x;\lambda;\mu;\nu)\frac{z^n}{n!} &= \left(\frac{2^\mu z^\nu}{\lambda e^z+1}\right)^\alpha e^{xz} \\
&= 2^{\mu\alpha}z^{\nu\alpha}\left(1+\lambda e^z\right)^{-\alpha}e^{xz} \\
&= 2^{\mu\alpha}z^{\nu\alpha}\sum_{k=0}^{\infty}\frac{(\alpha)_k}{k!}(-\lambda)^k e^{(k+x)z} \\
&= \sum_{n=0}^{\infty}\left[2^{\mu\alpha}\sum_{k=0}^{\infty}\frac{(\alpha)_k}{k!}(-\lambda)^k(k+x)^n\right]\frac{z^{n+\nu\alpha}}{n!} \\
&= \sum_{n=\nu\alpha}^{\infty}\left[2^{\mu\alpha}(\nu\alpha)!\binom{n}{\nu\alpha}\sum_{k=0}^{\infty}\frac{(\alpha)_k}{k!}\frac{(-\lambda)^k}{(k+x)^{\nu\alpha-n}}\right]\frac{z^n}{n!}.
\end{aligned} \tag{4.11}$$

Noting (4.9), (4.10) follows. □

Below we see that (4.10) implies some well-known results.

Let $\lambda \longmapsto -\lambda$, taking $\mu = 0$ and $\nu = 1$ in (4.10) and noting (1.19), we can obtain an explicit relation between the Apostol-Bernoulli polynomials and the Hurwitz-Lerch zeta function:

$$\mathcal{B}_n^{(l)}(x;\lambda) = (-n)_l\Phi_l(\lambda,l-n,x) \quad \left(n,l \in \mathbb{N}; n \geqq l; |\lambda|<1; x \in \mathbb{C}\setminus\mathbb{Z}_0^-\right). \tag{4.12}$$

The above result is just one of the main results of Garg *et al.* (see [29], p.809).

Clearly, we have the following relation between the Apostol-Bernoulli polynomials and the Hurwitz-Lerch zeta function:

$$\mathcal{B}_n(x;\lambda) = -n\Phi(\lambda,1-n,x) \quad \left(n \in \mathbb{N}; |\lambda| \leqq 1; x \in \mathbb{C}\setminus\mathbb{Z}_0^-\right), \tag{4.13}$$

which is just the result of Apostol (see [3]).

Taking $\lambda = 1$ in (4.13), we obtain the following well-known relationship between the Bernoulli polynomials and Hurwitz zeta function (see [26], p.264, Theorem 12.13):

$$B_n(x) = -n\zeta(1-n,x) \quad (n \in \mathbb{N}). \tag{4.14}$$

Taking $x = 0$ in (4.14), we obtain the following the well-known relationship between the Bernoulli numbers and the Riemann zeta function (see [26], p.266, Theorem 12.16):

$$B_n = -n\zeta(1-n) \quad (n \in \mathbb{N}). \tag{4.15}$$

Taking $\mu = 1$ and $\nu = 0$ in (4.10) and noting (1.20), we can obtain the following result of Luo (see [30], p.339, Theorem 2.1):

$$\mathcal{E}_n^{(\alpha)}(x;\lambda) = 2^\alpha \Phi_\alpha(-\lambda, -n, x). \tag{4.16}$$

Further taking $\alpha = 1$ in (4.16), we have the following relation between the Apostol-Euler polynomials and the Hurwitz-Lerch zeta function:

$$\mathcal{E}_n(x;\lambda) = 2\Phi(-\lambda, -n, x) \quad \left(n \in \mathbb{N}; -1 < \lambda \leqq 1; x \in \mathbb{C} \setminus \mathbb{Z}_0^-\right). \tag{4.17}$$

Taking $\lambda = 1$ in (4.17), we can obtain the following well-known relation between the Euler polynomials and the L-function:

$$E_n(x) = 2L(-n, x), \tag{4.18}$$

where the L-function is defined by

$$L(s,x) := \sum_{n=0}^{\infty} \frac{(-1)^n}{(n+x)^s} \quad \left(\Re(s) > 1; x \in \mathbb{C} \setminus \mathbb{Z}_0^-\right). \tag{4.19}$$

From (4.19), we further obtain the following well-known relation between the Euler numbers and the l-function:

$$E_n = 2l(-n), \tag{4.20}$$

where the l-function is defined by

$$l(s) := \sum_{n=1}^{\infty} \frac{(-1)^n}{n^s}, \quad \Re(s) > 0. \tag{4.21}$$

Taking $\mu = \nu = 1$ in (4.10), and noting (1.21), we can deduce the following relation between the Apostol-Genocchi polynomials and Hurwitz-Lerch zeta function (see [31], p.124, Corollary 4.2):

$$\mathcal{G}_n^{(l)}(x;\lambda) = \{n\}_l 2^l \Phi_l(-\lambda, l-n, x) \quad \left(n, l \in \mathbb{N}; n \geqq l; |\lambda| \leqq 1;\ x \in \mathbb{C} \setminus \mathbb{Z}_0^-\right) \tag{4.22}$$

and

$$\mathcal{G}_n(x;\lambda) = 2n\Phi(-\lambda, 1-n, x) \quad \left(n \in \mathbb{N}; |\lambda| \leqq 1; x \in \mathbb{C} \setminus \mathbb{Z}_0^-\right). \tag{4.23}$$

Taking $\lambda = -(\frac{\beta}{a})^b$, $\mu = 1-\kappa$, $\nu = \kappa$ in (4.10), and noting (1.23), we can deduce the following relations between the polynomials $\mathcal{Y}_{n,\beta}^{(\alpha)}(x;\kappa,a,b)$, $\mathcal{Y}_{n,\beta}(x;\kappa,a,b)$, and the (generalized)

Hurwitz zeta functions [13, 14, 18]:

$$\mathcal{Y}_{n,\beta}^{(\alpha)}(x;\kappa,a,b) = (-1)^{\alpha}\frac{2^{(1-\kappa)\alpha}(\kappa\alpha)!}{a^{b\alpha}}\binom{n}{\kappa\alpha}\Phi_{\alpha}\left(\left(\frac{\beta}{a}\right)^{b},\kappa\alpha-n,x\right) \tag{4.24}$$

and

$$\mathcal{Y}_{n,\beta}(x;\kappa,a,b) = -\frac{2^{(1-\kappa)}(\kappa)!}{a^{b}}\binom{n}{\kappa}\Phi\left(\left(\frac{\beta}{a}\right)^{b},\kappa-n,x\right). \tag{4.25}$$

Competing interests

The authors declare that they have no competing interests.

Authors' contributions

The authors contributed equally in writing this paper, and they read and approved the final manuscript.

Author details

[1]Department of Mathematics and Information, Taizhou University, Taizhou, Zhejiang 317000, People's Republic of China. [2]Department of Mathematics, Chongqing Normal University, Chongqing Higher Education Mega Center, Huxi Campus, Chongqing, 401331, People's Republic of China.

Acknowledgements

The authors sincerely thank the referees for their valuable suggestions and comments. The present investigation was supported by *Natural Science Foundation Project of Chongqing, China* under Grant CSTC2011JJA00024, *Research Project of Science and Technology of Chongqing Education Commission, China* under Grant KJ120625, *Fund of Chongqing Normal University, China* under Grants 10XLR017 and 2011XLZ07.

References

1. Sándor, J, Crstici, B: Handbook of Number Theory, vol. II. Kluwer Academic, Dordrecht (2004)
2. Srivastava, HM, Choi, J: Series Associated with the Zeta and Related Functions. Kluwer Academic, Dordrecht (2001)
3. Apostol, TM: On the Lerch zeta function. Pac. J. Math. **1**, 161-167 (1951)
4. Srivastava, HM: Some formulas for the Bernoulli and Euler polynomials at rational arguments. Math. Proc. Camb. Philos. Soc. **129**, 77-84 (2000)
5. Luo, Q-M, Srivastava, HM: Some generalizations of the Apostol-Bernoulli and Apostol-Euler polynomials. J. Math. Anal. Appl. **308**, 290-320 (2005)
6. Luo, Q-M: Apostol-Euler polynomials of higher order and Gaussian hypergeometric functions. Taiwan. J. Math. **10**, 917-925 (2006)
7. Horadam, AF: Genocchi polynomials. In: Proceedings of the Fourth International Conference on Fibonacci Numbers and Their Applications, pp. 145-166. Kluwer Academic, Dordrecht (1991)
8. Horadam, AF: Negative order Genocchi polynomials. Fibonacci Q. **30**, 21-34 (1992)
9. Horadam, AF: Generation of Genocchi polynomials of first order by recurrence relations. Fibonacci Q. **30**, 239-243 (1992)
10. Jang, L-C, Kim, T: On the distribution of the q-Euler polynomials and the q-Genocchi polynomials of higher order. J. Inequal. Appl. **2008**, Article ID 723615 (2008)
11. Kim, T, Jang, L-C, Pak, HK: A note on q-Euler numbers and Genocchi numbers. Proc. Jpn. Acad., Ser. A, Math. Sci. **77**, 139-141 (2001)
12. Luo, Q-M: Extension for the Genocchi polynomials and its Fourier expansions and integral representations. Osaka J. Math. **48**, 291-309 (2011)
13. Ozden, H, Simsek, Y, Srivastava, HM: A unified presentation of the generating functions of the generalized Bernoulli, Euler and Genocchi polynomials. Comput. Math. Appl. **60**, 2779-2787 (2010)
14. Özarslan, MA: Unified Apostol-Bernoulli, Euler and Genocchi polynomials. Comput. Math. Appl. **62**, 2452-2462 (2011)
15. Luo, Q-M, Srivastava, HM: Some generalizations of the Apostol-Genocchi polynomials and the Stirling numbers of the second kind. Appl. Math. Comput. **217**, 5702-5728 (2011)
16. Luo, Q-M: Elliptic extensions of the Apostol-Bernoulli and Apostol-Euler polynomials. Appl. Math. Comput. **261**, 156-166 (2015)
17. Mansour, T, Ozden, H, Simsek, Y: Modified Apostol-type polynomials arising from umbral calculus. Adv. Stud. Contemp. Math. **24**, 175-182 (2014)
18. Ozden, H: Unification of generating function of the Bernoulli, Euler and Genocchi numbers and polynomials. AIP Conf. Proc. **1281**, 1125-1128 (2010)
19. Tremblay, R, Gaboury, S, Fugére, B-J: Some new classes of generalized Apostol-Euler and Apostol-Genocchi polynomials. Int. J. Math. Math. Sci. **2012**, Article ID 182785 (2012)
20. Lin, S-D, Srivastava, HM, Wang, P-Y: Some expansion formulas for a class of generalized Hurwitz-Lerch zeta functions. Integral Transforms Spec. Funct. **17**(11), 817-827 (2006)
21. Lu, D-Q, Srivastava, HM: Some series identities involving the generalized Apostol-type and related polynomials. Comput. Math. Appl. **62**, 3591-3602 (2011)

22. Luo, Q-M, Srivastava, HM: Some relationships between the Apostol-Bernoulli and Apostol-Euler polynomials. Comput. Math. Appl. **51**, 631-642 (2006)
23. Comtet, L: Advanced Combinatorics: The Art of Finite and Infinite Expansions. Reidel, Dordrecht (1974) (Translated from the French by JW Nienhuys)
24. Luo, Q-M: The multiplication formulas for the Apostol-Bernoulli and Apostol-Euler polynomials of higher order. Integral Transforms Spec. Funct. **20**, 377-391 (2009)
25. Abramowitz, M, Stegun, IA: Handbook of Mathematical Functions with Formulas, Graphs, and Mathematical Tables, 4th edn. Applied Mathematics Series, vol. 55. National Bureau of Standards, Washington (1965)
26. Apostol, TM: Introduction to Analytic Number Theory. Springer, New York (1976)
27. Lin, S-D, Srivastava, HM: Some families of the Hurwitz-Lerch zeta functions and associated fractional derivative and other integral representations. Appl. Math. Comput. **154**, 725-733 (2004)
28. Goyal, SP, Laddha, RK: On the generalized Riemann zeta functions and the generalized Lambert transform. Ganita Sandesh **11**, 99-108 (1997)
29. Garg, M, Jain, K, Srivastava, HM: Some relationships between the generalized Apostol-Bernoulli polynomials and Hurwitz-Lerch zeta functions. Integral Transforms Spec. Funct. **17**, 803-815 (2006)
30. Luo, Q-M: Some formulas for Apostol-Euler polynomials associated with Hurwitz zeta function at rational arguments. Appl. Anal. Discrete Math. **3**, 336-346 (2009)
31. Luo, Q-M: q-Extensions for the Apostol-Genocchi polynomials. Gen. Math. **17**(2), 113-125 (2009)

9

Caputo type fractional difference operator and its application on discrete time scales

Mohamad Rafi Segi Rahmat[1*] and Mohd Salmi Md Noorani[2]

*Correspondence:
Mohd.Rafi@nottingham.edu.my
[1]School of Applied Mathematics, The University of Nottingham Malaysia Campus, Jalan Broga, Semenyih, Selangor Darul Ehsan 43500, Malaysia
Full list of author information is available at the end of the article

Abstract

In this paper, we introduce the analogue of Caputo type fractional derivatives on a (q,h)-discrete time scale which can be reduced to Caputo type fractional differences studied by Abdeljawad (Comput. Math. Appl. 62:1602-1611, 2011) and Caputo type fractional q-differences studied by Atici and Eloe via the choice $q = h = 1$ and $h = 0$, respectively. Then, we solve linear fractional difference equations involving Caputo type (q,h)-derivatives and give the general solutions in terms of discrete Mittag-Leffler functions introduced by Cermak *et al.* In addition, we also apply the (q,h)-Laplace transform method to solve these linear fractional order difference equations.

Keywords: fractional difference equations; discrete time scales; fractional calculus; (q,h)-calculus

1 Introduction

Fractional calculus deals with the study of fractional order integrals and derivatives and their applications [1–3]. Riemann-Liouville and Caputo are kinds of fractional derivatives which generalize the ordinary integral and differential operators. Differential equations with fractional derivative provided a natural framework for the discussion of various kinds of real problems modeled by the aid of fractional derivative, such as viscoelastic system, signal processing, diffusion processes, control processing, fractional stochastic system, allometry in biology and ecology (see [1, 4–7] and the references therein).

The study of fractional calculus in discrete settings has been initiated in [8–10]. While the papers [8, 10] present the introduction to fractional q-derivatives and q-integrals, the paper [9, 11] discusses the basics of fractional difference calculus. Among other significant papers dealing with these problems, we can mention, *e.g.*, [4, 12–15] or [16], where discrete analogues of some topics of continuous fractional calculus have been developed. For some recent contributions on fractional differential/difference equations, see [17–22] or [23] and the references therein.

The extension of basic notions of fractional calculus to other discrete settings was performed in [24, 25], where fractional sums and differences have been introduced and studied in the framework of (q,h)-calculus, which can be reduced to ordinary difference calculus and q-difference calculus via the choice $q = h = 1$ and $h = 0$, respectively. This extension follows recent trends in continuous and discrete analysis, characterized by a unification and generalization resulting into development of the time scales theory [26, 27]. Some interesting results concerning (q,h)-discrete time scales can be found in [28, 29].

The aim of this paper is to introduce Caputo type nabla (q,h)-fractional difference operators and investigate their basic properties. Here, we solve some linear fractional difference equations involving Caputo type (q,h)-derivatives and provide the general solutions in terms of discrete Mittag-Leffler functions introduced in [25]. In addition, we also apply the (q,h)-Laplace transform method (introduced in [28]) to solve these linear fractional order difference equations.

2 Preliminaries

For the convenience of readers, we provide some basic concepts concerning the nabla calculus on time scales. The readers are referred to [26, 27, 30] for further details of the time scale theory. By a time scale $\mathbb{T}$ we understand any nonempty, closed subset of reals with the ordering inherited from reals. Thus the reals $\mathbb{R}$, the integers $\mathbb{Z}$, the natural numbers $\mathbb{N}$, the nonnegative integers $\mathbb{N}_0$, the h-numbers $h\mathbb{Z} = \{hk : k \in \mathbb{Z}\}$ with fixed $h > 0$, and the q-numbers $q^{\mathbb{N}_0} = \{q^k : k \in \mathbb{N}_0\}$ with fixed $q > 1$ are examples of time scales.

For any $t \in \mathbb{T}$, we define the forward and backward jump operators as $\sigma(t) := \inf\{s \in \mathbb{T} : s > t\}$ and $\rho(t) := \sup\{s \in \mathbb{T} : s < t\}$, respectively. The forward and backward graininess functions are defined as $\mu(t) := \sigma(t) - t$ and $\mu(t) := t - \rho(t)$, respectively. By convention, $\inf\{\emptyset\} = \sup\{\mathbb{T}\}$ and $\sup\{\emptyset\} = \inf\{\mathbb{T}\}$ for these operators. We say that a point is left-dense if $\rho(t) = t$ and left-scattered if $\rho(t) \neq t$. The right-dense and right-scattered points are defined in a similar manner. A point which is both left- and right-scattered is discrete. If $\inf\{\mathbb{T}\} = a_0 > -\infty$, we define $\mathbb{T}_\kappa := \mathbb{T}\backslash a_0$, otherwise $\mathbb{T}_\kappa := \mathbb{T}$.

For a function $f : \mathbb{T} \to \mathbb{R}$ and a point $t \in \mathbb{T}_\kappa$, we define $f^\nabla(t)$ to be a number such that for $\epsilon > 0$ there exists a neighborhood $U \subset \mathbb{T}$ of t which satisfies

$$\left|f(\rho(t)) - f(\tau) - f^\nabla(t)[\tau - \rho(t)]\right| \leq \epsilon\left|\tau - \rho(t)\right| \quad \text{for all } \tau \in U.$$

If $f^\nabla(t)$ is defined for all $t \in \mathbb{T}_\kappa$, then the function obtained is called the ∇-derivative of f.

A function $f : \mathbb{T} \to \mathbb{R}$ is called *left-dense continuous* or *ld-continuous* provided it is continuous at every left-dense point in $\mathbb{T}$, and $f(t+)$ exists for every right-dense point in $\mathbb{T}$, and the set of ld-continuous functions is denoted by $C_{ld}(\mathbb{T})$. If $f \in C_{ld}(\mathbb{T})$, then there is a function F such that $F^\nabla(t) = f(t)$. In this case, we define the ∇-integral as

$$\int_a^t f(\tau)\nabla\tau = F(t) - F(a) \quad \text{for all } t \in \mathbb{T}.$$

Clearly, if $f : \mathbb{T} \to \mathbb{R}$ is ld-continuous and $t \in \mathbb{T}_\kappa$, then

$$\int_{\rho(t)}^t f(\tau)\nabla\tau = \nu(t)f(t).$$

A function $f \in C_{ld}(\mathbb{T})$ is called *ν-regressive* if $1 + f\nu \neq 0$ on $\mathbb{T}_\kappa$, and $f \in C_{ld}(\mathbb{T})$ is called *positively regressive* if $1+f\nu > 0$ on $\mathbb{T}_\kappa$. The set of *ν-regressive* functions and the set of *positively ν-regressive* functions are denoted by $\mathcal{R}_\nu(\mathbb{T})$ and $\mathcal{R}_\nu^+(\mathbb{T})$, respectively. For simplicity, we denote by ${}_c\mathcal{R}_\nu(\mathbb{T})$ the set of ν-regressive constants.

Let $\lambda \in {}_c\mathcal{R}_\nu(\mathbb{T})$ and $s \in \mathbb{T}$, then the generalized exponential function $\widehat{e}_\lambda(\cdot, s)$ on time scale $\mathbb{T}$ is denoted by the unique solution of the initial value problem

$$\begin{cases} x^\nabla(t) = \lambda x(t), & t \in \mathbb{T}_\kappa; \\ x(s) = 1. \end{cases}$$

For $p \in {}_c\mathcal{R}_\nu(\mathbb{T})$, define circle minus p by

$$\ominus_\nu p := -\frac{p}{1 - p\nu}.$$

Then the unique solution of the initial value problem

$$\begin{cases} x^\nabla(t) = -\lambda x(t), & t \in \mathbb{T}_\kappa; \\ x(s) = 1 \end{cases}$$

takes the form $\widehat{e}_{\ominus_\nu \lambda}(\cdot, s)$. It is known that the exponential function $\widehat{e}_f(\cdot, s)$ is strictly positive on $[s, \infty)_\mathbb{T}$ provided that $f \in \mathcal{R}^+([s, \infty)_\mathbb{T})$.

The definition of the generalized monomials $\widehat{h}_n : \mathbb{T} \times \mathbb{T} \to \mathbb{R}$ $(n \in \mathbb{N}_0)$ is given by

$$\widehat{h}_n(t, s) = \begin{cases} 1, & n = 0; \\ \int_s^t \widehat{h}_{n-1}(\tau, s) \nabla\tau, & n \in \mathbb{N} \end{cases}$$

for $s, t \in \mathbb{T}$ (see [27]). If we let $\widehat{h}_n^\nabla(t, s)$ denote for each fixed $s \in \mathbb{T}$ the derivatives of $\widehat{h}_n^\nabla(t, s)$ with respect to t, then

$$\widehat{h}_n^\nabla(t, s) = \widehat{h}_{n-1}(t, s) \quad \text{for } n \in \mathbb{N}, t \in \mathbb{T}_\kappa.$$

An important relation between the generalized exponential function and the monomials is given by

$$\widehat{e}_\lambda(t, s) = \sum_{k=0}^{\infty} \lambda^k \widehat{h}_k(t, s) \quad \text{for } s, t \in \mathbb{T} \text{ with } t \geq s,$$

where $\lambda \in \mathcal{R}_c(\mathbb{T})$.

(First mean value theorem for integrals [27].) Let f and g be bounded and ∇-integrable functions on $[a, b]_\mathbb{T}$, and let g be nonnegative (or nonpositive) on $[a, b]_\mathbb{T}$. Set $m = \inf\{f(t) : t \in [a, b)_\mathbb{T}\}$ and $M = \sup\{f(t) : t \in [a, b)_\mathbb{T}\}$. Then there exists a real number η satisfying $m \leq \eta \leq M$ such that

$$\int_a^b f(t)g(t)\nabla t = \eta \int_a^b g(t)\nabla t.$$

For example, the function $\widehat{h}_n(t, s) \geq 0$ for all $t \geq s$ and $n \in \mathbb{N}_0$. Let f be an integrable function on $[t, s] \subset \mathbb{T}$, and let m and M be the infimum and the supremum, respectively, of f on $[t, s]$. Then there exists a number K between m and M such that

$$\int_s^t f(\tau)\widehat{h}_n(\tau, s)\nabla\tau = K \int_s^t \widehat{h}_n(\tau, s)\nabla\tau = K\widehat{h}_{n+1}(t, s).$$

The nabla Laplace transform of a function $f : \mathbb{T} \to \mathbb{R}$ is defined by [21]

$$\mathcal{L}\{f\}(z,s) := \int_s^{\infty} \widehat{e}_{\ominus_\nu z}(\rho(\tau),s)f(\tau)\nabla\tau \quad \text{for } z \in \mathcal{D},$$

where $\mathcal{D}$ consists of all complex numbers $z \in \mathbb{C}$ for which the improper ∇-integral exists.

The convolution of two functions $f, g : \mathbb{T} \to \mathbb{R}$ is defined by

$$(f * g)(t) := \int_a^t \widehat{f}(t,\rho(s))g(s)\nabla s, \quad t \in \mathbb{T},$$

where $\widehat{f}$ is the shift of f introduced in [31]. Then

$$\mathcal{L}\{f * g\}(z) = \mathcal{L}\{f\}(z) \cdot \mathcal{L}\{g\}(z).$$

The following result is the nabla version of the result obtained in [20]: (change of integration order) Let $f \in C_{ld}(\mathbb{T}^2)$, then

$$\int_s^t \int_s^{\eta} f(\eta,\zeta)\nabla\zeta\nabla\eta = \int_s^t \int_{\rho(\zeta)}^t f(\eta,\zeta)\nabla\eta\nabla\zeta \quad \text{for } s,t \in \mathbb{T}.$$

3 (q,h)-Fractional calculus

Consider the following (q,h)-time scale (for details, see [24, 25]):

$$\mathbb{T}^{t_0}_{(q,h)} = \{t_0 q^k + [k]_q h : k \in \mathbb{Z}\} \cup \left\{\frac{h}{1-q}\right\}$$

for $t_0 > 0$, $q \geq 1$, $h \geq 0$ and $q + h > 1$. Note that if $q = 1$, then the cluster point $h/(1-q) = -\infty$ is not involved in $\mathbb{T}$. The forward and backward jump operator is the linear function $\sigma(t) = qt + h$ and $\rho(t) = q^{-1}(t-h)$, respectively. Similarly, the forward and backward graininess is given by $\mu(t) = (q-1)t + h$ and $\nu(t) = q^{-1}\mu(t)$, respectively. Observe that

$$\sigma^k(t) = q^k t + [k]_q h \quad \text{and} \quad \rho^k(t) = q^{-k}(t - [k]_q h).$$

The following relation

$$\nu(\rho^k(t)) = q^{-k}\nu(t)$$

holds for $t \in \mathbb{T}$.

The nabla (q,h)-derivative of the function $f : \mathbb{T}^{t_0}_{(q,h)} \to \mathbb{R}$ is defined by

$$\nabla_{(q,h)} f(t) := \frac{f(t) - f(\rho(t))}{\nu(t)} = \frac{f(t) - f(\tilde{q}(t-h))}{(1-\tilde{q})t + \tilde{q}h},$$

where $\tilde{q} = q^{-1}$. Let $t, a \in \mathbb{T}^{t_0}_{(q,h)}$ such that $h/(1-q) \le a \le t$ and $f : \mathbb{T} \to \mathbb{R}$. Then the nabla (q,h)-integral exists and can be calculated (provided $t > a$) via the formula

$$
\begin{aligned}
{}_a\nabla^{-1}_{(q,h)} f(t) &:= \int_a^t f(\tau)\nabla\tau = \sum_{k=0}^{n-1} \nu(\rho^k(t)) f(\rho^k(t)) \\
&= \left((1-q^{-1})t + q^{-1}h\right) \sum_{k=0}^{n-1} q^{-k} f(q^{-k}t + [-k]_q h).
\end{aligned}
$$

The Taylor monomials and the power functions on $\mathbb{T}^{t_0}_{(q,h)}$ have the forms

$$
\widehat{h}_n(t,s) = \frac{\prod_{j=0}^{n-1}(\sigma^j(t) - s)}{[n]_q!} = \frac{\prod_{j=0}^{n-1}(t, \rho^j(s))}{[n]_{\tilde{q}}!},
$$

$$
(t-s)^{(n)}_{(\tilde{q},h)} = \prod_{j=0}^{n-1}(t - \rho^j(s)),
$$

respectively, and the extension of the monomials $\widehat{h}_n(t,s)$ corresponding to $\mathbb{T}^{t_0}_{(q,h)}$ takes the form

$$
\widehat{h}_\alpha(t,s) = \frac{(t-s)^{(\alpha)}_{(\tilde{q},h)}}{\Gamma_{\tilde{q}}(\alpha+1)}, \quad \alpha \in \mathbb{R}.
$$

Another (equivalent) expression of $\widehat{h}_\alpha(t,s)$ is provided by the following assertion.

Proposition 3.1 *Let $\alpha \in \mathbb{R}$, $s, t \in \mathbb{T}^{t_0}_{(q,h)}$ and $n \in \mathbb{N}_0$ be such that $t = \sigma^n(s)$. Then*

$$
\widehat{h}_\alpha(t,s) = \left(\nu(t)\right)^\alpha \begin{bmatrix} \alpha + n - 1 \\ n-1 \end{bmatrix}_{\tilde{q}} \tag{1}
$$

$$
= \left(\nu(t)\right)^\alpha \begin{bmatrix} -\alpha - 1 \\ n-1 \end{bmatrix}_{\tilde{q}} (-1)^{n-1} \tilde{q}^{\alpha(n-1)+\binom{n}{2}}. \tag{2}
$$

Proposition 3.2 *Let $\alpha \in \mathbb{R}$, $s, t \in \mathbb{T}^{t_0}_{(q,h)}$ and $k, n \in \mathbb{N}_0$ be such that $t = \sigma^n(s)$ and $k \le n$. Then*

$$
\widehat{h}_\alpha\left(\rho^k(t), s\right) = \tilde{q}^{\alpha k}\left(\nu(t)\right)^\alpha \begin{bmatrix} \alpha + n - k - 1 \\ n-k-1 \end{bmatrix}_{\tilde{q}} \tag{3}
$$

$$
= \tilde{q}^{\alpha k}\left(\nu(t)\right)^\alpha \begin{bmatrix} -\beta \\ n-k-1 \end{bmatrix}_{\tilde{q}} (-1)^{n-k-1} \tilde{q}^{(\beta-1)(n-k-1)+\binom{n-k}{2}}. \tag{4}
$$

Proof The proof is similar to the proof of Proposition 2.2 in [25].

Let $q > 1$. Using the relations

$$
[t] = \frac{\nu(t)}{(1-\tilde{q})} \quad \Rightarrow \quad \left[\rho^k(t)\right] = \tilde{q}^k \frac{\nu(t)}{(1-\tilde{q})} \quad \text{and} \quad \frac{[s]}{[\rho^k(t)]} = \tilde{q}^{n-k},
$$

we can derive that

$$\begin{aligned}
\widehat{h}_\alpha\left(\rho^k(t),s\right) &= \frac{[\rho^k(t)]^\alpha([s]/[\rho^k(t)],\tilde{q})_\infty}{\Gamma_{\tilde{q}}(\alpha+1)(\tilde{q}^\alpha[s]/[\rho^k(t)],\tilde{q})_\infty} \\
&= \frac{\tilde{q}^{\alpha k}(\nu(t))^\alpha(1-\tilde{q})^{-\alpha}(\tilde{q}^{n-k},\tilde{q})_\infty}{\Gamma_{\tilde{q}}(\alpha+1)(\tilde{q}^{\alpha+n-k},\tilde{q})_\infty} \\
&= \tilde{q}^{\alpha k}\left(\nu(t)\right)^\alpha \frac{\Gamma_{\tilde{q}}(\alpha+n-k)}{\Gamma_{\tilde{q}}(\alpha+1)\Gamma_{\tilde{q}}(n-k)} \\
&= \tilde{q}^{\alpha k}\left(\nu(t)\right)^\alpha \begin{bmatrix} \alpha+n-k-1 \\ n-k-1 \end{bmatrix}_{\tilde{q}} \\
&= \tilde{q}^{\alpha k}\left(\nu(t)\right)^\alpha \begin{bmatrix} -\beta \\ n-k-1 \end{bmatrix}_{\tilde{q}} (-1)^{n-k-1}\tilde{q}^{(\beta-1)(n-k-1)+\binom{n-k}{2}}.
\end{aligned}$$

Here, we use the identities

$$\begin{bmatrix} x \\ k \end{bmatrix}_{\tilde{q}} = \frac{\Gamma_{\tilde{q}}(x+1)}{\Gamma_{\tilde{q}}(k+1)\Gamma_{\tilde{q}}(x-k+1)}, \quad x\in\mathbb{R}, k\in\mathbb{Z},$$

$$\frac{\Gamma_{\tilde{q}}(x+m)}{\Gamma_{\tilde{q}}(x)} = (-1)^m\tilde{q}^{xm+\binom{m}{2}}\frac{\Gamma_{\tilde{q}}(1-x)}{\Gamma_{\tilde{q}}(1-x-m)}, \quad x\in\mathbb{R}, m\in\mathbb{Z}^+,$$

where $\Gamma_{\tilde{q}}$ is the q-gamma function defined as

$$\Gamma_{\tilde{q}}(x) = \frac{(\tilde{q},\tilde{q})_\infty(1-\tilde{q})^{1-x}}{(\tilde{q}^x,\tilde{q})_\infty}, \quad 0<\tilde{q}<1. \qquad \square$$

The following theorem is an extension of the result provided in [32] in $\mathbb{T}^{t_0}_{(q,h)}$.

Theorem 3.3 *Let $n\in\mathbb{N}$ and $t,a\in\mathbb{T}^{t_0}_{(q,h)}$. Suppose $f:\mathbb{T}^{t_0}_{(q,h)}\to\mathbb{R}$, then*

$$f(t) = \sum_{k=0}^{n-1}\widehat{h}_k(t,a)\nabla^k_{(q,h)}f(a) + \int_a^t \widehat{h}_{n-1}\left(t,\rho(\tau)\right)\nabla^n_{(q,h)}f(\tau)\nabla\tau. \tag{5}$$

Lemma 3.4 ([25]) *Let $m\in\mathbb{Z}^+$, $\alpha\in\mathbb{R}$, $s,t\in\mathbb{T}^{t_0}_{(q,h)}$ and $n\in\mathbb{Z}^+$, $n\geq m$ be such that $t=\sigma^n(s)$. Then*

$$\nabla^m_{(q,h)}\widehat{h}_\alpha(t,s) = \begin{cases} \widehat{h}_{\alpha-m}(t,s), & \alpha\notin\{0,1,\ldots,m-1\}, \\ 0, & \alpha\in\{0,1,\ldots,m-1\}. \end{cases} \tag{6}$$

Let $a\in\mathbb{T}^{t_0}_{(q,h)}$, $a>h/(1-q)$ be fixed. We consider the following restricted (q,h)-time scale:

$$\widetilde{\mathbb{T}}^{\sigma^i(a)}_{(q,h)} = \left\{t\in\mathbb{T}, t\geq\sigma^i(a)\right\}, \quad i=0,1,2,\ldots, \tag{7}$$

where the symbol σ^i stands for the ith iterate of σ (analogously, we use the symbol ρ^i).

Now we can continue with the introduction of (q,h)-fractional integral and (q,h)-fractional derivative of a function $f:\widetilde{\mathbb{T}}^{\sigma^i(a)}_{(q,h)}\to\mathbb{R}$. Let $t\in\widetilde{\mathbb{T}}^{\sigma^i(a)}_{(q,h)}$.

Definition 3.5 ([25]) The nabla (q,h)-fractional integral (in the sense of Riemann-Liouville) of order $\alpha \in \mathbb{R}^+$ over the time scale interval $[a,t] \cap \widetilde{\mathbb{T}}^a_{(q,h)}$ is defined by

$$ {}_a\nabla^{-\alpha}_{(q,h)} f(t) := \int_a^t \widehat{h}_{\alpha-1}(t,\rho(\tau)) f(\tau)\nabla\tau. \tag{8}$$

The nabla (q,h)-fractional derivative (in the sense of Caputo) of order $\alpha \in \mathbb{R}^+$ is defined by

$$ {}^C_a\nabla^{\alpha}_{(q,h)} f(t) := {}_a\nabla^{-(m-\alpha)} \nabla^m f(t) \tag{9}$$

$$ = \int_a^t \widehat{h}_{m-\alpha-1}(t,\rho(\tau)) \nabla^m f(\tau)\nabla\tau, \tag{10}$$

where $m \in \mathbb{Z}^+$ is such that $m-1 < \alpha \leq m$.

Lemma 3.6 ([25]) *Let $\alpha \in \mathbb{R}^+$, $\beta \in \mathbb{R}$ and $t \in \widetilde{\mathbb{T}}^{\sigma(a)}_{(q,h)}$. Then*

$$ {}_a\nabla^{-\alpha}_{(q,h)} \widehat{h}_\beta(t,a) = \widehat{h}_{\alpha+\beta}(t,a). \tag{11}$$

The following lemma is useful.

Lemma 3.7 *For all $\alpha, \beta \in \mathbb{R}$ and $t, s \in \widetilde{\mathbb{T}}^{\sigma(a)}_{(q,h)}$,*

$$ \int_a^t \widehat{h}_{\alpha-1}(t,\rho(\tau)) \widehat{h}_{\beta-1}(\tau,a)\nabla\tau = \widehat{h}_{\alpha+\beta-1}(t,a). \tag{12}$$

The proof is similar to the proof of Proposition 4.1 in [25]. For the sake of completeness, we give the following proof.

Proof Let $\widetilde{\mathbb{T}}^{\sigma(a)}_{(q,h)}$ be such that $t = \sigma^n(a)$ for some $n \in \mathbb{Z}^+$. We have

$$\begin{aligned}
&\int_a^t \widehat{h}_{\alpha-1}(t,\rho(\tau)) \widehat{h}_{\beta-1}(\tau,a)\nabla\tau \\
&= \sum_{k=0}^{n-1} \widehat{h}_{\alpha-1}(t,\rho^{k+1}(t)) \widehat{h}_{\beta-1}(\rho^k(t),a)\nu(\rho^k(t)) \\
&= \sum_{k=0}^{n-1} (\nu(t))^{\alpha-1} \begin{bmatrix} -\alpha \\ k \end{bmatrix}_{\tilde{q}} (-1)^k \tilde{q}^{(\alpha-1)k+\binom{k+1}{2}} \\
&\quad \times \tilde{q}^{(\beta-1)k} (\nu(t))^{\beta-1} \begin{bmatrix} -\beta \\ n-k-1 \end{bmatrix}_{\tilde{q}} (-1)^{n-k-1} \tilde{q}^{(\beta-1)(n-k-1)+\binom{n-k}{2}} \tilde{q}^k \nu(t) \\
&= (\nu(t))^{\alpha+\beta-1} \sum_{k=0}^{n-1} \begin{bmatrix} -\alpha \\ k \end{bmatrix}_{\tilde{q}} \begin{bmatrix} -\beta \\ n-k-1 \end{bmatrix}_{\tilde{q}} (-1)^{n-1} \tilde{q}^{k^2-k(n-1)+k\alpha+\binom{n}{2}+(\beta-1)(n-1)} \\
&= (\nu(t))^{\alpha+\beta-1} \sum_{k=0}^{n-1} \begin{bmatrix} -\alpha \\ n-k-1 \end{bmatrix}_{\tilde{q}} \begin{bmatrix} -\beta \\ k \end{bmatrix}_{\tilde{q}} \\
&\quad \times (-1)^{n-1} \tilde{q}^{(n-k-1)^2-(n-k-1)(n-1)+(n-k-1)\alpha+\binom{n}{2}+(\beta-1)(n-1)}
\end{aligned}$$

$$= \left(\nu(t)\right)^{\alpha+\beta-1} \sum_{k=0}^{n-1} \begin{bmatrix} -\alpha \\ n-k-1 \end{bmatrix}_{\tilde{q}} \begin{bmatrix} -\beta \\ k \end{bmatrix}_{\tilde{q}} (-1)^{n-1} \tilde{q}^{k^2-k(n-1)-k\alpha+(\alpha+\beta)(n-1)+\binom{n}{2}}$$

$$= \left(\nu(t)\right)^{\alpha+\beta-1} \sum_{k=0}^{n-1} \begin{bmatrix} -\alpha \\ (n-1)-k \end{bmatrix}_{\tilde{q}} \begin{bmatrix} -\beta \\ k \end{bmatrix}_{\tilde{q}} (-1)^{n-1} \tilde{q}^{k^2-k(n-1)-k\alpha+(\alpha+\beta-1)(n-1)+\binom{n}{2}}$$

$$= \left(\nu(t)\right)^{\alpha+\beta-1} \sum_{k=0}^{n-1} \begin{bmatrix} -(\alpha+\beta-1)-1 \\ n-1 \end{bmatrix}_{\tilde{q}} (-1)^{n-1} \tilde{q}^{(\alpha+\beta-1)(n-1)+\binom{n}{2}}$$

$$= \widehat{h}_{\alpha+\beta-1}(t,a),$$

where we used the q-Vandermonde identity [25]

$$\sum_{k=0}^{n} \begin{bmatrix} x \\ n-k \end{bmatrix}_{\tilde{q}} \begin{bmatrix} y \\ k \end{bmatrix}_{\tilde{q}} \tilde{q}^{k^2-nk+xk} = \begin{bmatrix} x+y \\ n \end{bmatrix}_{\tilde{q}}, \quad x, y \in \mathbb{R}, m \in \mathbb{N}_0.$$

The following relations are used in the above proof:

$$\nu\left(\rho^k(t)\right) = \tilde{q}^k \nu(t).$$

□

With the power rule stated in Lemma 3.4, we can perform its extension to Caputo type nabla (q,h)-fractional derivative of the function $\widehat{h}_\beta(t,a)$ as below.

Lemma 3.8 *Let $\alpha \in \mathbb{R}^+$, $\beta \in \mathbb{R}$, $t \in \widetilde{\mathbb{T}}_{(q,h)}^{\sigma^{m+1}(a)}$, where $m \in \mathbb{Z}^+$ satisfies $m-1 < \alpha \leq m$. Then*

$${}_a^C\nabla_{(q,h)}^{\alpha} \widehat{h}_\beta(t,a) = \begin{cases} \widehat{h}_{\beta-\alpha}(t,a), & \beta \notin \{0,1,\ldots,m-1\}; \\ 0, & \beta \in \{0,1,\ldots,m-1\}. \end{cases} \tag{13}$$

Proof The definition of Caputo nabla derivative and Lemma 3.4 yield

$$\begin{aligned} {}_a^C\nabla_{(q,h)}^{\alpha} \widehat{h}_\beta(t,a) &= {}_a\nabla^{-(m-\alpha)} \nabla^m \widehat{h}_\beta(t,a) \\ &= {}_a\nabla^{-(m-\alpha)} \begin{cases} \widehat{h}_{\beta-m}(t,a), & \beta \notin \{0,1,\ldots,m-1\}; \\ 0, & \beta \in \{0,1,\ldots,m-1\}, \end{cases} \\ &= \begin{cases} \widehat{h}_{\beta-\alpha}(t,a), & \beta \notin \{0,1,\ldots,m-1\}; \\ 0, & \beta \in \{0,1,\ldots,m-1\}. \end{cases} \end{aligned}$$

□

Theorem 3.9 *Let $\alpha \in \mathbb{R}$ and $n \in \mathbb{Z}^+$ so that $n-1 < \alpha \leq n$. Then*

$${}_a\nabla_{(q,h)}^{-\alpha} {}_a^C\nabla_{(q,h)}^{\alpha} f(t) = f(t) - \sum_{k=0}^{n-1} \widehat{h}_k(t,a) \nabla_{(q,h)}^k f(a). \tag{14}$$

Proof By deploying the definitions of fractional integral and Caputo nabla derivative, Lemma 3.7 and change of integration order, we obtain

$$\begin{aligned} {}_a\nabla_{(q,h)}^{-\alpha} {}_a^C\nabla_{(q,h)}^{\alpha} f(t) &= \int_a^t \widehat{h}_{\alpha-1}(t,\rho(\tau)) \left({}_a^C\nabla_{(q,h)}^{\alpha} f(\tau)\right) \nabla\tau \\ &= \int_a^t \widehat{h}_{\alpha-1}(t,\rho(\tau)) \left(\int_a^\tau \widehat{h}_{m-\alpha-1}(\tau,\rho(u)) \left(\nabla_{(q,h)}^m f(u)\right) \nabla u \right) \nabla\tau \end{aligned}$$

$$= \int_a^t \nabla^m_{(q,h)} f(u)\left(\int_{\rho(u)}^t \widehat{h}_{\alpha-1}(t,\rho(\tau))\widehat{h}_{m-\alpha-1}(\tau,\rho(u))\nabla\tau\right)\nabla u$$

$$= \int_a^t \widehat{h}_{m-1}(t,\rho(u))\nabla^m_{(q,h)} f(u))\nabla u$$

$$= f(t) - \sum_{k=0}^{m-1} \widehat{h}_k(t,a)\nabla^k_{(q,h)} f(a) \quad \text{(by Taylor's formula, (16))}. \qquad \square$$

In the following theorem we give the alternative definition of nabla (q,h)-derivative (Caputo).

Theorem 3.10 *Let $\alpha \in \mathbb{R}$ and $n \in \mathbb{Z}^+$ so that $n-1 < \alpha \leq n$. Then*

$$ {}_a^C\nabla^\alpha_{(q,h)} f(t) = {}_a\nabla^\alpha_{(q,h)} \left\{ f(t) - \sum_{k=0}^{n-1} \widehat{h}_k(t,a)\nabla^k_{(q,h)} f(a) \right\}. \tag{15}$$

Proof It follows immediately from relation (14) and the fact that ${}_a\nabla^\alpha({}_a\nabla^{-\alpha} f)(t) = f(t)$ [24]. □

4 Application

In this section we give solutions for fractional initial value problems on $\mathbb{T}^{t_0}_{(q,h)}$. The following definition and results are due to Čermák *et al.* [25].

Definition 4.1 Let $\alpha, \beta, \lambda \in \mathbb{R}$. The (q,h)-Mittag-Leffler function $E^{s,\lambda}_{\alpha,\beta}(t)$ is defined by

$$E^{s,\lambda}_{\alpha,\beta}(t) = \sum_{k=0}^{\infty} \lambda^k \widehat{h}_{\alpha k+\beta-1}(t,s) \quad \left(= \sum_{k=0}^{\infty} \lambda^k \frac{(t-s)^{(\alpha k+\beta-1)}_{(\tilde{q},h)}}{\Gamma_{\tilde{q}}(\alpha k+\beta)} \right) \tag{16}$$

for $s, t \in \widetilde{\mathbb{T}}^{\sigma(a)}_{(q,h)}$ and $t \geq s$.

It is easy to check that the series on the right-hand side converges (absolutely) if $|\lambda|(\nu(t))^\alpha < 1$.

Theorem 4.2 ([25]) *Let $\eta \in \mathbb{R}^+$ and $t \in \mathbb{T}^{\sigma(a)}_{(q,h)}$. Then*

$$ {}_a\nabla^{-\eta}_{(q,h)} E^{a,\lambda}_{\alpha,\beta}(t) = E^{a,\lambda}_{\alpha,\beta+\eta}(t). \tag{17}$$

Theorem 4.3 *Let $\gamma \in \mathbb{R}^+$, $m \in \mathbb{Z}^+$ be such that $m-1 < \gamma \leq m$, and let $\alpha k + \beta - 1 \notin \{1,2,\dots,m\}$ for all $k \in \mathbb{Z}^+$. If $t \in \mathbb{T}^{\sigma^{m+1}(a)}_{(q,h)}$, then*

$$ {}_a^C\nabla^\gamma_{(q,h)} E^{a,\lambda}_{\alpha,\beta}(t) = \begin{cases} E^{a,\lambda}_{\alpha,\beta-\gamma}(t), & \beta \notin \{1,2,\dots,m\}, \\ \lambda E^{a,\lambda}_{\alpha,\alpha+\beta-\gamma}(t), & \beta \in \{1,2,\dots,m\}. \end{cases} \tag{18}$$

Proof By definition of (q,h)-Mittag-Leffler function,

$$ {}_a^C\nabla^\gamma_{(q,h)} E^{a,\lambda}_{\alpha,\beta}(t) = {}_a^C\nabla^\gamma_{(q,h)} \sum_{k=0}^{\infty} \lambda^k \widehat{h}_\alpha(t,s)_{\alpha k+\beta-1}(t,a) = \sum_{k=0}^{\infty} \lambda^k {}_a^C\nabla^\gamma_{(q,h)} \widehat{h}_\alpha(t,s)_{\alpha k+\beta-1}(t,a) \tag{19}$$

due to the absolute convergence property.

If $k \in \mathbb{Z}^+$, then Lemma 3.8 implies the relation

$$ {}_a^C\nabla_{(q,h)}^{\gamma} \widehat{h}_\alpha(t,s)_{\alpha k+\beta-1}(t,a) = \widehat{h}_\alpha(t,s)_{\alpha k+\beta-\gamma-1}(t,a) \tag{20}$$

due to the assumption $\alpha k+\beta-1 \notin \{1,2,\dots,m\}$. If $k=0$, then two possibilities may occur. If $\beta \notin \{1,2,\dots,m\}$, we get that (20) with $k=0$ implies the validity of $(18)_1$. If $\beta \in \{1,2,\dots,m\}$, then ${}_a^C\nabla_{(q,h)}^{\gamma} \widehat{h}_\alpha(t,s)_{\alpha k+\beta-1}(t,a)=0$, and by shifting the index k, *i.e.*,

$$\sum_{k=1}^{\infty} \lambda^k {}_a^C\nabla_{(q,h)}^{\gamma} \widehat{h}_\alpha(t,s)_{\alpha k+\beta-1}(t,a) = \lambda \sum_{k=0}^{\infty} \lambda^k \widehat{h}_\alpha(t,s)_{\alpha k+\alpha+\beta-\gamma-1}(t,a), \tag{21}$$

we obtain $(18)_2$. This completes the proof. □

Let us consider the following Cauchy type fractional initial value problem (FIVP):

$$ {}_a^C\nabla_{(q,h)}^{\alpha} y(t) - \lambda y(t) = f(t), \quad t \in \widetilde{\mathbb{T}}_{(q,h)}^{\sigma^{n+1}(a)}, \tag{22}$$

$$\nabla^j y(t)|_{t=\sigma^{n-1}(a)} = b_j \in \mathbb{R}, \quad j=0,1,\dots,n-1, \tag{23}$$

where $f : \widetilde{\mathbb{T}}_{(q,h)}^{\sigma^{n+1}(a)} \to \mathbb{R}$, $\alpha \in \mathbb{R}^+$ and $n \in \mathbb{Z}^+$ are such that $n-1 < \alpha \leq n$. Here, we assume that ν-regressivity condition is ensured, that is, $\lambda(\nu(t))^\alpha \neq 1$.

It can be easily verified that the above FIVP can be reduced to the Volterra type integral equation

$$y(t) = \sum_{k=0}^{n-1} b_k \widehat{h}_k(t,a) + \lambda \int_a^t \widehat{h}_{\alpha-1}(t,\rho(\tau)) y(\tau)\nabla\tau + \int_a^t \widehat{h}_{\alpha-1}(t,\rho(\tau)) f(\tau)\nabla\tau. \tag{24}$$

We apply the method of successive approximations by setting

$$y_0(t) = \sum_{k=0}^{n-1} b_k \widehat{h}_k(t,a),$$

$$y_m(t) = y_0(t) + \lambda \int_a^t \widehat{h}_{\alpha-1}(t,\rho(\tau)) y_{m-1}(\tau)\nabla\tau + \int_a^t \widehat{h}_{\alpha-1}(t,\rho(\tau)) f(\tau)\nabla\tau.$$

Thus

$$\begin{aligned}
y_1(t) &= \sum_{k=0}^{n-1} b_k \widehat{h}_k(t,a) + \sum_{k=0}^{n-1} \lambda b_k \widehat{h}_{\alpha+k}(t,a) + \int_a^t \widehat{h}_{\alpha-1}(t,\rho(\tau)) f(\tau)\nabla\tau \\
&= \sum_{k=0}^{n-1} b_k \sum_{j=0}^{1} \lambda^j \widehat{h}_{\alpha j+k}(t,a) + \int_a^t \widehat{h}_{\alpha-1}(t,\rho(\tau)) f(\tau)\nabla\tau, \\
y_2(t) &= \sum_{k=0}^{n-1} b_k \sum_{j=0}^{2} \lambda^j \widehat{h}_{\alpha j+k}(t,a) + \int_a^t \left[\lambda \widehat{h}_{2\alpha-1}(t,\rho(\tau)) + \widehat{h}_{\alpha-1}(t,\rho(\tau))\right] f(\tau)\nabla\tau \\
&= \sum_{k=0}^{n-1} b_k \sum_{j=0}^{2} \lambda^j \widehat{h}_{\alpha j+k}(t,a) + \int_a^t \sum_{j=1}^{2} \lambda^{j-1} \widehat{h}_{\alpha j-1}(t,\rho(\tau)) f(\tau)\nabla\tau,
\end{aligned}$$

$$\vdots$$

$$y_m(t) = \sum_{k=0}^{n-1} b_k \sum_{j=0}^{m} \lambda^j \widehat{h}_{\alpha j+k}(t,a) + \int_a^t \sum_{j=1}^{m} \lambda^{j-1} \widehat{h}_{\alpha j-1}\big(t,\rho(\tau)\big) f(\tau) \nabla \tau.$$

Taking the limit as $m \to \infty$, we arrive at the approximate solution to equation (24), and taking into account (16), we obtain

$$y(t) = \sum_{k=0}^{n-1} b_k E_{\alpha,k+1}^{a,\lambda}(t) + \int_a^t E_{\alpha,\alpha}^{\rho(\tau),\lambda}(\tau) f(\tau) \nabla \tau. \tag{25}$$

This yields an explicit solution $y(t)$ of the Volterra integral equation (24) and hence a solution for (22)-(23).

Clearly, the Cauchy problem (22) involving the homogeneous equation

$${}_a^C\nabla_{(q,h)}^{\alpha} y(t) - \lambda y(t) = 0, \quad t \in \widetilde{\mathbb{T}}_{(q,h)}^{\sigma^{n+1}(a)}, n-1 < \alpha \leq n \tag{26}$$

has a unique solution of the form

$$y(t) = \sum_{k=0}^{n-1} b_k E_{\alpha,k+1}^{a,\lambda}(t). \tag{27}$$

Corollary 4.4 *The solution of the Cauchy problem*

$${}_a^C\nabla_{(q,h)}^{\alpha} y(t) - \lambda y(t) = f(t), \quad t \in \widetilde{\mathbb{T}}_{(q,h)}^{\sigma^2(a)}, 0 < \alpha < 1, \tag{28}$$

$$y(a) = 1 \in \mathbb{R} \tag{29}$$

has the form

$$y(t) = E_{\alpha,1}^{a,\lambda}(t) + \int_a^t E_{\alpha,\alpha}^{\rho(\tau),\lambda}(\tau) f(\tau) \nabla \tau, \tag{30}$$

while the solution to the problem

$${}_a^C\nabla_{(q,h)}^{\alpha} y(t) = \lambda y(t), \quad t \in \widetilde{\mathbb{T}}_{(q,h)}^{\sigma^2(a)}, 0 < \alpha < 1, \tag{31}$$

$$y(a) = 1 \in \mathbb{R} \tag{32}$$

is given by

$$y(t) = E_{\alpha,1}^{a,\lambda}(t). \tag{33}$$

Finally, we give the Laplace transforms of fractional nabla integral and Caputo nabla derivative on $\mathbb{T}_{(q,h)}^{t_0}$. For $t, s \in \mathbb{T}_{(q,h)}^{t_0}$ and $\alpha \in \mathbb{R}^+$, we have [21]

$$\widehat{\mathcal{L}}_{(q,h)}\left\{\widehat{h}_{\alpha}(t,s)\right\}(z) = \frac{1}{z^{\alpha+1}}.$$

From the definition of (q,h)-Mittag-Leffler function $E^{s,\lambda}_{\alpha,\beta}(t)$, we have

$$\begin{aligned}\widehat{\mathcal{L}}_{(q,h)}\left\{E^{a,\lambda}_{\alpha,\beta}\right\}(z) &= \widehat{\mathcal{L}}_{(q,h)}\left\{\sum_{k=0}^{\infty}\lambda^k\widehat{h}_{\alpha k+\beta-1}(\cdot,a)\right\}(z)\\ &= \sum_{k=0}^{\infty}\lambda^k\widehat{\mathcal{L}}_{(q,h)}\left\{\widehat{h}_{\alpha k+\beta-1}(\cdot,a)\right\}(z)\\ &= \sum_{k=0}^{\infty}\frac{\lambda^k}{z^{\alpha k-\beta}}\\ &= \frac{z^{\alpha-\beta}}{z^{\alpha}-\lambda} \quad \text{provided } \left|z^{-\alpha}\lambda\right| < 1. \end{aligned} \tag{34}$$

For $\alpha = 1$, $\beta = 1$ and $t \in \widetilde{\mathbb{T}}^{\sigma^i(a)}_{(q,h)}$, we have

$$E^{a,\lambda}_{1,1}(t) = \widehat{e}_{\lambda}(t,a). \tag{35}$$

Hence,

$$\widehat{\mathcal{L}}_{(q,h)}\left\{\widehat{e}_{\lambda}(t,a)\right\}(z) = \widehat{\mathcal{L}}_{(q,h)}\left\{E^{a,\lambda}_{1,1}(t)\right\} = \frac{1}{z-\lambda}. \tag{36}$$

Theorem 4.5 *The (q,h)-Laplace transform of fractional nabla integral is given by*

$$\widehat{\mathcal{L}}_{(q,h)}\left\{{}_a\nabla^{-\alpha}_{(q,h)}f\right\}(z) = z^{-\alpha}\widehat{\mathcal{L}}_{(q,h)}\{f\}(z). \tag{37}$$

Proof By convolution, one may write

$${}_a\nabla^{-\alpha}_{(q,h)}f(t) = \int_a^t \widehat{h}_{\alpha-1}(t,\rho(\tau))f(\tau)\nabla\tau = \left(\widehat{h}_{\alpha-1}(\cdot,a) * f\right)(t). \tag{38}$$

Thus,

$$\begin{aligned}\widehat{\mathcal{L}}_{(q,h)}\left\{{}_a\nabla^{-\alpha}_{(q,h)}f\right\}(z) &= \widehat{\mathcal{L}}_{(q,h)}\left\{\widehat{h}_{\alpha-1}(t,a)\right\}(z)\cdot\widehat{\mathcal{L}}_{(q,h)}\{f\}(z)\\ &= \frac{1}{z^{\alpha}}\widehat{\mathcal{L}}_{(q,h)}\{f\}(z). \end{aligned}$$

□

Theorem 4.6 *The (q,h)-Laplace transform of Caputo nabla derivative is given by*

$$\widehat{\mathcal{L}}_{(q,h)}\left\{{}^C_a\nabla^{\alpha}_{(q,h)}f\right\}(z) = z^{\alpha}\widehat{\mathcal{L}}_{(q,h)}\{f\}(z) - \sum_{k=0}^{m-1} z^{\alpha-k-1}\nabla^k_{(q,h)}f(a), \tag{39}$$

where $m-1 < \alpha \leq m$.

Proof Note that, in general (see Theorem 2.10 in [28]) for $n \in \mathbb{Z}^+$, we have

$$\widehat{\mathcal{L}}_{(q,h)}\left\{\nabla^n_{(q,h)}f\right\}(z) = z^n\widehat{\mathcal{L}}_{(q,h)}\{f\}(z) - \sum_{i=0}^{n-1} z^{n-1-i}\nabla^i_{(q,h)}f(a). \tag{40}$$

Thus, for $m-1<\alpha\leq m$, in view of the definition of fractional derivative, we have

$$\begin{aligned}\widehat{\mathcal{L}}_{(q,h)}\{{}_a^C\nabla^{\alpha}_{(q,h)}f\}(z) &= \widehat{\mathcal{L}}_{(q,h)}\{\widehat{h}_{m-\alpha-1}(\cdot,a)\}(z)\cdot\widehat{\mathcal{L}}_{(q,h)}\{\nabla^m_{(q,h)}f\}(z)\\ &= \frac{1}{z^{m-\alpha}}\left(z^m\widehat{\mathcal{L}}_{(q,h)}\{f\}(z)-\sum_{k=0}^{m-1}z^{m-1-k}\nabla^k_{(q,h)}f(a)\right)\\ &= z^{\alpha}\widehat{\mathcal{L}}_{(q,h)}\{f\}(z)-\sum_{k=0}^{m-1}z^{\alpha-k-1}\nabla^k_{(q,h)}f(a).\end{aligned}$$

□

In particular, if $0<\alpha\leq 1$, then

$$\widehat{\mathcal{L}}_{(q,h)}\{{}_a^C\nabla^{\alpha}_{(q,h)}f\}(z)=z^{\alpha}\widehat{\mathcal{L}}_{(q,h)}\{f\}(z)-z^{\alpha-1}f(a). \tag{41}$$

As an application, we apply the Laplace transform method to derive explicit solutions to the homogeneous equations of the form

$${}_a^C\nabla^{\alpha}_{(q,h)}y(t)-\lambda y(t)=0, \qquad d_k=\nabla^k y(a), \quad k=0,\dots,m-1, \tag{42}$$

where $t\in\widetilde{\mathbb{T}}^{\sigma^m(a)}_{(q,h)}$ $(m\in\mathbb{N})$, $m-1<\alpha\leq m$, $\lambda\in\mathbb{R}$, in terms of the (q,h)-Mittag-Leffler functions.

The following statement holds.

Theorem 4.7 *Let $m\in\mathbb{N}$ be given by $m-1<\alpha\leq m$ and $\lambda\in\mathbb{R}$. Then the functions*

$$y_i(t)=E^{a,\lambda}_{\alpha,i+1}(t), \quad i=0,1,\dots,m-1, \tag{43}$$

yield the fundamental system of solutions to equation (42).

Proof Applying the Laplace transform to (42) and taking (39) into account, we have

$$\widehat{\mathcal{L}}_{(q,h)}\{y\}(z)=\sum_{i=0}^{m-1}d_k\frac{z^{\alpha-i-1}}{z^{\alpha}-\lambda}. \tag{44}$$

Formula (34) with $\beta=i+1$ yields

$$\widehat{\mathcal{L}}_{(q,h)}\{E^{a,\lambda}_{\alpha,i+1}(t)\}(z)=\frac{z^{\alpha-i-1}}{z^{\alpha}-\lambda}. \tag{45}$$

Thus, relation (44) yields

$$y(t)=\sum_{i=0}^{m-1}d_iy_i(t), \quad \text{where } y_i(t)=E^{a,\lambda}_{\alpha,i+1}(t). \tag{46}$$

□

Corollary 4.8 *The equation*

$${}_a^C\nabla^{\alpha}_{(q,h)}y(t)-\lambda y(t)=0, \qquad y(a)=1, \quad 0<\alpha\leq 1, \tag{47}$$

has solution in the form

$$y(t) = E^{a,\lambda}_{\alpha,1}(t),$$

while the equation

$$^{C}_{a}\nabla^{\alpha}_{(q,h)}y(t) - \lambda y(t) = 0, \qquad \nabla^{k}y(a) = 1, \quad k = 0,1, 1 < \alpha \leq 2, \tag{48}$$

has its fundamental system of solutions given by

$$y_1(t) = E^{a,\lambda}_{\alpha,1}(t), \qquad y_2(t) = E^{a,\lambda}_{\alpha,2}(t).$$

Competing interests
The authors declare that they have no competing interests.

Authors' contributions
The main idea of this paper was proposed by SR and MS. SR prepared the manuscript initially and performed all the steps of the proofs in this research. All authors read and approved the final manuscript.

Author details
[1]School of Applied Mathematics, The University of Nottingham Malaysia Campus, Jalan Broga, Semenyih, Selangor Darul Ehsan 43500, Malaysia. [2]School of Mathematical Science, Faculty of Science and Technology, University Kebangsaan Malaysia, Bangi, Selangor 43600, Malaysia.

Acknowledgements
The authors would like to thank Professor J Čermák and the anonymous referees for their valuable comments and suggestions. The authors would like to acknowledge the grants: UKM Grant DIP-2014-034 and Ministry of Education, Malaysia, Grant FRGS/1/2014/ST06/UKM/01/1 for financial support.

References

1. Kilbas, A, Srivastava, MH, Trujillo, JJ: Theory and Application of Fractional Differential Equations. Elsevier, Amsterdam (2006)
2. Miller, KS, Ross, B: An Introduction to the Fractional Calculus and Fractional Differential Equations. Wiley, New York (1993)
3. Podlubny, I: Fractional Differential Equations. Academic Press, San Diego (1999)
4. Guy, J: Modeling fractional stochastic systems an non-random fractional dynamics driven by Brownian motions. Appl. Math. Model. **32**(5), 836-859 (2008)
5. Hilfer, R: Applications of Fractional Calculus in Physics. World Scientific, River Edge (2000)
6. Lakshmikantham, V, Leela, S, Devi, JV: Theory of Fractional Dynamical Systems. Cambridge Scientific Publishers, Cambridge (2009)
7. Li, C, Qian, D, Chen, YQ: On Riemann-Liouville and Caputo derivatives. Discrete Dyn. Nat. Soc. **2011**, Article ID 562494 (2011)
8. Agarwal, RP: Certain fractional *q*-integrals and *q*-derivatives. Proc. Camb. Philos. Soc. **66**, 365-370 (1969)
9. Díaz, JB, Osler, TJ: Differences of fractional order. Math. Comput. **28**, 185-202 (1974)
10. Al-Salam, WA: Some fractional *q*-integrals and *q*-derivatives. Proc. Camb. Philos. Soc. **15**, 135-140 (1966)
11. Grey, HL, Zhang, NF: On a new definition of the fractional difference. Math. Comput. **50**, 513-529 (1988)
12. Atici, FM, Eloe, PW: A transform method in discrete fractional calculus. Int. J. Differ. Equ. **2**, 165-176 (2007)
13. Atici, FM, Eloe, PW: Fractional *q*-calculus on a time scale. J. Nonlinear Math. Phys. **14**(3), 341-352 (2007)
14. Atici, FM, Eloe, PW: Discrete fractional calculus with the nabla operator. Electron. J. Qual. Theory Differ. Equ. **2009**, 3 (2009)
15. Fang, JP: Some applications of *q*-differential operator. J. Korean Math. Soc. **47**(2), 223-233 (2010)
16. Mansour, ZSI: Linear sequential *q*-difference equations of fractional order. Fract. Calc. Appl. Anal. **12**(2), 159-178 (2009)
17. Dassios, IK, Baleanu, DI: On a singular system of fractional nabla difference equations with boundary conditions. Bound. Value Probl. **2013**, Article ID 148 (2013)
18. Jagan Mohan, J, Deekshitulu, GVSR: Solutions of nabla fractional difference equations using *N*-transforms. Commun. Math. Stat. **2**, 1-16 (2014)
19. Jarad, F, Kaymakçalan, B, Teş, K: A new transform method in nabla discrete fractional calculus. Adv. Differ. Equ. **2012**, Article ID 190 (2012)
20. Karpuz, B: Volterra theory on time scales. Results Math. **65**, 263-292 (2014)
21. Segi Rahmat, MR: Integral transform method for solving fractional dynamic equations on time scales. Abstr. Appl. Anal. **2014**, Article ID 261348 (2014)
22. Wu, G-C, Baleanu, D: New applications of the variational iteration method- from differential equations to *q*-fractional difference equations. Adv. Differ. Equ. **2013**, Article ID 21 (2013)

23. Abdeljawad, T: On Riemann and Caputo fractional differences. Comput. Math. Appl. **62**, 1602-1611 (2011)
24. Čermák, J, Nechvátal, L: On (q, h)-analogue of fractional calculus. J. Nonlinear Math. Phys. **17**(1), 1-18 (2010)
25. Čermák, J, Kisela, T, Nechvátal, L: Discrete Mittag-Leffler functions in linear fractional difference equations. Abstr. Appl. Anal. **2011**, Article ID 565067 (2011)
26. Bohner, M, Peterson, A: Dynamic Equations on Time Scales: An Introduction with Applications. Birkhäuser, Boston (2001)
27. Bohner, M, Peterson, A (eds.): Advances in Dynamic Equations on Time Scales. Birkhäuser, Boston (2003)
28. Segi Rahmat, MR: The (q, h)-Laplace transform on discrete time scales. Comput. Math. Appl. **62**, 272-281 (2011)
29. Segi Rahmat, MR: On some (q, h)-analogues of integral inequalities on discrete time scales. Comput. Math. Appl. **62**, 1790-1797 (2011)
30. Agarwal, RP, Bohner, M: Basic calculus on time scales and some of its applications. Results Math. **35**(1-2), 3-22 (1999)
31. Kisela, T: Power functions and essentials of fractional calculus on isolated time scales. Adv. Differ. Equ. **2013**, Article ID 259 (2013)
32. Williams, PA: Fractional calculus on time scales with Taylor's theorem. Fract. Calc. Appl. Anal. **15**(4), 616-638 (2012)

10

Inverse spectral problem for Sturm-Liouville operator with discontinuous coefficient and cubic polynomials of spectral parameter in boundary condition

Aynur Çöl*

*Correspondence: aynurcol@sinop.edu.tr
Department of Mathematics Education, Sinop University, Sinop, 57100, Turkey

Abstract

In this paper, the inverse scattering problem for the Sturm-Liouville operator with discontinuous coefficient and cubic polynomials of the spectral parameter in the boundary condition is considered. The scattering data of the problem is defined, and its properties are investigated. The modified Marchenko main equation is obtained and it is shown that the potential is uniquely recovered by the scattering data.

MSC: 34A55; 34B40; 34B07; 34L25; 81U40

Keywords: inverse problem; scattering data; Sturm-Liouville operator; spectral parameter; discontinuous coefficient

1 Introduction

Consider the boundary value problem generated by the differential equation

$$-y'' + q(x)y = \lambda^2 \rho(x)y \quad (0 < x < \infty), \tag{1}$$

with the boundary condition

$$(\alpha_0 + i\alpha_1\lambda - \alpha_2\lambda^2 - i\alpha_3\lambda^3)y'(0) - (\beta_0 + i\beta_1\lambda - \beta_2\lambda^2 - i\beta_3\lambda^3)y(0) = 0, \tag{2}$$

where λ is a spectral parameter, $q(x)$ is real valued function with the condition

$$\int_0^\infty (1+x)|q(x)|\,dx < \infty, \tag{3}$$

and $\rho(x)$ is a piecewise constant function in the form

$$\rho(x) = \begin{cases} \alpha^2, & 0 \le x < a, \\ 1, & a \le x < \infty, \end{cases} \tag{4}$$

$1 \neq \alpha > 0$. Here $p_j(\lambda)$ $(j = 1, 2)$ is a polynomial

$$p_1(\lambda) \equiv \alpha_0 + i\alpha_1\lambda - \alpha_2\lambda^2 - i\alpha_3\lambda^3, \qquad p_2(\lambda) \equiv \beta_0 + i\beta_1\lambda - \beta_2\lambda^2 - i\beta_3\lambda^3$$

with the relations

$$\alpha_{i+1}\beta_i - \alpha_i\beta_{i+1} > 0, \qquad \alpha_{i+2}\beta_i - \alpha_i\beta_{i+2} < 0, \qquad \alpha_{i+3}\beta_i - \alpha_i\beta_{i+3} = 0 \tag{5}$$

for $\alpha_i, \beta_i \in \mathbb{R}$ $(i = \overline{0,3})$.

Inverse spectral problems of spectral analysis often appear in mathematics, mechanics, physics and other branches of natural sciences. The direct scattering problem consists of the determination the collection $\{S(\lambda), \{\lambda_j\}_{j=1}^N, \{m_j\}_{j=1}^N\}$ when $q(x)$ is known. The inverse scattering problem deals with the construction of $q(x)$ in terms of the scattering data. In this paper in the case of discontinuous coefficient (4), we consider the inverse scattering problem for Sturm-Liouville operator with cubic polynomials of spectral parameter in boundary condition (2).

Discontinuous inverse problems appear in electronics for constructing parameters of heterogeneous electronic lines with desirable technical characteristics and in geophysical models for oscillations of the earth (see [1] and the references therein). In the case that $\rho(x) \equiv 1$ and the boundary condition does not contain a spectral parameter, the inverse scattering problem for (1) was solved by Marchenko [2, 3] and Levitan [4, 5]. The inverse scattering problem of the Sturm-Liouville operator with discontinuous coefficient was studied in [6–8]. The problem was examined in [9] by using a new integral representation of the Jost solution of (1).

An important case in spectral theory is that containing the spectral parameter in equations and boundary conditions. Sturm-Liouville problems with spectral parameter-dependent boundary conditions arise in studies of heat conduction problems and vibrating string problems. Fulton and Pruess showed a kind of heat conduction problems in [10]. Problems with the dependence on spectral parameter can be found in [10–17]. The inverse scattering problem for (1) with a linear spectral parameter in the boundary condition was solved in [18].

The main result of this paper is that the potential $q(x)$ can be uniquely recovered from the given scattering data. The Marchenko method is applied to solving the boundary value problem when the boundary conditions depend on spectral parameter as nonlinear.

If $q(x) = 0$, the result is obtained that the function

$$f_0(x,\lambda) = \frac{1}{2}\left(1 + \frac{1}{\sqrt{\rho(x)}}\right)e^{i\lambda\mu^+(x)} + \frac{1}{2}\left(1 - \frac{1}{\sqrt{\rho(x)}}\right)e^{i\lambda\mu^-(x)}$$

is a solution of (1), where $\mu^{\pm}(x) = \pm x\sqrt{\rho(x)} + a(1 \mp \sqrt{\rho(x)})$.

As proven in [9], if the condition (3) is satisfied, (1) has a unique solution $f(x,\lambda)$, which satisfies the asymptotic behavior

$$\lim_{x\to+\infty} e^{-i\lambda x} f(x,\lambda) = 1$$

for $\operatorname{Im}\lambda \geq 0$ and can be expressed by

$$f(x,\lambda) = f_0(x,\lambda) + \int_{\mu^+(x)}^{\infty} K(x,t)e^{i\lambda t}\,dt, \tag{6}$$

which is called the Jost solution. For the kernel function $K(x,t)$, the inequality

$$\int_{\mu^+(x)}^{\infty} |K(x,t)|\,dt \leq c\left(\exp\left(\int_x^{\infty} t|q(t)|\,dt\right)\right), \quad 0 < c = \text{const},$$

is satisfied. Also, if $q(x)$ is differentiable, the kernel $K(x,t)$ is twice differentiable and satisfies both the equation

$$\frac{\partial^2 K(x,t)}{\partial x^2} - \rho(x)\frac{\partial^2 K(x,t)}{\partial t^2} = q(x)K(x,t), \quad 0 < x < \infty, t > \mu^+(x),$$

and the conditions

$$\frac{dK(x,\mu^+(x))}{dx} = -\frac{1}{4\sqrt{\rho(x)}}\left(1 + \frac{1}{\sqrt{\rho(x)}}\right)q(x), \tag{7}$$

$$\frac{d}{dx}\left\{K\left(x,\mu^-(x)+0\right) - K\left(x,\mu^-(x)-0\right)\right\} = \frac{1}{4\sqrt{\rho(x)}}\left(1 - \frac{1}{\sqrt{\rho(x)}}\right)q(x). \tag{8}$$

Denote by $\varphi(\lambda,x)$ the solution of (1) satisfying the conditions

$$\varphi(0,\lambda) = p_1(\lambda), \qquad \varphi'(0,\lambda) = p_2(\lambda).$$

This solution satisfies the condition (2). The function

$$S(\lambda) = \frac{p_1(\lambda)\overline{f'(0,\lambda)} - p_2(\lambda)\overline{f(0,\lambda)}}{p_1(\lambda)f'(0,\lambda) - p_2(\lambda)f(0,\lambda)} \tag{9}$$

is the scattering function of the boundary value problem (1)-(3). It is a meromorphic function in the upper half plane $\operatorname{Im}\lambda > 0$ with the poles at the zeros of the function $p_1(\lambda)f'(0,\lambda) - p_2(\lambda)f(0,\lambda)$. These poles are simple and lie on the imaginary axis. The norming numbers are defined as

$$m_k^{-2} \equiv \int_0^{\infty} f^2(x,i\mu_k)\rho(x)\,dx + \frac{f^2(0,i\mu_k)}{p_1^2(i\mu_k)}\left[\frac{1}{2}\sum_{m=0}^{2}(\alpha_{1+m}\beta_m - \alpha_m\beta_{1+m})\mu_k^{2m-1} + \sum_{m=0}^{1}(\alpha_m\beta_{2+m} - \alpha_{2+m}\beta_m)\mu_k^{2m}\right], \tag{10}$$

where $\lambda_k = i\mu_k$, $\mu_k > 0$. We show that the kernel $K(x,y)$ of the solution (6) satisfies the integral equation which is called the main equation,

$$K(x,y) + F(x,y) + \int_{\mu^+(x)}^{\infty} K(x,t)F_0(t+y)\,dt + \frac{1-\sqrt{\rho(x)}}{1+\sqrt{\rho(x)}}K(x,2a-y) = 0, \quad y > \mu^+(x), \tag{11}$$

where

$$F(x,y) = \frac{1}{2}\left(1 + \frac{1}{\sqrt{\rho(x)}}\right)F_0\left(y + \mu^+(x)\right) + \frac{1}{2}\left(1 - \frac{1}{\sqrt{\rho(x)}}\right)F_0\left(y + \mu^-(x)\right), \tag{12}$$

$$F_0(x)=\frac{1}{2\pi}\int_{-\infty}^{\infty}\left[S_\infty(\lambda)-S(\lambda)\right]e^{i\lambda x}d\lambda+\sum_{k=1}^{N}m_k^2e^{-\mu_k x},\quad \lambda_k=i\mu_k,$$

$$S_\infty(\lambda)=\begin{cases} e^{-2i\lambda a}\frac{\tau e^{-2i\lambda a\alpha}-1}{\tau-e^{-2i\lambda a\alpha}}, & \alpha_n\neq 0,\\ e^{-2i\lambda a}\frac{\alpha\alpha_2(-1+\tau e^{-2i\lambda a\alpha})-\beta_3(1+\tau e^{-2i\lambda a\alpha})}{\alpha\alpha_2(e^{-2i\lambda a\alpha}-\tau)-\beta_3(e^{-2i\lambda a\alpha}+\tau)}, & \alpha_n=0,\end{cases}$$

and $\tau=\frac{\alpha-1}{\alpha+1}$.

Obviously, from the given scattering data $\{S(\lambda),\{\lambda_j\}_{j=1}^N,\{m_j\}_{j=1}^N\}$, the function $F(x,y)$ is found by (12) which is then introduced in (11). If (11) has a unique solution $K(x,y)$, then the function $q(x)$ can be found from (7) and (8). The solvability of the integral equation (11) is examined and the algorithm of recovering the potential $q(x)$ is given.

This paper is organized as follows. In Section 2, the scattering data to the boundary value problem (1)-(3) is found by using a new integral representation for the solution of (1), and its properties are investigated. In Section 3, the main equation for the boundary value problem (1)-(3) is derived. Finally, the solvability of the main equation is proved and the unique recovery of the potential from the solution of the main equation is shown in Section 4.

Let $y(x,\lambda)$ and $z(x,\lambda)$ be solutions of (1). The expression

$$W\left[y(x,\lambda),z(x,\lambda)\right]=y'z-yz'$$

is called the Wronskian of the functions $y(x,\lambda)$ and $z(x,\lambda)$. It is clear that for all real $\lambda\neq 0$, $f(x,\lambda)$ and $\overline{f(x,\lambda)}$ constitute fundamental solutions of (1). The Wronksian of these functions does not depend on x and equals $2i\lambda$.

2 Scattering data

Lemma 1 *For all real $\lambda\neq 0$, the following identity is valid:*

$$2i\lambda\frac{\varphi(x,\lambda)}{E(\lambda)}=\overline{f(x,\lambda)}-S(\lambda)f(x,\lambda)\tag{13}$$

and $S(\lambda)$ possesses the following properties:

$$S(\lambda)=\overline{S(-\lambda)},\qquad \left|S(\lambda)\right|<1.$$

Proof Since $f(x,\lambda)$ and $\overline{f(x,\lambda)}$ constitute the fundamental solution system of (1) for real $\lambda\neq 0$, we have

$$\varphi(x,\lambda)=c_1(\lambda)f(x,\lambda)+c_2(\lambda)\overline{f(x,\lambda)},\tag{14}$$

where $c_1(\lambda)$ and $c_2(\lambda)$ are functions which we have to find. Taking account of the following equalities:

$$c_1(\lambda)f(0,\lambda)+c_2(\lambda)\overline{f(0,\lambda)}=p_1(\lambda),$$
$$c_1(\lambda)f'(0,\lambda)+c_2(\lambda)\overline{f'(0,\lambda)}=p_2(\lambda),$$

$c_1(\lambda)$ and $c_2(\lambda)$ are found and substituted in (14). Thus, we obtain

$$\varphi(x,\lambda) = -\frac{p_1(\lambda)\overline{f'(0,\lambda)} - p_2(\lambda)\overline{f(0,\lambda)}}{2i\lambda}f(x,\lambda) + \frac{p_1(\lambda)f'(0,\lambda) - p_2(\lambda)f(0,\lambda)}{2i\lambda}\overline{f(x,\lambda)}. \quad (15)$$

Let

$$E(\lambda) \equiv p_1(\lambda)f'(0,\lambda) - p_2(\lambda)f(0,\lambda).$$

Now, it is necessary to show $E(\lambda) \neq 0$ for all real $\lambda \neq 0$. Assume the contrary, then there exists $\lambda_0 \in \mathbb{R}$, $\lambda_0 \neq 0$, such that

$$p_1(\lambda_0)f'(0,\lambda_0) = p_2(\lambda_0)f(0,\lambda_0).$$

Also

$$W\left[f(0,\lambda_0),\overline{f(0,\lambda_0)}\right] = 2i\lambda_0$$

is satisfied. By using these relations, we obtain

$$\frac{|f(0,\lambda_0)|^2}{|p_1(\lambda_0)|^2}\left[\alpha_1\beta_0 - \alpha_0\beta_1 + (\alpha_2\beta_1 - \alpha_1\beta_2)|\lambda_0|^2 + (\alpha_3\beta_2 - \alpha_2\beta_3)|\lambda_0|^4\right] = -1$$

and this is a contradiction since the left-hand side is positive. Thus, by dividing the equality (15) by $\frac{1}{2i\lambda}E(\lambda)$, we obtain (13) where $S(\lambda)$ is defined with (9). Since $\overline{E(\lambda)} = E(-\lambda)$ and the numerator has the same property, it is clear that

$$\overline{S(\lambda)} = S(-\lambda).$$

Also $[p_1(\lambda)\overline{p_2(\lambda)} - p_2(\lambda)\overline{p_1(\lambda)}][f'(0,\lambda)\overline{f(0,\lambda)} - \overline{f'(0,\lambda)}f(0,\lambda)] < 0$ holds for all real $\lambda \neq 0$, and so the identity

$$\left|S(\lambda)\right|^2 < 1$$

is satisfied. Thus, the lemma is proved. □

Lemma 2 *The function $E(\lambda)$ has only a finite number of zeros on the half plane* ($\operatorname{Im}\lambda > 0$). *All the zeros are simple and lie on the imaginary axis.*

Proof By the proof of Lemma 1 we have $E(\lambda) \neq 0$ for all real $\lambda \neq 0$, the point $\lambda = 0$ is the possible zero of $E(\lambda)$. Since $E(\lambda)$ is analytic in the upper plane $\operatorname{Im}\lambda > 0$ and $f(0,\lambda)$ is bounded as $|\lambda| \to \infty$, it follows that the set of zeros of $E(\lambda)$ is bounded and forms at most countable set having as zero the only possible limit point.

Now we show that zeros of $E(\lambda)$ lie on the imaginary axis. Assume that λ_1 and λ_2 are zeros of $E(\lambda)$. Then they satisfy (1):

$$-f''(x,\lambda_1) + q(x)f(x,\lambda_1) = \lambda_1^2\rho(x)f(x,\lambda_1), \quad (16)$$

$$-\overline{f''(x,\lambda_2)} + q(x)\overline{f(x,\lambda_2)} = \overline{\lambda_2^2}\rho(x)\overline{f(x,\lambda_2)}. \quad (17)$$

We multiply (16) by $\overline{f(x,\lambda_2)}$ and (17) by $f(x,\lambda_1)$, subtract the latter from the former, and finally integrate this relation according to x from 0 to ∞. As a result,

$$\left(\lambda_1^2-\overline{\lambda_2^2}\right)\int_0^{\infty} f(x,\lambda_1)\overline{f(x,\lambda_2)}\rho(x)\,dx - W\left[f(x,\lambda_1),\overline{f(x,\lambda_2)}\right]\Big|_{x=0}=0. \tag{18}$$

On the other hand, since λ_j $(j=1,2)$ is a zero to $E(\lambda)$ we have

$$W\left[f(x,\lambda_1),\overline{f(x,\lambda_2)}\right]\Big|_{x=0}=\left(\frac{p_2(\lambda_1)}{p_1(\lambda_1)}-\frac{\overline{p_2(\lambda_2)}}{\overline{p_1(\lambda_2)}}\right)f(0,\lambda_1)\overline{f(0,\lambda_2)}.$$

If we take $\lambda_1=\lambda_2=\mu$ and substitute in (18), the result is obtained that

$$(\mu+\overline{\mu})\left[\frac{|f(0,\mu)|^2}{|p_1(\mu)|^2}\left[\sum_{k=0}^{2}(\alpha_{1+k}\beta_k-\alpha_k\beta_{1+k})|\mu|^{2k}+2\operatorname{Im}\mu\sum_{k=0}^{1}(\alpha_k\beta_{2+k}-\alpha_{2+k}\beta_k)|\mu|^{2k}\right]\right.$$
$$\left.+2\operatorname{Im}\mu\int_0^{\infty}|f(x,\mu)|^2\rho(x)\,dx\right]=0.$$

Since we have the condition (5), the expression in the parentheses is positive, and it implies that $\mu+\overline{\mu}=0$, *i.e.*, μ is pure imaginary.

Let us prove that there are only finitely many zeros. Let δ denote the infimum of the distances between two neighboring zeros of $E(\lambda)$, and show $\delta>0$. Let us assume the contrary and let $\{i\lambda_k\}$ and $\{i\hat{\lambda}_k\}$ be two sequences of zeros of the function $E(\lambda)$ such that

$$\lim_{k\to\infty}(\hat{\lambda}_k-\lambda_k)=0,\quad 0<\lambda_k<\hat{\lambda}_k,\qquad \max_k\hat{\lambda}_k<M.$$

For A large enough, the inequality

$$f(x,i\lambda_k)>\frac{1}{2}e^{-\lambda_k x}$$

holds uniformly with respect to $x\in[A,\infty)$ and $\lambda\in[0,\infty)$. Thus, we obtain

$$\int_A^{\infty} f(i\hat{\lambda}_k,x),\overline{f(i\lambda_k,x)}\rho(x)\,dx>\frac{1}{4}\frac{e^{-A(\hat{\lambda}_k+\lambda_k)}}{(\hat{\lambda}_k+\lambda_k)}>\frac{e^{-2AM}}{8M}. \tag{19}$$

On the other hand, the equality (18) yields

$$0=\frac{f(0,i\hat{\lambda}_k)\overline{f(0,i\lambda_k)}}{p_1(i\lambda_k)p_1(i\hat{\lambda}_k)}\left[\sum_{m=0}^{2}(\alpha_{1+m}\beta_m-\alpha_m\beta_{1+m})(\hat{\lambda}_k\lambda_k)^m\right.$$
$$\left.+\sum_{m=0}^{1}(\alpha_m\beta_{2+m}-\alpha_{2+m}\beta_m)(\hat{\lambda}_k\lambda_k)^m(\hat{\lambda}_k+\lambda_k)\right]$$
$$+(\hat{\lambda}_k+\lambda_k)\int_0^{\infty} f(x,i\hat{\lambda}_k)\overline{f(x,i\lambda_k)}\rho(x)\,dx$$
$$=(\hat{\lambda}_k+\lambda_k)\int_0^{A} f(x,i\hat{\lambda}_k)\left[\overline{f(x,i\lambda_k)}-\overline{f(x,i\hat{\lambda}_k)}\right]\rho(x)\,dx$$

$$+ (\hat{\lambda}_k + \lambda_k) \int_0^A f(x, i\hat{\lambda}_k)\overline{f(x, i\hat{\lambda}_k)}\rho(x)\,dx$$

$$+ (\hat{\lambda}_k + \lambda_k) \int_A^\infty f(x, i\hat{\lambda}_k)\overline{f(x, i\lambda_k)}\rho(x)\,dx$$

$$+ \frac{f(0, i\hat{\lambda}_k)\overline{f(0, i\lambda_k)}}{\overline{p_1(i\lambda_k)}p_1(i\hat{\lambda}_k)} \left[\sum_{m=0}^{2} (\alpha_{1+m}\beta_m - \alpha_m\beta_{1+m})(\hat{\lambda}_k\lambda_k)^m \right.$$

$$\left. + \sum_{m=0}^{1} (\alpha_m\beta_{2+m} - \alpha_{2+m}\beta_m)(\hat{\lambda}_k\lambda_k)^m(\hat{\lambda}_k + \lambda_k) \right]$$

and letting $k \to \infty$, we get

$$\lim_{k\to\infty} \int_A^\infty f(x, i\hat{\lambda}_k)\overline{f(x, i\lambda_k)}\rho(x)\,dx \leq 0. \tag{20}$$

Since

$$\lim_{k\to\infty} \left[\overline{f(x, i\lambda_k)} - \overline{f(x, i\hat{\lambda}_k)}\right] = 0$$

uniformly with respect to $x \in [0, A]$. Comparing (19) and (20) we reach a contradiction. We conclude that $\delta > 0$ and so the function $E(\lambda)$ has only a finite number of zeros.

Now, let us show that all zeros of the function $E(\lambda)$ are simple. By the derivation of the identity

$$-f''(x, \lambda) + q(x)f(x, \lambda) = \lambda^2\rho(x)f(x, \lambda) \tag{21}$$

with respect to λ, we get

$$-\dot{f}''(x, \lambda) + q(x)\dot{f}(x, \lambda) = \lambda^2\rho(x)\dot{f}(x, \lambda) + 2\lambda\rho(x)f(x, \lambda), \tag{22}$$

here $\dot{f}$ denotes differentiation with respect to λ. Multiplying (21) by $\dot{f}(x, \lambda)$ and (22) by $f(x, \lambda)$ and subtracting the second from the first and integrating this relation with respect to x over $(0, \infty)$, the result is obtained that

$$2\lambda \int_0^\infty f^2(x, \lambda)\rho(x)\,dx + W\left[f(x, \lambda), \dot{f}(x, \lambda)\right]\Big|_{x=0} = 0.$$

Let λ be a zero of the function $E(\lambda)$. By using the expression for the function $E(\lambda)$, it is found that

$$\frac{\dot{E}(\lambda)f(0, \lambda)}{p_1(\lambda)} = 2\lambda \int_0^\infty f^2(x, \lambda)\rho(x)\,dx + i\frac{f^2(0, \lambda)}{p_1^2(\lambda)} \left[\sum_{m=0}^{2} (\alpha_{1+m}\beta_m - \alpha_m\beta_{1+m})(i\lambda)^{2m} \right.$$

$$\left. + \sum_{m=0}^{1} (\alpha_{2+m}\beta_m - \alpha_m\beta_{2+m})(i\lambda)^{2m+1} \right]. \tag{23}$$

Substituting $\lambda_k = i\mu_k$, $\mu_k > 0$, in (23) and multiplying by $-i$ the result is obtained that the right side of the equality is positive. Thus $\dot{E}(i\mu_k) \neq 0$, *i.e.* the zeros of $E(\lambda)$ are simple. The lemma is proved. □

The numbers

$$m_k^{-2} \equiv \int_0^{\infty} f^2(x, i\mu_k)\rho(x)\,dx + \frac{f^2(0, i\mu_k)}{p_1^2(i\mu_k)}\left[\frac{1}{2}\sum_{m=0}^{2}(\alpha_{1+m}\beta_m - \alpha_m\beta_{1+m})\mu_k^{2m-1} + \sum_{m=0}^{1}(\alpha_m\beta_{2+m} - \alpha_{2+m}\beta_m)\mu_k^{2m}\right]$$

are called *norming numbers.*

The collection $\{S(\lambda), \{\lambda_k\}_{k=1}^N, \{m_k\}_{k=1}^N\}$ is called the *scattering data* for the boundary value problem (1)-(3).

Using (6) and substituting the related expressions into $S(\lambda)$, the following result is obtained:

$$S(\lambda) = e^{-2i\lambda a}\frac{\tau e^{-2i\lambda a\alpha} - 1}{e^{-2i\lambda a\alpha} - \tau} + O\left(\frac{1}{\lambda}\right)$$

if $\alpha_3 \neq 0$ as $|\lambda| \to \infty$, and

$$S(\lambda) = e^{-2i\lambda a}\frac{\alpha\alpha_2(-1 + \tau e^{-2i\lambda a\alpha}) - \beta_3(1 + \tau e^{-2i\lambda a\alpha})}{\alpha\alpha_2(e^{-2i\lambda a\alpha} - \tau) - \beta_3(e^{-2i\lambda a\alpha} + \tau)} + O\left(\frac{1}{\lambda}\right)$$

if $\alpha_3 = 0$ as $|\lambda| \to \infty$.

Let

$$S_\infty(\lambda) = \begin{cases} e^{-2i\lambda a}\frac{\tau e^{-2i\lambda a\alpha} - 1}{e^{-2i\lambda a\alpha} - \tau}, & \alpha_3 \neq 0, \\ e^{-2i\lambda a}\frac{\alpha\alpha_2(-1+\tau e^{-2i\lambda a\alpha}) - \beta_3(1+\tau e^{-2i\lambda a\alpha})}{\alpha\alpha_2(e^{-2i\lambda a\alpha} - \tau) - \beta_3(e^{-2i\lambda a\alpha} + \tau)}, & \alpha_3 = 0. \end{cases}$$

Hence $S_\infty(\lambda) - S(\lambda) \in L_2(-\infty, \infty)$ and so the function

$$F_{0S}(x) = \frac{1}{2\pi}\int_{-\infty}^{\infty}(S_\infty(\lambda) - S(\lambda))e^{i\lambda x}\,d\lambda$$

also belongs to the space $L_2(-\infty, \infty)$.

3 The main equation

The inverse scattering problem consists in recovering the coefficient $q(x)$ from the scattering data. It is clear that in order to determine $q(x)$ it is sufficient to know the kernel $K(x, t)$ of the solution (6). To derive the integral equation for $K(x, t)$, we use the equality (13), which was obtained in Lemma 1. Rewriting the identity (13) we get the following form:

$$\begin{aligned}
&\frac{2i\lambda\varphi(\lambda, x)}{E(\lambda)} - \overline{f_0(x, \lambda)} + S_\infty(\lambda)f_0(x, \lambda) \\
&= \int_{\mu^+(x)}^{\infty}[S_\infty(\lambda) - S(\lambda)]K(x, t)e^{i\lambda t}\,dt + [S_\infty(\lambda) - S(\lambda)]f_0(x, \lambda) \\
&\quad + \int_{\mu^+(x)}^{\infty}K(x, t)e^{-i\lambda t}\,dt - \int_{\mu^+(x)}^{\infty}S_\infty(\lambda)K(x, t)e^{i\lambda t}\,dt.
\end{aligned}$$

Multiplying both sides of this equality by $\frac{1}{2\pi}e^{i\lambda y}$ and integrating it to λ from $-\infty$ to ∞, it is found that

$$\begin{aligned}
&\frac{1}{2\pi}\int_{-\infty}^{\infty}\left[\frac{2i\lambda\varphi(\lambda,x)}{E(\lambda)}-\overline{f_0(x,\lambda)}+S_\infty(\lambda)f_0(x,\lambda)\right]e^{i\lambda y}\,d\lambda\\
&=\frac{1}{2\pi}\int_{-\infty}^{\infty}\left[S_\infty(\lambda)-S(\lambda)\right]f_0(x,\lambda)e^{i\lambda y}\,d\lambda\\
&+\frac{1}{2\pi}\int_{-\infty}^{\infty}\int_{\mu^+(x)}^{\infty}K(x,t)e^{-i\lambda(t-y)}\,dt\,d\lambda-\frac{1}{2\pi}\int_{-\infty}^{\infty}\int_{\mu^+(x)}^{\infty}S_\infty(\lambda)K(x,t)e^{i\lambda(t+y)}\,dt\,d\lambda\\
&+\frac{1}{2\pi}\int_{-\infty}^{\infty}\int_{\mu^+(x)}^{\infty}\left[S_\infty(\lambda)-S(\lambda)\right]K(x,t)e^{i\lambda(t+y)}\,dt\,d\lambda. \qquad (24)
\end{aligned}$$

To compute the third term on the right, we need to find new expression for $S_\infty(\lambda)$. After some calculations, in the case $\alpha_3 = 0$ we have

$$\begin{aligned}
S_\infty(\lambda)&=e^{-2i\lambda a}\frac{(\alpha\alpha_2+\beta_3)(\tau^2-1)e^{2i\lambda a\alpha}}{(\alpha\alpha_2-\beta_3)-(\alpha\alpha_2+\beta_3)\tau e^{2i\lambda a\alpha}}+\tau e^{-2i\lambda a}\\
&=e^{-2i\lambda a(1-\alpha)}\left(\tau^2-1\right)\sum_{k=0}^{\infty}\left(\frac{\alpha\alpha_2+\beta_3}{\alpha\alpha_2-\beta_3}\right)^{k+1}\tau^k e^{2i\lambda a\alpha k}+\tau e^{-2i\lambda a}
\end{aligned}$$

and then

$$\begin{aligned}
&\frac{1}{2\pi}\int_{-\infty}^{\infty}S_\infty(\lambda)e^{i\lambda(t+y)}\,d\lambda\\
&=\left(\tau^2-1\right)\sum_{k=0}^{\infty}\left(\frac{\alpha\alpha_2+\beta_3}{\alpha\alpha_2-\beta_3}\right)^{k+1}\tau^k\delta\left(t+y-2a(1-\alpha)+2a\alpha k\right)+\tau\delta(t+y-2a).
\end{aligned}$$

Hence, the right-hand side of (24) equals

$$\begin{aligned}
&K(x,y)+F_S(x,y)+\int_{\mu^+(x)}^{\infty}K(x,t)F_{0S}(t+y)\,dt-\tau K(x,2a-y)\\
&-\left(\tau^2-1\right)\sum_{k=0}^{\infty}\left(\frac{\alpha\alpha_2+\beta_3}{\alpha\alpha_2-\beta_3}\right)^{k+1}\tau^k K\left(x,2a(1-\alpha)-2a\alpha k-y\right),
\end{aligned}$$

where

$$\begin{aligned}
F_{0S}(x)&=\frac{1}{2\pi}\int_{-\infty}^{\infty}\left[S_\infty(\lambda)-S(\lambda)\right]e^{i\lambda x}\,d\lambda,\\
F_S(x,y)&=\frac{1}{2}\left(1+\frac{1}{\sqrt{\rho(x)}}\right)F_{0S}\left(y+\mu^+(x)\right)+\frac{1}{2}\left(1-\frac{1}{\sqrt{\rho(x)}}\right)F_{0S}\left(y+\mu^-(x)\right).
\end{aligned}$$

We note that $K(x,y)=0$ for $y<\mu^+(x)$. If $0<x<a$, then $\mu^+(x)=\alpha x-\alpha a+a$, and so we have for $k=0,1,\ldots,$

$$2a(1-\alpha)-2a\alpha k-y<2a(1-\alpha)-2a\alpha k-\alpha x+\alpha a-a=a-\alpha a-2a\alpha k-\alpha x\le\mu^+(x).$$

Hence, this shows that the last term is equal to zero. If $x\ge a$, then $\mu^+(x)=x$, and the inequality holds for this case.

Therefore, the right-hand side of (24) becomes

$$K(x,y)+F_S(x,y)+\int_{\mu^+(x)}^{\infty}K(x,t)F_{0S}(t+y)\,dt-\frac{1-\sqrt{\rho(x)}}{1+\sqrt{\rho(x)}}K(x,2a-y),\quad y>\mu^+(x).$$

In the case $\alpha_3\neq 0$,

$$S_\infty(\lambda)=e^{-2i\lambda a}\frac{(\tau^2-1)e^{2i\lambda a\alpha}}{1-\tau e^{2i\lambda a\alpha}}+\tau e^{-2i\lambda a}=e^{-2i\lambda a(1-\alpha)}\left(\tau^2-1\right)\sum_{k=0}^{\infty}\tau^k e^{2i\lambda a\alpha k}+\tau e^{-2i\lambda a}$$

and we arrive the same result of the right of (24).

On the other side, using the residue theorem and Jordan's lemma we have

$$\frac{1}{2\pi}\int_{-\infty}^{\infty}\left[\frac{2i\lambda\varphi(\lambda,x)}{E(\lambda)}-\overline{f_0(x,\lambda)}+S_\infty(\lambda)f_0(x,\lambda)\right]e^{i\lambda y}\,d\lambda=-\sum_{k=1}^{n}\frac{2i\lambda_k\varphi(i\lambda_k,x)}{\dot{E}(i\lambda_k)}e^{-\lambda_k y}.$$

Taking (23) into account we can transform this expression to the form

$$\begin{aligned}
&-\sum_{k=1}^{n}\frac{2i\lambda_k\varphi(i\lambda_k,x)}{\dot{E}(i\lambda_k)}e^{-\lambda_k y}\\
&=-\sum_{k=1}^{n}\frac{2i\lambda_k p_1(i\lambda_k)}{f(0,i\lambda_k)\dot{E}(i\lambda_k)}f(x,i\lambda_k)e^{-\lambda_k y}=-\sum_{k=1}^{n}m_k^2 f(x,i\lambda_k)e^{-\lambda_k y}\\
&=-\sum_{k=1}^{n}m_k^2\left[f_0(x,i\lambda_k)e^{-\lambda_k y}+\int_{\mu^+(x)}^{\infty}K(x,t)e^{-\lambda_k(t+y)}\,dt\right].
\end{aligned}$$

Substituting this value into the left side of (24), we obtain (11).

Thus we arrive at the following theorem.

Theorem 1 *For every fixed $x\geq 0$, the kernel $K(x,t)$ to the special solution* (6) *satisfies the integral equation* (11).

The integral equation (11) is called the *main equation* for the boundary value problem (1)-(3). The main equation does not have same form as the classical Marchenko equation and we call (11) the *modified Marchenko equation.*

4 Solvability of the main equation

We construct (11) only on the basis of the given scattering data. In this equation, we can take kernel $K(x,t)$ as unknown and regard it as a Fredholm-type equation for every fixed x. The main equation is rewritten in the form

$$\begin{aligned}
&K\left(x,y+\mu^+(x)\right)+F\left(x,y+\mu^+(x)\right)+\upsilon K\left(x,2a-y-\mu^+(x)\right)\\
&+\int_0^{\infty}K\left(x,t+\mu^+(x)\right)F_0\left(t+y+2\mu^+(x)\right)dt=0,\quad y>0.
\end{aligned}\tag{25}$$

Theorem 2 *For every fixed $x\geq 0$, the main equation* (11) *has a unique solution $K(x,\cdot)\in L_1(\mu^+(x),\infty)$.*

Proof Let

$$f(y) = K\left(x, y + \mu^+(x)\right).$$

For the proof of the solvability of the given main equation, it is enough to show that the homogeneous equation

$$f(y) + \upsilon K\left(x, 2a - y - \mu^+(x)\right) + \int_0^{\infty} f(t) F_0\left(t + y + 2\mu^+(x)\right) dt = 0 \tag{26}$$

has no nontrivial solution in the corresponding space.

If $f(y) \in L_1(0,\infty)$ is a solution of (26), then both $f(y) \in L_\infty(0,\infty)$ and $f(y) \in L_2(0,\infty)$. Hence $f(y) \in L_1(0,\infty) \cap L_\infty(0,\infty) \subset L_2(0,\infty)$ and it is sufficient to investigate (26) in $L_2(0,\infty)$. Equation (26) has the same properties as the fundamental equation of the problem in [18]. The proof of this fact is analog to Lemma 4.2 and Corollary 4.3 in [18].

Multiplying (26) by $\overline{f(y)}$ and integrating it from $-\infty$ to ∞ with respect to y, we obtain

$$\frac{1}{2\pi}\int_{\infty}^{\infty} \left|\widetilde{f(\lambda)}\right|^2 d\lambda + \frac{1}{2\pi}\int_{\infty}^{\infty} \upsilon e^{2i\lambda\mu^+(x) - 2i\lambda a} \widetilde{f(-\lambda)}\overline{\widetilde{f(\lambda)}}\, d\lambda + \sum_{k=1}^{n} m_k^2 e^{-2\lambda_k \mu^+(x)} \left|\widetilde{f(-i\lambda_k)}\right|^2$$
$$+ \frac{1}{2\pi}\int_{\infty}^{\infty} \left(S_\infty(\lambda) - S(\lambda)\right) e^{2i\lambda\mu^+(x)} \widetilde{f(-\lambda)}\overline{\widetilde{f(\lambda)}}\, d\lambda = 0$$

and hence

$$\frac{1}{2\pi}\int_{\infty}^{\infty} \left|\widetilde{f(\lambda)}\right|^2 d\lambda = \frac{1}{2\pi}\int_{\infty}^{\infty} S(\lambda) e^{2i\lambda\mu^+(x)} \widetilde{f(-\lambda)}\overline{\widetilde{f(\lambda)}}\, d\lambda - \sum_{k=1}^{n} m_k^2 e^{-2\lambda_k \mu^+(x)} \left|\widetilde{f(-i\lambda_k)}\right|^2$$
$$\leq \frac{1}{2\pi}\int_{\infty}^{\infty} \left|S(\lambda) e^{2i\lambda\mu^+(x)} \widetilde{f(-\lambda)}\overline{\widetilde{f(\lambda)}}\right| d\lambda \leq \frac{1}{2\pi}\int_{\infty}^{\infty} |S(\lambda)| \left|\widetilde{f(\lambda)}\right|^2 d\lambda,$$

i.e.,

$$\int_{\infty}^{\infty} \left(1 - |S(\lambda)|\right) \left|\widetilde{f(\lambda)}\right|^2 d\lambda \leq 0.$$

Since $1 - |S(\lambda)| > 0$ for all $\lambda \neq 0$, this implies that $\widetilde{f(\lambda)} \equiv 0$. Therefore, the homogeneous equation (26) has only the null solution, and this proves the theorem. □

Theorem 3 *The scattering data of the boundary value problem* (1)-(3) *determines the potential* $q(x)$ *in* (1) *uniquely.*

Proof Obviously, to form the main equation (11) it is sufficient to know the matrix function $F(x,y)$ and in its turn, to find $F(x,y)$ it is sufficient to know the scattering data $\{S(\lambda), \{\lambda_j\}_{j=1}^N, \{m_j\}_{j=1}^N\}$. It is seen in Theorem 2 that the main equation (11), constructed only on the basis of the scattering data, has a unique solution $K(x,y)$. Then the function $q(x)$ in (1) can be uniquely found according to (7) and (8). Equation (1) is constructed by the given algorithm. The theorem is proved. □

Competing interests

The author declares that she has no competing interests.

Author's contributions

The author carried out the research of this paper, and she read and approved the final version of the manuscript.

Acknowledgements

This work is supported by The Scientific and Technological Research Council of Turkey (TÜBİTAK).

References

1. Shieh, CT, Yurko, VA: Inverse nodal and inverse spectral problems for discontinuous boundary value problems. J. Math. Anal. Appl. **347**, 266-272 (2008)
2. Marchenko, VA: Sturm-Liouville Operators and Applications. Birkhäuser, Basel (1986)
3. Marchenko, VA: On reconstruction of the potential energy from phases of the scattered waves. Dokl. Akad. Nauk SSSR **104**, 695-698 (1955)
4. Levitan, BM: The inverse scattering problem of quantum theory. Mat. Zametki **17**(4), 611-624 (1975)
5. Levitan, BM: Inverse Sturm-Liouville Problems. VSP, Zeist (1987). Translated from the Russian by O Efimov
6. Gasymov, MG: Direct and inverse problems of spectral analysis for a class of equations with discontinuous coefficients. In: Proceedings of the International Conference on Non-Classical Methods in Geophysics, Novosibirsk, pp. 37-44 (1977) (in Russian)
7. Darwish, AA: The inverse scattering problem for a singular boundary value problem. N.Z. J. Math. **22**, 37-56 (1993)
8. Darwish, AA: On a direct and inverse scattering problem for a boundary value problem with discontinuous coefficient. N.Z. J. Math. **25**, 1-14 (1996)
9. Guseinov, IM, Pashaev, RT: On an inverse problem for a second-order differential equation. Russ. Math. Surv. **57**(3), 597-598 (2002)
10. Fulton, CT, Pruess, S: Numerical methods for a singular eigenvalue problem with eigenparameter in the boundary conditions. J. Math. Anal. Appl. **71**, 431-462 (1979)
11. Walter, J: Regular eigenvalue problems with eigenvalue parameter in the boundary conditions. Math. Z. **133**, 301-312 (1973)
12. Fulton, CT: Two-point boundary value problems with eigenvalue parameter contained in the boundary conditions. Proc. R. Soc. Edinb., Sect. A **77**, 293-308 (1977)
13. Binding, PA, Browne, PJ, Watson, BA: Inverse spectral problems for Sturm-Liouville equations with eigenparameter dependent boundary conditions. J. Lond. Math. Soc. **62**, 161-182 (2000)
14. Chugunova, MV: Inverse spectral problem for the Sturm-Liouville operator with eigenvalue parameter dependent boundary conditions. Oper. Theory, Adv. Appl. **123**, 187-194 (2001)
15. Chernozhukova, A, Freiling, G: A uniqueness theorem for the boundary value problems with non-linear dependence on the spectral parameter in the boundary conditions. Inverse Probl. Sci. Eng. **17**(6), 777-785 (2009)
16. Yurko, VA: An inverse problem for pencils of differential operators. Mat. Sb. **191**(10), 137-160 (2000)
17. Mamedov, KR, Kosar, NP: Inverse scattering problem for Sturm-Liouville operator with nonlinear dependence on the spectral parameter in the boundary condition. Math. Methods Appl. Sci. **34**, 231-241 (2011)
18. Mamedov, KR: On an inverse scattering problem for a discontinuous Sturm-Liouville equation with a spectral parameter in the boundary condition. Bound. Value Probl. **2010**, 171967 (2010)

Anti-periodic solutions for high-order cellular neural networks with mixed delays and impulses

Changjin Xu[1*] and Yusen Wu[2]

*Correspondence: xcj403@126.com
[1]Guizhou Key Laboratory of Economics System Simulation, Guizhou University of Finance and Economics, Guiyang, 550004, P.R. China
Full list of author information is available at the end of the article

Abstract

In this paper, delayed high-order cellular neural networks with impulses are investigated. Some sufficient conditions on the existence and exponential stability of anti-periodic solutions are established. An example with its numerical simulations is presented. Our results are new and complement previously known results.

MSC: 34C25; 34K13; 34K25

Keywords: high-order cellular neural networks; anti-periodic solution; exponentially stability; celay; impulse

1 Introduction

During the past decades, high-order cellular neural networks (HCNNs) have been extensively investigated due to their immense potential of application perspective in various fields such as signal and image processing, pattern recognition, optimization, and many other subjects. Many results on the problem of global stability of equilibrium points and periodic solutions of HCNNs have been reported (see [1–9]). In applied sciences, the existence of anti-periodic solutions plays a key role in characterizing the behavior of nonlinear differential equations [10–13]. In recent years, there have been some papers which deal with the problem of existence and stability of anti-periodic solutions. For example, Gong [14] investigated the existence and exponential stability of anti-periodic solutions for a class of Cohen-Grossberg neural networks; Peng and Huang [15] studied the anti-periodic solutions for shunting inhibitory cellular neural networks with continuously distributed delays, Zhang [16] focused on the existence and exponential stability of anti-periodic solutions for HCNNs with time-varying leakage delays. For details, we refer readers to [15, 17–35]. We know that many evolutionary processes exhibit impulsive effects [33, 36–43]. Thus, it is worthwhile to investigate the existence and stability of anti-periodic solutions for HCNNs with impulses. To the best of our knowledge, very few scholars have considered the problem of anti-periodic solutions for such impulsive systems. In this paper, we study the anti-periodic solution of the following high-order cellular neural network with mixed delays and impulses modeled by

$$\begin{cases} \dot{x}_i(t) = -b_i(t)x_i(t) + \sum_{j=1}^n c_{ij}(t)g_j(x_j(t-\tau_{ij}(t))) \\ \qquad + \sum_{j=1}^n d_{ij}(t)\int_0^\sigma k_{ij}(s)g_j(x_j(t-s))\,ds \\ \qquad + \sum_{j=1}^n \sum_{l=1}^n e_{ijl}(t)g_j(x_j(t-\alpha_{jl}(t)))g_l(x_l(t-\beta_{jl}(t))) + I_i(t), \quad t \neq t_k, \\ x_i(t_k^+) = (1+\gamma_{ik})x_i(t_k), \quad k = 1,2,\ldots, \end{cases} \tag{1.1}$$

where $i = 1,2,\ldots,n$, $x_i(t)$ denotes the state of the ith unit, $b_i(t) > 0$ denotes the passive decay, c_{ij}, d_{ij}, e_{ijl} are the synaptic connections strengths, $\tau_{ij}(t) \geq 0$, $\alpha_{jl}(t) \geq 0$ and $\beta_{jl}(t) \geq 0$ correspond to the delays, $I_i(t)$ stands for the external inputs, g_j is the activation function of signal transmission, the delay kernels k_{ij} is a real-valued negative continuous function defined on $R^+ := [0,\infty)$, t_k is the impulsive moment, and γ_{ik} characterizes the impulsive jump at time t_k for the ith unit.

For convenience, we introduce some notations as follows.

$$\bar{c}_{ij} = \sup_{t\in R}|c_{ij}(t)|, \qquad \bar{d}_{ij} = \sup_{t\in R}|d_{ij}(t)|, \qquad \bar{e}_{ijl} = \sup_{t\in R}|e_{ijl}(t)|, \qquad \bar{I}_i = \sup_{t\in R}|I_i(t)|,$$

$$\underline{b}_i = \inf_{t\in R}|b_i(t)|, \qquad \tau = \sup_{t\in R}\max_{1\leq i,j,l\leq n}\{\tau_{ij}(t), \alpha_{jl}(t), \beta_{jl}(t), \sigma\}.$$

Throughout this paper, we assume that

(H1) For $i,j,l = 1,2,\ldots,n$, $b_i, c_{ij}, d_{ij}, e_{ijl}, I_i(t), g_j : R \to R$, $k_{ij} : R^+ \to R^+$, $\alpha_{jl}, \beta_{jl} : R \to R^+$ are continuous functions, and there exists a constant $T > 0$ such that

$$b_i(t+T) = b_i(t), \qquad I_i(t+T) = -I_i(t), \qquad \tau_{ij}(t+T) = \tau_{ij}(t), \qquad \alpha_{jl}(t+T) = \alpha_{jl}(t),$$

$$c_{ij}(t+T)g_j(u) = -c_{ij}(t)g_j(-u), \qquad d_{ij}(t+T)g_j(u) = -d_{ij}(t)g_j(-u),$$

$$\beta_{jl}(t+T) = \beta_{jl}(t), \qquad e_{ijl}(t+T)g_j(u)g_l(u) = -e_{ijl}(t)g_j(-u)g_l(-u).$$

(H2) The sequence of times $\{t_k\}$ $(k \in N)$ satisfies $t_k < t_{k+1}$ and $\lim_{k\to+\infty} t_k = +\infty$, and γ_{ik} satisfies $-2 \leq \gamma_{ik} \leq 0$ for $i \in \{1,2,\ldots,n\}$ and $k \in N$.

(H3) There exists $q \in N$ such that $\gamma_{i(k+q)} = \gamma_{ik}$, $t_{k+q} = t_k + T$, $k \in N$.

(H4) For each $j \in \{1,2,\ldots,n\}$, the activation function $g_j : R \to R$ is continuous, and there exist nonnegative constants L_g^j and M_g such that, for all $u, v \in R$,

$$g_j(0) = 0, \qquad |g_j(u)| \leq M_g, \qquad |g_j(u) - g_j(v)| \leq L_g^j|u-v| \quad \text{for all } u,v \in R.$$

(H5) There exist constants $\eta > 0$, $\lambda > 0$, $i = 1,2,\ldots,n$, such that

$$(\lambda - \underline{b}_i) + \sum_{j=1}^n \bar{c}_{ij}L_g^j e^{\lambda\tau} + \sum_{j=1}^n |\bar{d}_{ij}| \int_0^\sigma |k_{ij}(s)|L_g^j e^{\lambda s}\,ds$$
$$+ \sum_{j=1}^n \sum_{l=1}^n \bar{e}_{ijl}\left(M_g L_g^l e^{\lambda\tau} + M_g L_g^j e^{\lambda\tau}\right) < -\eta < 0.$$

(H6) For $i = 1,2,\ldots,n$, the following condition holds:

$$\begin{cases} -b_i + \sum_{j=1}^n \bar{c}_{ij}L_g^j + \sum_{j=1}^n L_g^j\bar{d}_{ij}\int_0^\sigma |k_{ij}(s)|\,ds < 0, \\ (-b_i + \sum_{j=1}^n \bar{c}_{ij}L_g^j + \sum_{j=1}^n L_g^j\bar{d}_{ij}\int_0^\sigma |k_{ij}(s)|\,ds)^2 - 4\bar{I}_i\sum_{j=1}^n\sum_{l=1}^n \bar{e}_{ijl}L_g^jL_g^l > 0. \end{cases}$$

Let $x = (x_1, x_2, \ldots, x_n)^T \in R^n$, in which '$T$' denotes the transposition. We define $|x| = (|x_1|, |x_2|, \ldots, |x_n|)^T$ and $\|x\| = \max_{1 \le i \le n} |x_i|$. Obviously, the solution $x(t) = (x_1(t), x_2(t), \ldots, x_n(t))^T$ of (1.1) has components $x_i(t)$ piece-wise continuous on $(-\tau, +\infty)$, $x(t)$ is differentiable on the open intervals (t_{k-1}, t_k) and $x(t_k^+)$ exists.

Definition 1.1 Let $u(t) : R \to R$ be a piece-wise continuous function having a countable number of discontinuous $\{t_k\}|_{k=1}^{+\infty}$ of the first kind. It is said to be T-anti-periodic on R if

$$\begin{cases} u(t + T) = -u(t), & t \neq t_k, \\ u((t_k + T)^+) = -u(t_k^+), & k = 1, 2, \ldots. \end{cases}$$

Definition 1.2 Let $x^*(t) = (x_1^*(t), x_2^*(t), \ldots, x_n^*(t))^T$ be an anti-periodic solution of (1.1) with initial value $\varphi^* = (\varphi_1^*(t), \varphi_2^*(t), \ldots, \varphi_n^*(t))^T$. If there exist constants $\lambda > 0$ and $M > 1$ such that for every solution $x(t) = (x_1(t), x_2(t), \ldots, x_n(t))^T$ of (1.1) with an initial value $\varphi = (\varphi_1(t), \varphi_2(t), \ldots, \varphi_n(t))^T$,

$$|x_i(t) - x_i^*(t)| \le M \|\varphi - \varphi^*\| e^{-\lambda t} \quad \text{for all } t > 0, i = 1, 2, \ldots, n,$$

where

$$\|\varphi - \varphi^*\| = \sup_{-\tau \le s \le 0} \max_{1 \le i \le n} |\varphi_i(s) - \varphi_i^*(s)|.$$

Then $x^*(t)$ is said to be globally exponentially stable.

The purpose of this paper is to present sufficient conditions of existence and exponential stability of anti-periodic solution of system (1.1). Not only can our results be applied directly to many concrete examples of cellular neural networks, but they also extend, to a certain extent, the results in some previously known ones. In addition, an example with its numerical simulations is presented to illustrate the effectiveness of our main results.

The rest of this paper is organized as follows. In the next section, we give some preliminary results. In Section 3, we derive the existence of T-anti-periodic solution, which is globally exponential stable. In Section 4, we present an example to illustrate the effectiveness of our main results.

2 Preliminary results

In this section, we present two important lemmas which are used to prove our main results in Section 3.

Lemma 2.1 *Let* (H1)-(H6) *hold. Suppose that* $x(t) = (x_1(t), x_2(t), \ldots, x_n(t))^T$ *is a solution of* (1.1) *with initial conditions*

$$x_i(s) = \varphi_i(s), \quad |\varphi_i(s)| < \delta, s \in [-\tau, 0], i = 1, 2, \ldots, n. \tag{2.1}$$

Then

$$|x_i(t)| < \delta \quad \text{and} \quad |x_i(t_k^+)| < \delta \quad \text{for all } t \ge 0, i = 1, 2, \ldots, n, \tag{2.2}$$

where δ *satisfies*

$$-\underline{b}_i\delta + \sum_{j=1}^{n} \overline{c}_{ij}L_g^j\delta + \sum_{j=1}^{n} L_g^j\overline{d}_{ij}\delta \int_0^{\sigma} |k_{ij}(s)|\,ds + \sum_{j=1}^{n}\sum_{l=1}^{n} \overline{e}_{ijl}L_g^jL_g^l\delta^2 + \overline{I}_i < 0. \tag{2.3}$$

Proof For any given initial condition, hypothesis (H4) guarantees the existence and uniqueness of $x(t)$, the solution to (1.1) in $[-\tau, +\infty)$. Consider the following polynomial $ax^2 + bx + c$, where a, b, c are all real numbers. If $a < 0$ and $b^2 - 4ac < 0$, then $ax^2 + bx + c > 0$. In view of (H6), we know that there exists a positive constant δ which satisfies (2.3). By way of contradiction, we assume that (2.2) does not hold. Notice that $x_i(t_k^+) = (1 + \gamma_{ik})x_i(t_k)$ and by assumption (H2), $-2 \leq \gamma_{ik} \leq 0$, then $|x_i(t_k^+)| = |(1 + \gamma_{ik})||x_i(t_k)| \leq |x_i(t_k)|$. Then, if $|x_i(t_k^+)| \geq \delta$, then $|x_i(t_k)| \geq \delta$. Thus we may assume that there must exist $i \in \{1, 2, \ldots, n\}$ and $\widetilde{t} \in (t_k, t_{k+1}]$ such that

$$|x_i(\widetilde{t})| = \delta \quad \text{and} \quad |x_j(\widetilde{t})| < \delta \quad \text{for all } t \in (-\tau, \widetilde{t}), j = 1, 2, \ldots, n. \tag{2.4}$$

By directly computing the upper left derivative of $|x_i(t)|$, together with assumptions (2.3), (H4) and (2.4), we deduce that

$$\begin{aligned}
0 &\leq D^+\left(|x_i(\widetilde{t})|\right) \\
&\leq -b_i(t)|x_i(\widetilde{t})| + \Bigg|\sum_{j=1}^{n} c_{ij}(\widetilde{t})g_j(x_j(\widetilde{t} - \tau_{ij}(\widetilde{t}))) + \sum_{j=1}^{n} d_{ij}(\widetilde{t}) \int_0^{\sigma} k_{ij}(s)g_j(x_j(\widetilde{t} - s))\,ds \\
&\quad + \sum_{j=1}^{n}\sum_{l=1}^{n} e_{ijl}(\widetilde{t})g_j(x_j(\widetilde{t} - \alpha_{jl}(\widetilde{t})))g_l(x_l(\widetilde{t} - \beta_{jl}(\widetilde{t}))) + I_i(\widetilde{t})\Bigg| \\
&\leq -b_i(t)|x_i(\widetilde{t})| + \sum_{j=1}^{n} |c_{ij}(\widetilde{t})||g_j(x_j(\widetilde{t} - \tau_{ij}(\widetilde{t})))| + \sum_{j=1}^{n} |d_{ij}(\widetilde{t})| \int_0^{\sigma} |k_{ij}(s)||g_j(x_j(\widetilde{t} - s))|\,ds \\
&\quad + \sum_{j=1}^{n}\sum_{l=1}^{n} |e_{ijl}(\widetilde{t})||g_j(x_j(\widetilde{t} - \alpha_{jl}(\widetilde{t})))||g_l(x_l(\widetilde{t} - \beta_{jl}(\widetilde{t})))| + |I_i(\widetilde{t})| \\
&\leq -b_i(t)|x_i(\widetilde{t})| + \sum_{j=1}^{n} \overline{c}_{ij}L_g^j|x_j(\widetilde{t} - \tau_{ij}(\widetilde{t}))| + \sum_{j=1}^{n} L_g^j\overline{d}_{ij} \int_0^{\sigma} |k_{ij}(s)||x_j(\widetilde{t} - s)|\,ds \\
&\quad + \sum_{j=1}^{n}\sum_{l=1}^{n} \overline{e}_{ijl}L_g^jL_g^l|x_j(\widetilde{t} - \alpha_{jl}(\widetilde{t}))||x_l(\widetilde{t} - \beta_{jl}(\widetilde{t}))| + \overline{I}_i \\
&\leq -\underline{b}_i\delta + \sum_{j=1}^{n} \overline{c}_{ij}L_g^j\delta + \sum_{j=1}^{n} L_g^j\overline{d}_{ij}\delta \int_0^{\sigma} |k_{ij}(s)|\,ds \\
&\quad + \sum_{j=1}^{n}\sum_{l=1}^{n} \overline{e}_{ijl}L_g^jL_g^l\delta^2 + \overline{I}_i < 0,
\end{aligned} \tag{2.5}$$

which is a contradiction and implies that (2.2) holds. This completes the proof. □

Lemma 2.2 *Suppose that* (H1)-(H6) *hold. Let* $x^*(t) = (x_1^*(t), x_2^*(t), \ldots, x_n^*(t))^T$ *be the solution of* (1.1) *with initial value* $\varphi^* = (\varphi_1^*(t), \varphi_2^*(t), \ldots, \varphi_n^*(t))^T$, *and* $x(t) = (x_1(t), x_2(t),$

$\ldots, x_n(t))^T$ be the solution of (1.1) with initial value $\varphi = (\varphi_1(t), \varphi_2(t), \ldots, \varphi_n(t))^T$. Then there exist constants $\lambda > 0$ and $M > 1$ such that

$$|x_i(t) - x_i^*(t)| \le M\|\varphi - \varphi^*\|e^{-\lambda t} \quad \textit{for all } t > 0, i = 1, 2, \ldots, n.$$

Proof Let $y(t) = \{y_j(t)\} = \{x_j(t) - x_j^*(t)\} = x(t) - x^*(t)$. Then

$$\begin{aligned}
y_i'(t) &= -b_i(t)[x_i(t) - x_i^*(t)] + \sum_{j=1}^n c_{ij}(t)[g_j(x_j(t-\tau_{ij}(t))) - g_j(x_j^*(t-\tau_{ij}(t)))] \\
&\quad + \sum_{j=1}^n d_{ij}(t)\int_0^\sigma k_{ij}(s)[g_j(x_j(t-s)) - g_j(x_j^*(t-s))]\,ds \\
&\quad + \sum_{j=1}^n\sum_{l=1}^n e_{ijl}(t)[g_j(x_j(t-\alpha_{jl}(t)))g_l(x_l(t-\beta_{jl}(t))) \\
&\quad - g_j(x_j^*(t-\alpha_{jl}(t)))g_l(x_l^*(t-\beta_{jl}(t)))] + I_i(t) \\
&= -b_i(t)[x_i(t) - x_i^*(t)] + \sum_{j=1}^n c_{ij}(t)[g_j(x_j(t-\tau_{ij}(t))) - g_j(x_j^*(t-\tau_{ij}(t)))] \\
&\quad + \sum_{j=1}^n d_{ij}(t)\int_0^\sigma k_{ij}(s)[g_j(x_j(t-s)) - g_j(x_j^*(t-s))]\,ds \\
&\quad + \sum_{j=1}^n\sum_{l=1}^n e_{ijl}(t)[g_j(x_j(t-\alpha_{jl}(t)))g_l(x_l(t-\beta_{jl}(t))) \\
&\quad - g_j(x_j(t-\alpha_{jl}(t)))g_l(x_l^*(t-\beta_{jl}(t))) \\
&\quad + g_j(x_j(t-\alpha_{jl}(t)))g_l(x_l^*(t-\beta_{jl}(t))) \\
&\quad - g_j(x_j^*(t-\alpha_{jl}(t)))g_l(x_l^*(t-\beta_{jl}(t)))], \quad t \ne t_k, \qquad (2.6)
\end{aligned}$$

$$y_i(t_k^+) = (1+\gamma_{ik})y_i(t_k), \quad k = 1, 2, \ldots, \qquad (2.7)$$

where $i = 1, 2, \ldots, n$. Next, define a Lyapunov functional as

$$V_i(t) = |y_i(t)|e^{\lambda t}, \quad i = 1, 2, \ldots, n. \qquad (2.8)$$

It follows from (2.6), (2.7) and (2.8) that

$$\begin{aligned}
D^+(V_i(t)) &\le D^+(|y_i(t)|)e^{\lambda t} + \lambda|y_i(t)|e^{\lambda t} \\
&\le (\lambda - b_i(t))|y_i(t)|e^{\lambda t} + \Bigg[\sum_{j=1}^n |c_{ij}(t)||g_j(x_j(t-\tau_{ij}(t))) - g_j(x_j^*(t-\tau_{ij}(t)))| \\
&\quad + \sum_{j=1}^n |d_{ij}(t)|\int_0^\sigma |k_{ij}(s)||g_j(x_j(t-s)) - g_j(x_j^*(t-s))|\,ds \\
&\quad + \sum_{j=1}^n\sum_{l=1}^n |e_{ijl}(t)||g_j(x_j(t-\alpha_{jl}(t)))g_l(x_l(t-\beta_{jl}(t)))
\end{aligned}$$

$$
\begin{aligned}
&-g_j(x_j(t-\alpha_{jl}(t)))g_l(x_l^*(t-\beta_{jl}(t)))| \\
&+|g_j(x_j(t-\alpha_{jl}(t)))g_l(x_l^*(t-\beta_{jl}(t))) \\
&-g_j(x_j^*(t-\alpha_{jl}(t)))g_l(x_l^*(t-\beta_{jl}(t)))|\Big]e^{\lambda t} \\
\le\ & (\lambda - b_i(t))|y_i(t)|e^{\lambda t} + \Bigg\{\sum_{j=1}^{n}|c_{ij}(t)|L_g^j|y_j(x_j(t-\tau_{ij}(t))| \\
&+\sum_{j=1}^{n}|d_{ij}(t)|\int_0^{\sigma}|k_{ij}(s)|L_g^j|y_j(t-s)|\,ds \\
&+\sum_{j=1}^{n}\sum_{l=1}^{n}|e_{ijl}(t)|[M_gL_g^l|y_l(t-\beta_{jl}(t))| \\
&+M_gL_g^j|y_j(t-\alpha_{jl}(t))|]\Bigg\}e^{\lambda t}, \quad t\neq t_k,
\end{aligned}
\tag{2.9}
$$

and

$$
V_i(t_k^+) = |y_i(t_k^+)|e^{\lambda t_k} = |x_i(t_k^+) - x_i^*(t_k^+)|e^{\lambda t_k} = |1+\gamma_{ik}||y_i(t_k)|e^{\lambda t_k}, \tag{2.10}
$$

where $i = 1,2,\dots,n$. Let $M > 1$ denote an arbitrary real number and set

$$
\|\varphi - \varphi^*\| = \sup_{-\tau\le s\le 0}\max_{1\le j\le n}|\varphi_j(s) - \varphi_j^*(s)| > 0, \quad j = 1,2,\dots,n.
$$

Then, by (2.8), we have

$$
V_i(t) = |y_i(t)|e^{\lambda t} < M\|\varphi - \varphi^*\| \quad \text{for all } t\in[-\tau,0], i = 1,2,\dots,n.
$$

Thus we can claim that

$$
V_i(t) = |y_i(t)|e^{\lambda t} < M\|\varphi - \varphi^*\| \quad \text{for all } t\in[-\tau,t_1], i = 1,2,\dots,n. \tag{2.11}
$$

Otherwise, there must exist $i\in\{1,2,\dots,n\}$ and $\sigma\in(-\tau,t_1]$ such that

$$
V_i(\sigma) = M\|\varphi - \varphi^*\|, \qquad V_j(t) < M\|\varphi - \varphi^*\| \quad \text{for all } t\in[-\tau,\sigma), j = 1,2,\dots,n. \tag{2.12}
$$

Combining (2.9), (2.10) with (2.12), we obtain

$$
\begin{aligned}
0 &\le D^+(V_i(\sigma) - M\|\varphi - \varphi^*\|) \\
&= D^+(V_i(\sigma)) \\
&\le (\lambda - b_i(\sigma))|y_i(\sigma)|e^{\lambda\sigma} + \Bigg\{\sum_{j=1}^{n}|c_{ij}(\sigma)|L_g^j|y_j(x_j(\sigma-\tau_{ij}(\sigma))| \\
&\quad+\sum_{j=1}^{n}|d_{ij}(\sigma)|\int_0^{\sigma}|k_{ij}(s)|L_g^j|y_j(\sigma-s)|\,ds
\end{aligned}
$$

$$+\sum_{j=1}^{n}\sum_{l=1}^{n}|e_{ijl}(\sigma)|\left[M_g L_g^l|y_l(\sigma-\beta_{jl}(\sigma))|+M_g L_g^j|y_j(\sigma-\alpha_{jl}(\sigma))|\right]\Bigg\}e^{\lambda\sigma}$$

$$=(\lambda-b_i(\sigma))|y_i(\sigma)|e^{\lambda\sigma}+\sum_{j=1}^{n}|c_{ij}(\sigma)|L_g^j|y_j(x_j(\sigma-\tau_{ij}(\sigma))|e^{\lambda(\sigma-\tau_{ij}(\sigma))}e^{\lambda\tau_{ij}(\sigma)}$$

$$+\sum_{j=1}^{n}|d_{ij}(\sigma)|\int_0^{\sigma}|k_{ij}(s)|L_g^j|y_j(\sigma-s)|e^{\lambda(\sigma-s)}e^{\lambda s}\,ds$$

$$+\sum_{j=1}^{n}\sum_{l=1}^{n}|e_{ijl}(\sigma)|\left[M_g L_g^l|y_l(\sigma-\beta_{jl}(\sigma))|e^{\lambda(\sigma-\beta_{jl}(\sigma))}e^{\lambda\beta_{jl}(\sigma)}\right.$$

$$\left.+M_g L_g^j|y_j(\sigma-\alpha_{jl}(\sigma))|e^{\lambda(\sigma-\alpha_{jl}(\sigma))}e^{\lambda\alpha_{jl}(\sigma)}\right]$$

$$\le(\lambda-b_i(\sigma))M\|\varphi-\varphi^*\|+\sum_{j=1}^{n}|c_{ij}(\sigma)|L_g^j M\|\varphi-\varphi^*\|e^{\lambda\tau_{ij}(\sigma)}$$

$$+\sum_{j=1}^{n}|d_{ij}(\sigma)|\int_0^{\sigma}|k_{ij}(s)|L_g^j M\|\varphi-\varphi^*\|e^{\lambda s}\,ds$$

$$+\sum_{j=1}^{n}\sum_{l=1}^{n}|e_{ijl}(\sigma)|\left[M_g L_g^l M\|\varphi-\varphi^*\|e^{\lambda\beta_{jl}(\sigma)}+M_g L_g^j M\|\varphi-\varphi^*\|e^{\lambda\alpha_{jl}(\sigma)}\right]$$

$$=\Bigg[(\lambda-b_i(\sigma))+\sum_{j=1}^{n}|c_{ij}(\sigma)|L_g^j e^{\lambda\tau_{ij}(\sigma)}+\sum_{j=1}^{n}|d_{ij}(\sigma)|\int_0^{\sigma}|k_{ij}(s)|L_g^j e^{\lambda s}\,ds$$

$$+\sum_{j=1}^{n}\sum_{l=1}^{n}|e_{ijl}(\sigma)|\left(M_g L_g^l e^{\lambda\beta_{jl}(\sigma)}+M_g L_g^j M e^{\lambda\alpha_{jl}(\sigma)}\right)\Bigg]M\|\varphi-\varphi^*\|$$

$$\le\Bigg[(\lambda-\underline{b}_i)+\sum_{j=1}^{n}\bar{c}_{ij}L_g^j e^{\lambda\tau}+\sum_{j=1}^{n}|\bar{d}_{ij}|\int_0^{\sigma}|k_{ij}(s)|L_g^j e^{\lambda s}\,ds$$

$$+\sum_{j=1}^{n}\sum_{l=1}^{n}\bar{e}_{ijl}\left(M_g L_g^l e^{\lambda\tau}+M_g L_g^j e^{\lambda\tau}\right)\Bigg]M\|\varphi-\varphi^*\|. \tag{2.13}$$

Then

$$(\lambda-\underline{b}_i)+\sum_{j=1}^{n}\bar{c}_{ij}L_g^j e^{\lambda\tau}+\sum_{j=1}^{n}|\bar{d}_{ij}|\int_0^{\sigma}|k_{ij}(s)|L_g^j e^{\lambda s}\,ds$$

$$+\sum_{j=1}^{n}\sum_{l=1}^{n}\bar{e}_{ijl}\left(M_g L_g^l e^{\lambda\tau}+M_g L_g^j e^{\lambda\tau}\right)>0,$$

which contradicts (H5). Then (2.11) holds. In view of (2.11), we know that

$$V_i(t_1)=|y_i(t_1)|e^{\lambda t_1}<M\|\varphi-\varphi^*\|,\quad i=1,2,\ldots$$

and

$$V_i(t_1^+)=|1+\gamma_{i1}||y_i(t_1)|e^{\lambda t_1}\le|y_i(t_1)|e^{\lambda t_1}.$$

Then

$$V_i(t_1^+) < M\|\varphi - \varphi^*\|. \tag{2.14}$$

Thus, for $t \in [t_1, t_2]$, we can repeat the above procedure and obtain

$$V_i(t) = |y_i(t)|e^{\lambda t} < M\|\varphi - \varphi^*\| \quad \text{for all } t \in [t_1, t_2], i = 1, 2, \dots.$$

Similarly, we have

$$V_i(t) = |y_i(t)|e^{\lambda t} < M\|\varphi - \varphi^*\| \quad \text{for all } t > 0, i = 1, 2, \dots.$$

Namely,

$$|x_i(t) - x_i^*(t)| = |y_i(t)| < M\|\varphi - \varphi^*\| \quad \text{for all } t > 0, i = 1, 2, \dots.$$

This completes the proof. □

Remark 2.1 If $x^*(t) = (x_1^*(t), x_2^*(t), \dots, x_n^*(t))^T$ is a T-anti-periodic solution of (1.1), it follows from Lemma 2.2 and Definition 1.2 that $x^*(t)$ is globally exponentially stable.

3 Main result

In this section,we present our main result that there exists the exponentially stable anti-periodic solution of (1.1).

Theorem 3.1 *Assume that* (H1)-(H6) *are satisfied. Then* (1.1) *has exactly one T-anti-periodic solution $x^*(t)$. Moreover, this solution is globally exponentially stable.*

Proof Let $v(t) = (v_1(t), v_2(t), \dots, v_n(t))^T$ be a solution of (1.1) with initial conditions

$$v_i(s) = \varphi_i^v(s), \quad |\varphi_i^v(s)| < \delta, s \in (-\tau, 0], i = 1, 2, \dots, n. \tag{3.1}$$

Thus, according to Lemma 2.1, the solution $v(t)$ is bounded and

$$|v_i(t)| < \delta \quad \text{for all } t \in R, i = 1, 2, \dots, n. \tag{3.2}$$

For $p \in N$, if $t \notin t_k$, then $t + (p+1)T \notin \{t_k\}$; if $t \in t_k$, then $t + (p+1)T \in \{t_k\}$. From (1.1), we obtain

$$\begin{aligned}
&\left((-1)^{p+1}v_i(t + (p+1)T)\right)' \\
&\quad = (-1)^{p+1}\Bigg\{-b_i(t + (p+1)T)v_i(t + (k+1)T) \\
&\qquad + \sum_{j=1}^{n} c_{ij}(t + (p+1)T)g_j(v_j(t + (p+1)T - \tau_{ij}(t + (p+1)T))) \\
&\qquad + \sum_{j=1}^{n} d_{ij}(t + (p+1)T)\int_0^{\sigma} k_{ij}(s)g_j(v_j(t + (p+1)T - s))\,ds
\end{aligned}$$

$$+\sum_{j=1}^{n}\sum_{l=1}^{n} e_{ijl}\big(t+(p+1)T\big)g_j\big(v_j\big(t+(p+1)T-\alpha_{jl}\big(t+(p+1)T\big)\big)\big)$$

$$\times g_l\big(v_l\big(t+(p+1)T-\beta_{jl}\big(t+(p+1)T\big)\big)\big)+I_i\big(t+(p+1)T\big)\Big\}$$

$$=-b_i(t)(-1)^{p+1}v_i\big(t+(p+1)T\big)$$

$$+\sum_{j=1}^{n} c_{ij}(t)g_j\big((-1)^{p+1}v_j\big(t+(p+1)T-\tau_{ij}(t)\big)\big)$$

$$+\sum_{j=1}^{n} d_{ij}(t)\int_0^{\sigma} k_{ij}(s)g_j\big((-1)^{p+1}v_j\big(t+(p+1)T-s\big)\big)\,ds$$

$$+\sum_{j=1}^{n}\sum_{l=1}^{n} e_{ijl}(t)g_j\big((-1)^{p+1}v_j\big(t+(p+1)T-\alpha_{jl}(t)\big)\big)$$

$$\times g_l\big((-1)^{p+1}v_l\big(t+(p+1)T-\beta_{jl}(t)\big)\big)+I_i(t),\quad t\neq t_k \tag{3.3}$$

and

$$\begin{aligned}(-1)^{p+1}v_i\big(\big(t_k+(p+1)T\big)^{+}\big) &= (-1)^{p+1}\big(1+\gamma_{i(k+(p+1)q)}v_i\big(t_k+(p+1)T\big)\big)\\ &= (-1)^{p+1}(1+\gamma_{ik})v_i\big(t_k+(p+1)T\big)\\ &= (1+\gamma_{ik})\big((-1)^{p+1}v_i\big(t_k+(p+1)T\big)\big),\quad k=1,2,\ldots,\end{aligned} \tag{3.4}$$

where $i=1,2,\ldots,n$. Thus $(-1)^{p+1}v(t+(p+1)T)$ are solutions of (1.1) on R^{+} for any natural number p. Then, from Lemma 2.2, there exists a constant $M>1$ such that

$$\begin{aligned}&\big|(-1)^{p+1}v_i\big(t+(p+1)T\big)-(-1)^{k}v_i(t+pT)\big|\\ &\leq Me^{-\lambda(t+pT)}\sup_{-\tau\leq s\leq 0}\max_{1\leq i\leq n}\big|v_i\big(s+(p+1)T\big)+v_i(s+pT)\big|\\ &\leq 2e^{-\lambda(t+pT)}M\delta,\end{aligned} \tag{3.5}$$

and

$$\begin{aligned}&\big|(-1)^{p+1}v_i\big(\big(t_k+(p+1)T\big)^{+}\big)-(-1)^{p}v_i\big((t_k+pT)^{+}\big)\big|\\ &=\big|x_i\big(\big(t_k+(p+1)T\big)^{+}\big)+x_i\big((t_k+pT)^{+}\big)\big|\\ &=|1+\gamma_{ik}|\big|x_i\big(t_k+(p+1)T\big)+x_i(t_k+pT)\big|\leq 2M\delta e^{-\lambda(pT+t_k)},\end{aligned} \tag{3.6}$$

where $k\in N$, $i=1,2,\ldots,n$. Thus, for any natural number m, we have

$$\begin{aligned}&(-1)^{m+1}v_i\big(t+(m+1)T\big)\\ &=v_i(t)+\sum_{k=0}^{m}\big[(-1)^{k+1}v_i\big(t+(k+1)T\big)-(-1)^{k}v_i(t+kT)\big],\quad t\neq t_k.\end{aligned} \tag{3.7}$$

Hence

$$
\begin{aligned}
&\left|(-1)^{m+1}v_i\big(t+(m+1)T\big)\right| \\
&\quad \le \left|v_i(t)\right| + \sum_{k=0}^{m}\left|(-1)^{k+1}v_i\big(t+(k+1)T\big)-(-1)^k v_i(t+kT)\right|, \quad t\neq t_k,
\end{aligned} \tag{3.8}
$$

and

$$
\begin{aligned}
\left|(-1)^{m+1}v_i\big(\big(t_k+(m+1)T\big)^+\big)\right| &= \left|(1+\gamma_{ik})(-1)^{m+1}v_i\big(t_k+(m+1)T\big)\right| \\
&\le \left|(-1)^{m+1}v_i\big(t_k+(m+1)T\big)\right|,
\end{aligned} \tag{3.9}
$$

where $i=1,2,\ldots,n$. It follows from (3.5)-(3.9) that $(-1)^{m+1}v_i(t+(m+1)T)$ is a fundamental sequence on any compact set of R^+. Obviously, $\{(-1)^m v(t+mT)\}$ uniformly converges to a piece-wise continuous function $x^*(t)=(x_1^*(t),x_2^*(t),\ldots,x_n^*(t))^T$ on any compact set of R^+.

Now we show that $x^*(t)$ is T-anti-periodic solution of (1.1). Firstly, $x^*(t)$ is T-anti-periodic, since

$$
\begin{aligned}
x^*(t+T) &= \lim_{m\to\infty}(-1)^m v(t+T+mT) \\
&= -\lim_{(m+1)\to\infty}(-1)^{m+1}v\big(t+(m+1)T\big) = -x^*(t), \quad t\neq t_k,
\end{aligned} \tag{3.10}
$$

and

$$
\begin{aligned}
x^*\big((t+T)^+\big) &= \lim_{m\to\infty}(-1)^m v\big((t+T+mT)^+\big) \\
&= -\lim_{(m+1)\to\infty}(-1)^{m+1}v\big(\big(t+(m+1)T\big)^+\big) = -x^*(t_k)^+.
\end{aligned} \tag{3.11}
$$

In the sequel, we prove that $x^*(t)$ is a solution of (1.1). Noting that the right-hand side of (1.1) is piece-wise continuous, (3.3) and (3.4) imply that $\{((-1)^{m+1}v(t+(m+1)T))'\}$ uniformly converges to a piece-wise continuous function on any compact subset of R^+. Thus, letting $m\to\infty$ on both sides of (3.3) and (3.4), we can easily obtain

$$
\begin{cases}
\dot{x}_i^*(t) = -b_i(t)x_i^*(t) + \sum_{j=1}^n c_{ij}(t)g_j(x_j^*(t-\tau_{ij}(t))) \\
\qquad + \sum_{j=1}^n d_{ij}(t)\int_0^\sigma k_{ij}(s)g_j(x_j^*(t-s))\,ds \\
\qquad + \sum_{j=1}^n\sum_{l=1}^n e_{ijl}(t)g_j(x_j^*(t-\alpha_{jl}(t)))g_l(x_l^*(t-\beta_{jl}(t))) + I_i(t), \quad t\neq t_k, \\
x_i{}^*(t_k^+) = (1+\gamma_{ik})x_i^*(t_k), \quad k=1,2,\ldots,
\end{cases} \tag{3.12}
$$

where $i=1,2,\ldots,n$. Therefore, $x^*(t)$ is a solution of (1.1). Finally, by applying Lemma 2.2, it is easy to check that $x^*(t)$ is globally exponentially stable. The proof of Theorem 3.1 is completed. □

Remark 3.1 In [10–12, 14, 15, 20–24, 44], although authors consider the existence and exponential stability of anti-periodic solutions of neural networks, they do not consider the impulsive case. In this paper, we consider the high-order cellular neural networks with impulses. The obtained results show that impulses play a certain role in the existence and exponential stability of anti-periodic solutions of cellular neural networks. If the $\gamma_{ik}=0$

(*i.e.*, there is no impulse), then Theorem 3.1 is still valid if we delete the condition on the impulse. All the results in [10–12, 14, 15, 20–24, 44] cannot be applicable to system (1.1) to obtain the existence and exponential stability of anti-periodic solutions. This implies that the results of this paper are essentially new. Our results complement the previous work.

4 An example

In this section, we give an example to illustrate our main results obtained in previous sections. Consider the high-order cellular neural network with delays and impulses

$$\begin{aligned}
\dot{x}_1(t) &= -x_1(t) + \frac{1}{12}g_1\big(x_1\big(t-\sin^2 t\big)\big) + \frac{1}{14}g_2\big(x_2\big(t-7\sin^2 t\big)\big) \\
&\quad + \frac{1}{36}|\sin t|\int_0^{20} e^{-s}g_1\big(x_1(t-s)\big)\,ds + \frac{1}{20}|\cos t|\int_0^{20} e^{-s}g_2\big(x_2(t-s)\big)\,ds \\
&\quad + \frac{1}{80}\sin t g_1\big(x_1\big(t-3\sin^2 t\big)\big)g_2\big(x_2\big(t-2\sin^2 t\big)\big) + 2\sin t, \\
\dot{x}_2(t) &= -x_2(t) + \frac{1}{16}g_1\big(x_1\big(t-\cos^2 t\big)\big) + \frac{1}{16}g_2\big(x_2\big(t-4\sin^2 t\big)\big) \\
&\quad + \frac{1}{20}|\cos t|\int_0^{2C} e^{-s}g_1\big(x_1(t-s)\big)\,ds + \frac{1}{20}|\sin t|\int_0^{20} e^{-s}g_2\big(x_2(t-s)\big)\,ds \\
&\quad + \frac{1}{15}\sin t g_1\big(x_1\big(t-\sin^2 t\big)\big)g_2\big(x_2\big(t-2\sin^2 t\big)\big) + \sin t, \quad t \neq t_k, \\
x_1\big(t_k^+\big) &= 0.4x_1(t_k), \quad k = 1,2,\ldots, \\
x_2\big(t_k^+\big) &= 0.4x_2(t_k), \quad k = 1,2,\ldots,
\end{aligned} \tag{4.1}$$

where $g_1(u) = g_2(u) = \frac{1}{2}(|u+1| - |u-1|)$ and

$$\begin{bmatrix} c_{11}(t) & c_{12}(t) \\ c_{21}(t) & c_{22}(t) \end{bmatrix} = \begin{bmatrix} \frac{1}{11} & \frac{1}{14} \\ \frac{1}{16} & \frac{1}{16} \end{bmatrix}, \quad \begin{bmatrix} k_{11}(s) & k_{12}(s) \\ k_{21}(s) & k_{22}(s) \end{bmatrix} = \begin{bmatrix} e^{-s} & e^{-s} \\ e^{-s} & e^{-s} \end{bmatrix},$$

$$\begin{bmatrix} \tau_{11}(t) & \tau_{12}(t) \\ \tau_{21}(t) & \tau_{22}(t) \end{bmatrix} = \begin{bmatrix} \sin^2 t & 7\sin^2 t \\ \cos^2 t & 4\sin^2 t \end{bmatrix}, \quad \begin{bmatrix} d_{11}(t) & d_{12}(t) \\ d_{21}(t) & d_{22}(t) \end{bmatrix} = \begin{bmatrix} \frac{1}{36}|\sin t| & \frac{1}{20}|\cos t| \\ \frac{1}{20}|\cos t| & \frac{1}{20}|\sin t| \end{bmatrix},$$

$$\begin{bmatrix} u_{1l}(t) & u_{2l}(t) \\ v_{1l}(t) & v_{2l}(t) \end{bmatrix} = \begin{bmatrix} 3\sin^2 t & \sin^2 t \\ 2\sin^2 t & 2\sin^2 t \end{bmatrix}, \quad \begin{bmatrix} e_{11l}(t) & e_{12l}(t) \\ e_{21l}(t) & e_{22l}(t) \end{bmatrix} = \begin{bmatrix} \frac{1}{80}\sin t & \frac{1}{80}\sin t \\ \frac{1}{15}\sin t & \frac{1}{15}\sin t \end{bmatrix},$$

$$\begin{bmatrix} b_1(t) & b_2(t) \\ I_1(t) & I_2(t) \end{bmatrix} = \begin{bmatrix} 1 & 1 \\ 2\sin t & \sin t \end{bmatrix},$$

where $l = 1,2$. By a simple calculation, we get

$$\underline{b}_i = L_i^g = M_g = 1 \quad (i = 1,2), \qquad \bar{c}_{11} = \frac{1}{11}, \qquad \bar{c}_{12} = \frac{1}{14}, \qquad \bar{c}_{21} = \frac{1}{16}, \qquad \bar{c}_{22} = \frac{1}{16},$$

$$\bar{d}_{11} = \frac{1}{36}, \qquad \bar{d}_{12} = \frac{1}{20}, \qquad \bar{d}_{21} = \frac{1}{20}, \qquad \bar{d}_{22} = \frac{1}{20},$$

$$\bar{e}_{1jl} = \frac{1}{80}, \qquad \bar{e}_{2jl} = \frac{1}{15} \quad (j,l = 1,2),$$

$$\bar{\tau}_{11} = 1, \qquad \bar{\tau}_{12} = 7, \qquad \bar{\tau}_{21} = 1, \qquad \bar{\tau}_{22} = 4, \qquad \bar{I}_1 = 2, \qquad \bar{I}_2 = 1.$$

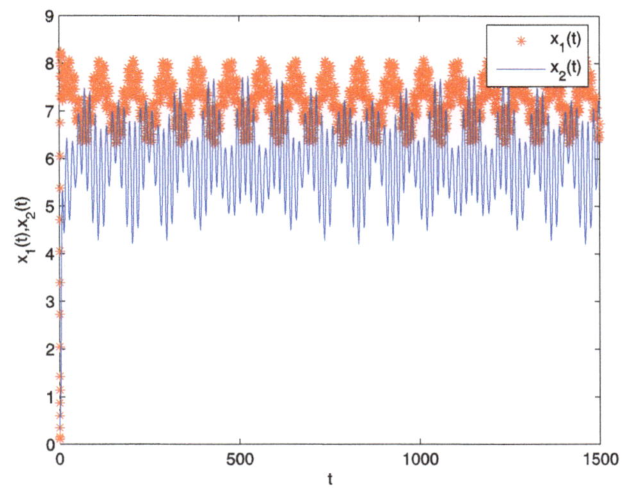

Figure 1 Times series of $x_1(t)$ and $x_2(t)$ of system (4.1).

Let $\eta = 0.2$, $\lambda = 0.001$. Then

$$
\begin{aligned}
&(\lambda - \underline{b}_i) + \sum_{j=1}^{2} \bar{c}_{ij} L_g^j e^{\lambda\tau} + \sum_{j=1}^{2} |\bar{d}_{ij}| \int_0^{\sigma} |k_{ij}(s)| L_g^j e^{\lambda s}\, ds \\
&\quad + \sum_{j=1}^{2} \sum_{l=1}^{2} \bar{e}_{ijl} \left(M_g L_g^l e^{\lambda\tau} + M_g L_g^j e^{\lambda\tau} \right) \\
&\quad \le (0.001 - 1) + \left(\frac{1}{11} + \frac{1}{14} + \frac{1}{16} + \frac{1}{16} \right) e^{0.007} + \left(\frac{1}{11} + \frac{1}{36} + \frac{3}{20} \right) \\
&\quad + \left(\frac{1}{80} + \frac{1}{80} + \frac{1}{15} + \frac{1}{15} \right) \times 2 \times e^{0.001} = -0.2149 < -0.2 < 0
\end{aligned}
$$

and

$$
\begin{aligned}
&-b_i + \sum_{j=1}^{2} \bar{c}_{ij} L_g^j + \sum_{j=1}^{2} L_g^j \bar{d}_{ij} \int_0^{\sigma} |k_{ij}(s)|\, ds \\
&\quad < -1 + \frac{1}{11} + \frac{1}{14} + \frac{1}{16} \times \frac{1}{36} + \frac{1}{10} \times \frac{1}{20} = -0.8309 < 0, \\
&\left(-b_i + \sum_{j=1}^{2} \bar{c}_{ij} L_g^j + \sum_{j=1}^{2} L_g^j \bar{d}_{ij} \int_0^{\sigma} |k_{ij}(s)|\, ds \right)^2 - 4\bar{I}_i \sum_{j=1}^{2} \sum_{l=1}^{2} \bar{e}_{ijl} L_g^j L_g^l \\
&\quad < \left(-1 + \frac{1}{11} + \frac{1}{14} + \frac{1}{16} \times \frac{1}{36} + \frac{1}{10} \times \frac{1}{20} \right)^2 - 4 \times \left(\frac{1}{40} + \frac{1}{10} \right) = 0.14 > 0,
\end{aligned}
$$

which implies that system (4.1) satisfies all the conditions in Theorem 3.1. Thus, (4.1) has exactly one π-anti-periodic solution. Moreover, this solution is globally exponentially stable. The results are shown in Figure 1.

Competing interests

The authors declare that they have no competing interests.

Authors' contributions

The authors have made the same contribution. All authors read and approved the final manuscript.

Author details

[1]Guizhou Key Laboratory of Economics System Simulation, Guizhou University of Finance and Economics, Guiyang, 550004, P.R. China. [2]School of Mathematics and Statistics, Henan University of Science and Technology, Luoyang, 471023, P.R. China.

Acknowledgements

The first author was supported by the National Natural Science Foundation of China (No. 11261010) and Governor Foundation of Guizhou Province ([2012]53). The second author was supported by the National Natural Science Foundation of China (No. 11101126). The authors would like to thank the referees and the editor for helpful suggestions incorporated into this paper.

References

1. Cao, J, Wang, J: Global exponent al stability and periodicity of recurrent neural networks with time delays. IEEE Trans. Circuits Syst. I **52**(5), 920-931 (2005)
2. Chen, Z, Zhao, DH, Ruan, J: Dynamic analysis of high order Cohen-Grossberg neural networks with time delay. Chaos Solitons Fractals **32**(4), 1538-1546 (2007)
3. Lou, XY, Cui, BT: Novel global stability criteria for high-order Hopfield-type neural networks with time-varying delays. J. Math. Anal. Appl. **330**(1), 144-158 (2007)
4. Wu, C, Ruan, J, Lin, W: On the existence and stability of the periodic solution in the Cohen-Grossberg neural network with time delay and high-order terms. Appl. Math. Comput. **177**(1), 194-210 (2006)
5. Xiang, H, Yan, KM, Wang, BY: Existence and global exponential stability of periodic solution for delayed high-order Hopfield-type neural networks. Phys. Lett. A **352**(4-5), 341-349 (2006)
6. Xu, BJ, Liu, XZ, Liao, XX: Global exponential stability of high order Hopfield type neural networks. Appl. Math. Comput. **174**(1), 98-116 (2006)
7. Xu, BJ, Liu, XZ, Liao, XX: Global exponential stability of high order Hopfield type neural networks with time delays. Comput. Math. Appl. **45**(10-11), 1729-1737 (2003)
8. Wu, W, Cui, BT: Global robust exponential stability of delayed neural networks. Chaos Solitons Fractals **35**(4), 747-754 (2008)
9. Zeng, ZG, Wang, J: Improved conditions for global exponential stability of recurrent neural networks with time-varying delays. IEEE Trans. Neural Netw. **17**(3), 623-635 (2006)
10. Li, YK, Yang, L: Anti-periodic solutions for Cohen-Grossberg neural networks with bounded and unbounded delays. Commun. Nonlinear Sci. Numer. Simul. **14**(7), 3134-3140 (2009)
11. Shao, JY: Anti-periodic solutions for shunting inhibitory cellular neural networks with time-varying delays. Phys. Lett. A **372**(30), 5011-5016 (2008)
12. Fan, QY, Wang, WT, Yi, XJ: Anti-periodic solutions for a class of nonlinear *n*th-order differential equations with delays. J. Comput. Appl. Math. **230**(2), 762-769 (2009)
13. Li, YK, Xu, EL, Zhang, TW: Existence and stability of anti-periodic solution for a class of generalized neural networks with impulses and arbitrary delays on time scales. J. Inequal. Appl. **2010**, Article ID 132790 (2010)
14. Gong, SH: Anti-periodic solutions for a class of Cohen-Grossberg neural networks. Comput. Math. Appl. **58**(2), 341-347 (2009)
15. Peng, GQ, Huang, LH: Anti-periodic solutions for shunting inhibitory cellular neural networks with continuously distributed delays. Nonlinear Anal., Real World Appl. **10**(40), 2434-2440 (2009)
16. Zhang, AP: Existence and exponential stability of anti-periodic solutions for HCNNs with time-varying leakage delays. Adv. Differ. Equ. **2013**, Article ID 162 (2013). doi:10.1186/1687-1847-2013-162
17. Aftabizadeh, AR, Aizicovici, S, Pavel, NH: On a class of second-order anti-periodic boundary value problems. J. Math. Anal. Appl. **171**(2), 301-320 (1992)
18. Aizicovici, S, McKibben, M, Reich, S: Anti-periodic solutions to nonmonotone evolution equations with discontinuous nonlinearities. Nonlinear Anal. TMA **43**(2), 233-251 (2001)
19. Liu, BW: An anti-periodic LaSalle oscillation theorem for a class of functional differential equations. J. Comput. Appl. Math. **223**(2), 1081-1086 (2009)
20. Huang, ZD, Peng, LQ, Xu, M: Anti-periodic solutions for high-order cellular neural networks with time-varying delays. Electron. J. Differ. Equ. **2010**, 59 (2010)
21. Li, YK, Yang, L, Wu, WQ: Anti-periodic solutions for a class of Cohen-Grossberg neural networks with time-varying on time scales. Int. J. Syst. Sci. **42**(7), 1127-1132 (2011)
22. Pan, LJ, Cao, JD: Anti-periodic solution for delayed cellular neural networks with impulsive effects. Nonlinear Anal., Real World Appl. **12**(6), 3014-3027 (2011)
23. Li, YK: Anti-periodic solutions to impulsive shunting inhibitory cellular neural networks with distributed delays on time scales. Commun. Nonlinear Sci. Numer. Simul. **16**(8), 3326-3336 (2011)
24. Fan, QY, Wang, WT, Yi, XJ, Huang, LH: Anti-periodic solutions for a class of third-order nonlinear differential equations with deviating argument. Electron. J. Qual. Theory Differ. Equ. **2010**, 8 (2010)
25. Wang, W, Shen, J: Existence of solutions for anti-periodic boundary value problems. Nonlinear Anal. **70**(2), 598-605 (2009)
26. Chen, YQ, Nieto, JJ, O'Regan, D: Anti-periodic solutions for fully nonlinear first-order differential equations. Math. Comput. Model. **46**(9-10), 1183-1190 (2007)
27. Peng, L, Wang, WT: Anti-periodic solutions for shunting inhibitory cellular neural networks with time-varying delays in leakage terms. Neurocomputing **111**, 27-33 (2013)
28. Park, JY, Ha, TG: Existence of anti-periodic solutions for quasilinear parabolic hemivariational inequalities. Nonlinear Anal. TMA **71**(7-8), 3203-3217 (2009)

29. Yu, YH, Shao, JY, Yue, GX: Existence and uniqueness of anti-periodic solutions for a kind of Rayleigh equation with two deviating arguments. Nonlinear Anal. TMA **71**(10), 4689-4695 (2009)
30. Lv, X, Yan, P, Liu, DJ: Anti-periodic solutions for a class of nonlinear second-order Rayleigh equations with delays. Commun. Nonlinear Sci. Numer. Simul. **15**(11), 3593-3598 (2010)
31. Li, YQ, Huang, LH: Anti-periodic solutions for a class of Liénard-type systems with continuously distributed delays. Nonlinear Anal., Real World Appl. **10**(4), 2127-2132 (2009)
32. Liu, BW: Anti-periodic solutions for forced Rayleigh-type equations. Nonlinear Anal., Real World Appl. **10**(5), 2850-2856 (2009)
33. Shi, PL, Dong, LZ: Existence and exponential stability of anti-periodic solutions of Hopfield neural networks with impulses. Appl. Math. Comput. **216**(2), 623-630 (2010)
34. Wang, Q, Fang, YY, Li, H, Su, LJ, Dai, BX: Anti-periodic solutions for high-order Hopfield neural networks with impulses. Neurocomputing **138**, 339-346 (2014)
35. Liu, Y, Yang, YQ, Liang, T, Li, L: Existence and global exponential stability of anti-periodic solutions for competitive neural networks with delays in the leakage terms on time scales. Neurocomputing **133**, 471-482 (2014)
36. Zheng, QQ, Shen, JW: Bifurcations and dynamics of cancer signaling network regulated by micro RNA. Discrete Dyn. Nat. Soc. **2013**, Article ID 176956 (2013)
37. Shen, JW, Liu, ZR, Zheng, WX, Xu, FD, Chen, LN: Oscillatory dynamics in a simple gene regulatory network mediated by small RNAs. Physica A **388**(14), 2995-3000 (2009)
38. Tang, XH, Zou, XF: The existence and global exponential stability of a periodic solution of a class of delay differential equations. Nonlinearity **22**(10), 2423-2442 (2009)
39. Xu, CJ, Tang, XH, Liao, MX: Bifurcation analysis of a delayed predator-prey model of prey migration and predator switching. Bull. Korean Math. Soc. **50**, 353-373 (2013)
40. Ferrara, M, Munteanu, F, Udriste, C, Zugravescu, D: Controllability of a nonholonomic macroeconomic system. J. Optim. Theory Appl. **154**, 1036-1054 (2012)
41. Ferrara, M, Bianca, C, Guerrini, L: High-order moments conservation in thermostatted kinetic models. J. Glob. Optim. **58**(2), 389-404 (2014)
42. Ferrara, M, Khademloob, S, Heidarkhani, S: Multiplicity results for perturbed fourth-order Kirchhoff type problems. Appl. Math. Comput. **234**, 316-325 (2014)
43. Ferrara, M, Heidarkhani, S: Multiple solutions for perturbed p-Laplacian boundary-value problems with impulsive effects. Electron. J. Differ. Equ. **2014**, 106 (2014)
44. Ou, CX: Anti-periodic solutions for high-order Hopfield neural networks. Comput. Math. Appl. **56**(7), 1838-1844 (2008)

12

A note on the initial value problem of fractional evolution equations

Pengyu Chen[1*], Xuping Zhang[1,2] and Yongxiang Li[1]

*Correspondence: chpengyu123@163.com
[1] Department of Mathematics, Northwest Normal University, Lanzhou, 730070, P.R. China
Full list of author information is available at the end of the article

Abstract

In this paper, we discuss the continuous dependence of mild solutions on initial values and orders for the initial value problem of fractional evolution equations in infinite dimensional spaces. The results obtained in this paper improve and extend some related conclusions on this topic. This paper can be considered as a contribution to this emerging field.

MSC: 35R´ 1; 47J35

Keywords: fractional evolution equation; initial value problem; continuous dependence

1 Introduction

Fractional differential equations have recently come to be considered to be a very powerful tool to help scientists explore the hidden properties of the dynamics of complex systems in various fields of sciences and engineering. In recent years, fractional differential equations played the key role of a fundamental, efficient, and convenient theoretical framework for more adequate modeling of complex dynamic processes. Indeed, we can find numerous applications in viscoelasticity, electromagnetism, diffusion, control, mechanics, physics, signal processing, chemistry, bioengineering, medicine, and in many other areas. For more details as regards fractional differential equations we refer to the monographs by Miller and Ross [1], Podlubny [2] and Kilbas *et al.* [3], the papers by Eidelman and Kochubei [4] and Lakshmikantham and Vatsala [5], and the survey by Agarwal *et al.* [6].

In this paper, we are concerned with the continuous dependence of mild solutions on initial values and orders for the initial value problem (IVP) of the fractional evolution equation

$$\begin{cases} {}_0^C D_t^\alpha u(t) + Au(t) = f(t, u(t)), & t \geq 0, \\ u(0) = u_0 \end{cases} \tag{1.1}$$

in Banach space E, where ${}_0^C D_t^\alpha$ is the Caputo fractional derivative of order $0 < \alpha < 1$, $A : D(A) \subset E \to E$ is a closed linear operator and $-A$ generates a uniformly bounded C_0-semigroup $T(t)$ $(t \geq 0)$ in E, $f : [0, \infty) \times E \to E$ is a continuous nonlinear mapping, $u_0 \in E$.

In recent years, there has been a significant development in the theory of fractional evolution equations. Due to fractional semilinear evolution equations being abstract formulations for many problems arising in engineering and physics, fractional evolution equations have attracted much attention in recent years; see [7–17] and the references cited therein. In addition, some numerical methods have been used to solve space-time or space fractional evolution equations; see [18–23].

However, we observed that all of the existing articles are only devoted to the study of the existence, uniqueness, and controllability of mild solutions for fractional evolution equations; up to now the continuous dependence of mild solutions on parameters for fractional evolution equations has not been considered in the literature. In order to fill this gap, we are concerned with the continuous dependence of mild solutions on the initial values and orders for IVP (1.1).

2 Preliminaries

In this section, we introduce some notations, definitions, and preliminary facts which are used in the sequel.

Let E be a Banach space with the norm $\|\cdot\|$ and let $a > 0$ be a constant. We denote by $C([0,a],E)$ the Banach space of all continuous E-value functions on interval $[0,a]$ with the supnorm $\|u\|_C = \sup_{t\in[0,a]} \|u(t)\|$. Throughout this paper, we assume that $A : D(A) \subset E \to E$ is a closed linear operator and $-A$ generates a uniformly bounded C_0-semigroup $T(t)$ $(t \geq 0)$ on E. Let $M = \sup_{t\in[0,+\infty)} \|T(t)\|_{\mathcal{L}(E)}$, where $\mathcal{L}(E)$ stands for the Banach space of all linear and bounded operators in E. For more details of the theory of operator semigroups, see [24].

Let $0 < \alpha < 1$. The Caputo fractional order derivative of order α with the lower limit 0 for a function $u \in C^1[0,a]$ is defined as

$$ {}_0^C D_t^\alpha u(t) = J_t^{1-\alpha} u'(t) = \frac{1}{\Gamma(1-\alpha)} \int_0^t (t-s)^{-\alpha} u'(s)\,ds, \quad t > 0, \tag{2.1}$$

where Γ stands for the gamma function, and the fractional integral of order $\alpha > 0$ with the lower limit 0 for a function u is defined as

$$ J_t^\alpha u(t) = \frac{1}{\Gamma(\alpha)} \int_0^t (t-s)^{\alpha-1} u(s)\,ds. \tag{2.2}$$

If u is an abstract function with values in E, then the integrals which appear in (2.1) and (2.2) are taken in Bochner's sense. For the details as regards the definitions of fractional derivative and integral, please see [1–3].

Definition 2.1 By a mild solution of IVP (1.1) on $[0,\infty)$, we mean a continuous function u defined from $[0,\infty)$ into E satisfying

$$ u(t) = \mathcal{T}_\alpha(t) u_0 + \int_0^t (t-s)^{\alpha-1} \mathcal{S}_\alpha(t-s) f\big(s,u(s)\big)\,ds, \quad t \in [0,\infty), \tag{2.3}$$

where

$$ \begin{aligned} \mathcal{T}_\alpha(t) &= \int_0^\infty h_\alpha(s) T\big(t^\alpha s\big)\,ds, \qquad \mathcal{S}_\alpha(t) = \alpha \int_0^\infty s h_\alpha(s) T\big(t^\alpha s\big)\,ds, \\ h_\alpha(s) &= \frac{1}{\pi\alpha} \sum_{n=1}^\infty (-s)^{n-1} \frac{\Gamma(n\alpha+1)}{n!} \sin(n\pi\alpha), \quad s \in (0,\infty), \end{aligned} \tag{2.4}$$

is a function of Wright type defined on $(0, \infty)$ which satisfies

$$h_\alpha(s) \geq 0, \quad s \in (0, \infty), \qquad \int_0^\infty h_\alpha(s)\,ds = 1$$

and

$$\int_0^\infty s^\nu h_\alpha(s)\,ds = \frac{\Gamma(1+\nu)}{\Gamma(1+\alpha\nu)}, \quad \nu \in [0,1].$$

The following lemma will be of fundamental importance in what follows (see [7, 9]).

Lemma 2.2 *The operators $\mathcal{T}_q(t)$ $(t \geq 0)$ and $\mathcal{S}_\alpha(t)$ $(t \geq 0)$ have the following properties:*

(1) *For every fixed $t \geq 0$, $\mathcal{T}_\alpha(t)$ and $\mathcal{S}_\alpha(t)$ are linear bounded operators, i.e.,*

$$\|\mathcal{T}_\alpha(t)u\| \leq M\|u\|; \qquad \|\mathcal{S}_\alpha(t)u\| \leq \frac{M}{\Gamma(\alpha)}\|u\|, \quad u \in E.$$

(2) *The operators $\mathcal{T}_\alpha(t)$ $(t \geq 0)$ and $\mathcal{S}_\alpha(t)$ $(t \geq 0)$ are strongly continuous on $[0, \infty)$.*

In what follows, we recall the following Gronwall-Bellman type inequalities, which can be used in fractional differential equations and integral equations with singular kernel.

Lemma 2.3 ([25]) *Suppose that $m > 0$, $n \geq 0$, and $q > 0$. If $u(t)$ is nonnegative and locally integrable on $0 \leq t < C$ (some $C \leq +\infty$) with*

$$u(t) \leq m + n\int_0^t (t-s)^{q-1}u(s)\,ds \tag{2.5}$$

on this interval. Then

$$u(t) \leq mE_q\big(n\Gamma(q)t^q\big), \quad 0 \leq t < C, \tag{2.6}$$

where E_q is the Mittag-Leffler function defined by

$$E_q(z) = \sum_{n=0}^{\infty} \frac{z^n}{\Gamma(nq+1)}, \quad z \in \mathbb{C}. \tag{2.7}$$

Lemma 2.4 ([26]) *Suppose $q > 0$, $m(t)$ is a nonnegative function locally integrable on $0 \leq t < C$ and $n(t)$ is a nonnegative, nondecreasing, continuous function defined on $0 \leq t < C$, $n(t) \leq K$ (constant), and suppose $u(t)$ is nonnegative and locally integrable on $0 \leq t < C$ with*

$$u(t) \leq m(t) + n(t)\int_0^t (t-s)^{q-1}u(s)\,ds \tag{2.8}$$

on this interval. Then

$$u(t) \leq m(t) + \int_0^t \left[\sum_{n=1}^{\infty} \frac{(n(t)\Gamma(q))^n}{\Gamma(nq)}(t-s)^{nq-1}m(s)\right]ds, \quad 0 \leq t < C. \tag{2.9}$$

3 Main results

In this section, we state and prove the main results of this paper. We suppose that the nonlinear term f satisfies the following local Lipschitz condition:

(H_f) For any $\bar{t} > 0$ and constant $r > 0$, there exists a positive constant $L = L(r, \bar{t})$ such that

$$\|f(t, u(t)) - f(t, v(t))\| \leq L\|u(t) - v(t)\|$$

for every $t \in [0, \bar{t}]$ and all $u, v \in E$ satisfying $\|u(t)\| \leq r$, $\|v(t)\| \leq r$.

Theorem 3.1 *Assume that the condition* (H_f) *is satisfied, then there exists a constant $a > 0$ such that IVP* (1.1) *has a unique mild solution u on $[0, a]$. Moreover, for any $\epsilon > 0$, there exists a constant $\delta > 0$ such that for arbitrary $\bar{u}_0 \in E$ satisfying $\|\bar{u}_0 - u_0\| < \delta$, the initial value problem (IVP) of the fractional evolution equation*

$$\begin{cases} {}_0^C D_t^{\alpha} u(t) + Au(t) = f(t, u(t)), \quad t \geq 0, \\ u(0) = \bar{u}_0 \end{cases} \tag{3.1}$$

has a unique mild solution $\bar{u}$ on $[0, a]$ and $\|\bar{u}(t) - u(t)\| < \epsilon$ for every $t \in [0, a]$.

Proof We first prove the local existence and uniqueness of mild solutions for IVP (1.1) on interval $[0, a_1]$ for an appropriately small constant $a_1 > 0$, which will be given later. Consider the operator $\mathcal{F} : C([0, a_1], E) \to C([0, a_1], E)$ defined by

$$(\mathcal{F}u)(t) = \mathcal{T}_{\alpha}(t)u_0 + \int_0^t (t-s)^{\alpha-1} \mathcal{S}_{\alpha}(t-s) f(s, u(s))\, ds, \quad t \in [0, a_1]. \tag{3.2}$$

From the continuity of nonlinear term f and Lemma 2.2(2) one can easily see that the operator $\mathcal{F} : C([0, a_1], E) \to C([0, a_1], E)$ is continuous. By Definition 2.1, the mild solution of IVP (1.1) on $[0, a_1]$ is equivalent to the fixed point of the operator $\mathcal{F}$ defined by (3.2). Let $R = 2M\|u_0\|$, $L = L(R, 1)$, $N = \max_{0 \leq t \leq 1} \|f(t, \theta)\|$. Denote $\Omega_R = \{u \in C([0, a_1], E) : \|u(t)\| \leq R, t \in [0, a_1]\}$, then Ω_R is a closed ball in $C([0, a_1], E)$ with center θ and radius R. Set

$$a_1 := \min\left\{1, \left(\frac{\Gamma(\alpha+1)\|u_0\|}{LR + N}\right)^{\frac{1}{\alpha}}, \left(\frac{\Gamma(\alpha+1)}{ML+1}\right)^{\frac{1}{\alpha}}\right\}. \tag{3.3}$$

For any $u \in \Omega_R$ and $t \in [0, a_1]$, by Lemma 2.2(1), (3.2), (3.3), and the condition (H_f), we have

$$\begin{aligned} \|(\mathcal{F}u)(t)\| &\leq M\|u_0\| + \frac{M}{\Gamma(\alpha)} \int_0^t (t-s)^{\alpha-1} \left(\|f(s, u(s)) - f(s, \theta)\| + \|f(s, \theta)\|\right) ds \\ &\leq M\|u_0\| + \frac{(LR+N)Ma_1^{\alpha}}{\Gamma(\alpha+1)} \\ &\leq 2M\|u_0\| \\ &= R. \end{aligned} \tag{3.4}$$

Therefore, $\mathcal{F}u \in \Omega_R$. Thus, we proved that $\mathcal{F} : \Omega_R \to \Omega_R$ is a continuous operator. Now we show that the operator $\mathcal{F}$ is a contraction in Ω_R. For any $u, v \in C([0, a_1], E)$, and $t \in [0, a_1]$, it follows from Lemma 2.2(1), (3.2), (3.3), and the condition (H_f) that

$$\begin{aligned}\|(\mathcal{F}u)(t)-(\mathcal{F}v)(t)\| &\leq \frac{M}{\Gamma(\alpha)}\int_0^t (t-s)^{\alpha-1}\|f(s,u(s))-f(s,v(s))\|\,ds\\ &\leq \frac{MLa_1^{\alpha}}{\Gamma(\alpha+1)}\|u-v\|_C\\ &< \|u-v\|_C.\end{aligned} \tag{3.5}$$

Thus, by (3.5) we have

$$\|\mathcal{F}u-\mathcal{F}v\|_C < \|u-v\|_C,$$

which means that $\mathcal{F}$ has a unique fixed point $u \in \Omega_R$, which is in turn a unique mild solution of IVP (1.1) on $[0,a_1]$. Using a similar method we can prove that IVP (3.1) has a unique mild solution $\overline{u}$ on $[0,a_2]$ for appropriate constant $a_2 > 0$. Let $a = \min\{a_1,a_2\}$. Then for any $t \in [0,a]$, $u(t)$ and $\overline{u}(t)$ are the unique mild solutions of IVP (1.1) and IVP (3.1), respectively, and they are given by

$$u(t) = \mathcal{T}_\alpha(t)u_0 + \int_0^t (t-s)^{\alpha-1}\mathcal{S}_\alpha(t-s)f(s,u(s))\,ds \tag{3.6}$$

and

$$\overline{u}(t) = \mathcal{T}_\alpha(t)\overline{u}_0 + \int_0^t (t-s)^{\alpha-1}\mathcal{S}_\alpha(t-s)f(s,\overline{u}(s))\,ds. \tag{3.7}$$

Next, we prove the continuous dependence of the mild solutions on the initial values. For any $\epsilon > 0$, choosing $\delta = \frac{\Gamma(\alpha)\epsilon}{ME_\alpha(ML(R,a)a^\alpha)+1}$ such that $\|u_0-\overline{u}_0\| \leq \delta$, where E_α is the Mittag-Leffler function defined by (2.7). From (3.6), (3.7), Lemma 2.2(1), and the condition (H_f), we get

$$\begin{aligned}\|u(t)-\overline{u}(t)\| &\leq \frac{M}{\Gamma(\alpha)}\|u_0-\overline{u}_0\| + \frac{M}{\Gamma(\alpha)}\int_0^t (t-s)^{\alpha-1}\|f(s,u(s))-f(s,\overline{u}(s))\|\,ds\\ &\leq \frac{M\delta}{\Gamma(\alpha)} + \frac{ML(R,a)}{\Gamma(\alpha)}\int_0^t (t-s)^{\alpha-1}\|u(s)-\overline{u}(s)\|\,ds\end{aligned} \tag{3.8}$$

for any $t \in [0,a]$. Therefore, by (3.8) and Lemma 2.3 one gets

$$\|u(t)-\overline{u}(t)\| \leq \frac{M\delta}{\Gamma(\alpha)}E_\alpha(ML(R,a)a^\alpha) < \epsilon.$$

This completes the proof of Theorem 3.1. □

Next, we discuss the continuous dependence of the mild solutions for IVP (1.1) on the orders.

Theorem 3.2 *Assume that the condition* (H_f) *is satisfied, then there exists a constant* $a > 0$ *such that for any* $0 < \beta < 1$, *for IVP* (1.1), *and for the IVP of the fractional evolution equation*

$$\begin{cases} {}_0^C D_t^\beta u(t) + Au(t) = f(t,u(t)), & t \geq 0,\\ u(0) = u_0 \end{cases} \tag{3.9}$$

there exist unique mild solutions u and $\tilde{u}$ on $[0,a]$, respectively, and $\|\tilde{u}(t)-u(t)\| \to 0$ as $\alpha \to \beta$ for every $t \in [0,a]$.

Proof By using a similar method to the proof of Theorem 3.1, we can prove that for IVP (1.1) and IVP (3.9) there exist unique mild solutions u and $\tilde{u}$ on $[0,a]$ for an appropriate constant $a > 0$, respectively, and u and $\tilde{u}$ are given by (3.6) and the following integral equation:

$$\tilde{u}(t) = \mathcal{T}_\beta(t)u_0 + \int_0^t (t-s)^{\beta-1}\mathcal{S}_\beta(t-s)f\big(s,\tilde{u}(s)\big)\,ds. \tag{3.10}$$

Without loss of generality, we may assume that $\alpha > \beta$. Then by (2.4), (3.6), (3.10), Lemma 2.2(1), and the condition (H_f) one has

$$\begin{aligned}
\|u(t)-\tilde{u}(t)\| &\le \|\mathcal{T}_\alpha(t)u_0 - \mathcal{T}_\beta(t)u_0\| + \Big\|\int_0^t (t-s)^{\alpha-1}\mathcal{S}_\alpha(t-s)f\big(s,u(s)\big)\,ds \\
&\quad - \int_0^t (t-s)^{\beta-1}\mathcal{S}_\beta(t-s)f\big(s,\tilde{u}(s)\big)\,ds\Big\| \\
&\le \int_0^\infty \big\|h_\alpha(\tau)T(t^\alpha\tau)u_0 - h_\beta(\tau)T(t^\beta\tau)u_0\big\|\,d\tau \\
&\quad + \frac{MF}{\alpha\Gamma(\beta)}t^\alpha - \frac{MF}{\Gamma(\beta+1)}t^\beta \\
&\quad + F\int_0^t (t-s)^{\alpha-1}\int_0^\infty \tau\big\|\alpha h_\alpha(\tau)T\big((t-s)^\alpha\tau\big) \\
&\quad - \beta h_\beta(\tau)T\big((t-s)^\beta\tau\big)\big\|\,d\tau\,ds \\
&\quad + \frac{ML(R,a)}{\Gamma(\beta)}\int_0^t (t-s)^{\alpha-1}\|u(s)-\tilde{u}(s)\|\,ds
\end{aligned} \tag{3.11}$$

for every $t \in [0,a]$, where $F = \max_{0\le t\le a}\{\|f(t,u(t))\| : \|u(t)\| \le R\}$. By (3.11), the properties of the Wright type functions $h_\alpha(\cdot)$, $h_\beta(\cdot)$, and the semigroup $T(t)$ $(t \ge 0)$, we know that for any $t \in [0,a]$,

$$\begin{aligned}
m(t) &:= \int_0^\infty \big\|h_\alpha(\tau)T(t^\alpha\tau)u_0 - h_\beta(\tau)T(t^\beta\tau)u_0\big\|\,d\tau + \frac{MF}{\alpha\Gamma(\beta)}t^\alpha - \frac{MF}{\Gamma(\beta+1)}t^\beta \\
&\quad + F\int_0^t (t-s)^{\alpha-1}\int_0^\infty \tau\big\|\alpha h_\alpha(\tau)T\big((t-s)^\alpha\tau\big) \\
&\quad - \beta h_\beta(\tau)T\big((t-s)^\beta\tau\big)\big\|\,d\tau\,ds \\
&\to 0 \quad \text{as } \alpha \to \beta.
\end{aligned} \tag{3.12}$$

Therefore, by (3.11), (3.12), and Lemma 2.4, we get

$$\begin{aligned}
\|u(t)-\tilde{u}(t)\| &\le m(t) + \int_0^t \left[\sum_{n=1}^\infty \frac{(ML(R,a)\Gamma(\alpha))^n}{\Gamma(\beta)^n\Gamma(n\alpha)}(t-s)^{n\alpha-1}m(s)\right]ds \\
&\to 0 \quad \text{as } \alpha \to \beta
\end{aligned}$$

for every $t \in [0,a]$. This completes the proof of Theorem 3.2. □

Remark 3.3 The results obtained in this paper can be applied to all kinds of time-fractional partial differential equations. Using Theorems 3.1 and 3.2 one can obtain the continuous dependence results for the concrete fractional partial differential equations. Based on Theorem 3.2, obtained in this paper, we can approximate the solutions of time-fractional differential equations by the solutions of integer order differential equations, which are easier to obtain. Therefore, Theorem 3.2 gives an approach to obtain an approximate solution of fractional differential equations.

Competing interests
The authors declare that they have no competing interests.

Authors' contributions
All authors contributed equally to the writing of this paper. All authors read and approved the final manuscript.

Author details
[1]Department of Mathematics, Northwest Normal University, Lanzhou, 730070, P.R. China. [2]Department of Mathematics, Zhixing College of Northwest Normal University, Lanzhou, 730070, P.R. China.

Acknowledgements
The authors are very grateful to the anonymous referees for their valuable suggestions. This work is supported by NWNU-LKQN-14-3 and NNSF of China (11261053).

References

1. Miller, KS, Ross, B: An Introduction to the Fractional Calculus and Differential Equations. Wiley, New York (1993)
2. Podlubny, I: Fractional Differential Equations. Academic Press, San Diego (1999)
3. Kilbas, AA, Srivastava, HM, Trujillo, JJ: Theory and Applications of Fractional Differential Equations. North-Holland Mathematics Studies, vol. 204. Elsevier, Amsterdam (2006)
4. Eidelman, SD, Kochubei, AN: Cauchy problem for fractional diffusion equations. J. Differ. Equ. **199**, 211-255 (2004)
5. Lakshmikantham, V, Vatsala, AS: Basic theory of fractional differential equations. Nonlinear Anal. **69**, 1677-1682 (2008)
6. Agarwal, RP, Benchohra, M, Hamani, S: A survey on existence results for boundary value problems of nonlinear fractional differential equations and inclusions. Acta Appl. Math. **109**, 973-1033 (2010)
7. El-Borai, MM: Some probability densities and fundamental solutions of fractional evolution equations. Chaos Solitons Fractals **14**, 433-440 (2002)
8. El-Borai, MM, El-Nadi, KS, El-Akabawy, EG: On some fractional evolution equations. Comput. Math. Appl. **59**, 1352-1355 (2010)
9. Wang, J, Zhou, Y: A class of fractional evolution equations and optimal controls. Nonlinear Anal., Real World Appl. **12**, 262-272 (2011)
10. Wang, J, Zhou, Y, Fečkan, M: Abstract Cauchy problem for fractional differential equations. Nonlinear Dyn. **74**, 685-700 (2013)
11. Wang, R, Xiao, TJ, Liang, J: A note on the fractional Cauchy problems with nonlocal conditions. Appl. Math. Lett. **24**, 1435-1442 (2011)
12. Chen, P, Li, Y: Existence of mild solutions for fractional evolution equations with mixed monotone nonlocal conditions. Z. Angew. Math. Phys. **65**, 711-728 (2014)
13. Chen, P, Li, Y, Chen, Q, Feng, B: On the initial value problem of fractional evolution equations with noncompact semigroup. Comput. Math. Appl. **67**, 1108-1115 (2014)
14. Chen, P, Li, Y, Li, Q: Existence of mild solutions for fractional evolution equations with non-local initial conditions. Ann. Pol. Math. **110**, 13-24 (2014)
15. Herzallah, MAE, Baleanu, D: On fractional order hybrid differential equations. Abstr. Appl. Anal. **2014**, Article ID 389386 (2014)
16. Herzallah, MAE, El-Shahed, M, Baleanu, D: Mild and strong solutions for a fractional nonlinear Neumann boundary value problem. J. Comput. Anal. Appl. **15**, 341-352 (2013)
17. Kerboua, M, Debbouche, A, Baleanu, D: Approximate controllability of Sobolev type nonlocal fractional stochastic dynamic systems in Hilbert spaces. Abstr. Appl. Anal. **2013**, Article ID 262191 (2013)
18. Bhrawy, AH, Zaky, MA, Baleanu, D: New numerical approximations for space-time fractional Burgers' equations via a Legendre spectral-collocation method. Rom. Rep. Phys. **67**(2), 1-13 (2015)
19. Bhrawy, AH, Baleanu, D: A spectral Legendre-Gauss-Lobatto collocation method for a space-fractional advection diffusion equations with variable coefficients. Rep. Math. Phys. **72**, 219-233 (2013)
20. Bhrawy, AH, Al-Zahrani, AA, Alhamed, YA, Baleanu, D: A new generalized Laguerre-Gauss collocation scheme for numerical solution of generalized fractional pantograph equations. Rom. J. Phys. **59**, 646-657 (2014)
21. Bhrawy, AH, Zaky, MK: A method based on the Jacobi tau approximation for solving multi-term time-space fractional partial differential equations. J. Comput. Phys. **281**, 876-895 (2015)
22. Bhrawy, AH, Doha, EH, Baleanu, D, Ezz-Eldien, SS: A spectral tau algorithm based on Jacobi operational matrix for numerical solution of time fractional diffusion-wave equations. J. Comput. Phys. (2014). doi:10.1016/j.jcp.2014.03.039
23. Abdelkawy, MA, Zaky, MA, Bhrawy, AH, Baleanu, D: Numerical simulation of time variable fractional order mobile-immobile advection-dispersion model. Rom. Rep. Phys. **67**(3), 1-19 (2015)

24. Pazy, A: Semigroups of Linear Operators and Applications to Partial Differential Equations. Springer, Berlin (1983)
25. Henry, D: Geometric Theory of Semilinear Parabolic Equations. Lecture Notes in Math., vol. 840. Springer, New York (1981)
26. Ye, H, Gao, J, Ding, Y: A generalized Gronwall inequality and its application to a fractional differential equation. J. Math. Anal. Appl. **328**, 1075-1081 (2007)

Some algebraic identities on quadra Fibona-Pell integer sequence

Arzu Özkoç*

*Correspondence:
arzuozkoc@duzce.edu.tr
Department of Mathematics,
Faculty of Arts and Science, Düzce
University, Düzce, Turkey

Abstract

In this work, we define a quadra Fibona-Pell integer sequence $W_n = 3W_{n-1} - 3W_{n-3} - W_{n-4}$ for $n \geq 4$ with initial values $W_0 = W_1 = 0, W_2 = 1, W_3 = 3$, and we derive some algebraic identities on it including its relationship with Fibonacci and Pell numbers.

Keywords: Fibonacci numbers; Lucas numbers; Pell numbers; Binet's formula; binary linear recurrences

1 Preliminaries

Let p and q be non-zero integers such that $D = p^2 - 4q \neq 0$ (to exclude a degenerate case). We set the sequences U_n and V_n to be

$$\begin{aligned} U_n &= U_n(p,q) = pU_{n-1} - qU_{n-2}, \\ V_n &= V_n(p,q) = pV_{n-1} - qV_{n-2} \end{aligned} \tag{1}$$

for $n \geq 2$ with initial values $U_0 = 0$, $U_1 = 1$, $V_0 = 2$, and $V_1 = p$. The sequences U_n and V_n are called the (first and second) Lucas sequences with parameters p and q. V_n is also called the companion Lucas sequence with parameters p and q.

The characteristic equation of U_n and V_n is $x^2 - px + q = 0$ and hence the roots of it are $x_1 = \frac{p+\sqrt{D}}{2}$ and $x_2 = \frac{p-\sqrt{D}}{2}$. So their Binet formulas are

$$U_n = \frac{x_1^n - x_2^n}{x_1 - x_2} \quad \text{and} \quad V_n = x_1^n + x_2^n$$

for $n \geq 0$. For the companion matrix $M = \left[\begin{smallmatrix} p & -q \\ 1 & 0 \end{smallmatrix}\right]$, one has

$$\begin{bmatrix} U_n \\ U_{n-1} \end{bmatrix} = M^{n-1} \begin{bmatrix} 1 \\ 0 \end{bmatrix} \quad \text{and} \quad \begin{bmatrix} V_n \\ V_{n-1} \end{bmatrix} = M^{n-1} \begin{bmatrix} p \\ 2 \end{bmatrix}$$

for $n \geq 1$. The generating functions of U_n and V_n are

$$U(x) = \frac{x}{1 - px + qx^2} \quad \text{and} \quad V(x) = \frac{2 - px}{1 - px + qx^2}. \tag{2}$$

Fibonacci, Lucas, Pell, and Pell-Lucas numbers can be derived from (1). Indeed for $p = 1$ and $q = -1$, the numbers $U_n = U_n(1,-1)$ are called the Fibonacci numbers (A000045 in

OEIS), while the numbers $V_n = V_n(1,-1)$ are called the Lucas numbers (A000032 in OEIS). Similarly, for $p = 2$ and $q = -1$, the numbers $U_n = U_n(2,-1)$ are called the Pell numbers (A000129 in OEIS), while the numbers $V_n = V_n(2,-1)$ are called the Pell-Lucas (A002203 in OEIS) (companion Pell) numbers (for further details see [1–6]).

2 Quadra Fibona-Pell sequence

In [7], the author considered the quadra Pell numbers $D(n)$, which are the numbers of the form $D(n) = D(n-2) + 2D(n-3) + D(n-4)$ for $n \geq 4$ with initial values $D(0) = D(1) = D(2) = 1$, $D(3) = 2$, and the author derived some algebraic relations on it.

In [8], the authors considered the integer sequence (with four parameters) $T_n = -5T_{n-1} - 5T_{n-2} + 2T_{n-3} + 2T_{n-4}$ with initial values $T_0 = 0$, $T_1 = 0$, $T_2 = -3$, $T_3 = 12$, and they derived some algebraic relations on it.

In the present paper, we want to define a similar sequence related to Fibonacci and Pell numbers and derive some algebraic relations on it. For this reason, we set the integer sequence W_n to be

$$W_n = 3W_{n-1} - 3W_{n-3} - W_{n-4} \tag{3}$$

for $n \geq 4$ with initial values $W_0 = W_1 = 0$, $W_2 = 1$, $W_3 = 3$ and call it a *quadra Fibona-Pell sequence*. Here one may wonder why we choose this equation and call it a quadra Fibona-Pell sequence. Let us explain: We will see below that the roots of the characteristic equation of W_n are the roots of the characteristic equations of both Fibonacci and Pell sequences. Indeed, the characteristic equation of (3) is $x^4 - 3x^3 + 3x + 1 = 0$ and hence the roots of it are

$$\alpha = \frac{1+\sqrt{5}}{2}, \qquad \beta = \frac{1-\sqrt{5}}{2}, \qquad \gamma = 1+\sqrt{2} \quad \text{and} \quad \delta = 1-\sqrt{2}. \tag{4}$$

(Here α, β are the roots of the characteristic equation of Fibonacci numbers and γ, δ are the roots of the characteristic equation of Pell numbers.) Then we can give the following results for W_n.

Theorem 1 *The generating function for* W_n *is*

$$W(x) = \frac{x^2}{x^4 + 3x^3 - 3x + 1}.$$

Proof The generating function $W(x)$ is a function whose formal power series expansion at $x = 0$ has the form

$$W(x) = \sum_{n=0}^{\infty} W_n x^n = W_0 + W_1 x + W_2 x^2 + \cdots + W_n x^n + \cdots.$$

Since the characteristic equation of (3) is $x^4 - 3x^3 + 3x + 1 = 0$, we get

$$\begin{aligned}(1 - 3x + 3x^3 + x^4)W(x) &= (1 - 3x + 3x^3 + x^4)(W_0 + W_1 x + \cdots + W_n x^n + \cdots) \\ &= W_0 + (W_1 - 3W_0)x + (W_2 - 3W_1)x^2\end{aligned}$$

$$+ (W_3 - 3W_2 + 3W_0)x^3 + \cdots$$
$$+ (W_n - 3W_{n-1} + 3W_{n-3} + W_{n-4})x^n + \cdots.$$

Notice that $W_0 = W_1 = 0$, $W_2 = 1$, $W_3 = 3$, and $W_n = 3W_{n-1} - 3W_{n-3} - W_{n-4}$. So $(1 - 3x + 3x^3 + x^4)W(x) = x^2$ and hence the result is obvious. □

Theorem 2 *The Binet formula for* W_n *is*

$$W_n = \left(\frac{\gamma^n - \delta^n}{\gamma - \delta}\right) - \left(\frac{\alpha^n - \beta^n}{\alpha - \beta}\right)$$

for $n \geq 0$.

Proof Note that the generating function is $W(x) = \frac{x^2}{x^4+3x^3-3x+1}$. It is easily seen that $x^4 + 3x^3 - 3x + 1 = (1 - x - x^2)(1 - 2x - x^2)$. So we can rewrite $W(x)$ as

$$W(x) = \frac{x}{1 - 2x - x^2} - \frac{x}{1 - x - x^2}. \tag{5}$$

From (2), we see that the generating function for Pell numbers is

$$P(x) = \frac{x}{1 - 2x - x^2} \tag{6}$$

and the generating function for the Fibonacci numbers is

$$F(x) = \frac{x}{1 - x - x^2}. \tag{7}$$

From (5), (6), (7), we get $W(x) = P(x) - F(x)$. So $W_n = (\frac{\gamma^n-\delta^n}{\gamma-\delta}) - (\frac{\alpha^n-\beta^n}{\alpha-\beta})$ as we wanted. □

The relationship with Fibonacci, Lucas, and Pell numbers is given below.

Theorem 3 *For the sequences* W_n, F_n, L_n, *and* P_n, *we have:*

(1) $W_n = P_n - F_n$ *for* $n \geq 0$.
(2) $W_{n+1} + W_{n-1} = (\gamma^n + \delta^n) - (\alpha^n + \beta^n)$ *for* $n \geq 1$.
(3) $\sqrt{5}F_n + 2\sqrt{2}P_n = (\gamma^n - \delta^n) + (\alpha^n - \beta^n)$ *for* $n \geq 1$.
(4) $L_n + P_{n+1} + P_{n-1} = \alpha^n + \beta^n + \gamma^n + \delta^n$ *for* $n \geq 1$.
(5) $2(W_{n+1} - W_n + F_{n-1}) = \gamma^n + \delta^n$ *for* $n \geq 1$.
(6) $\lim_{n\to\infty} \frac{W_n}{W_{n-1}} = \gamma$.

Proof (1) It is clear from the above theorem, since $W(x) = P(x) - F(x)$.

(2) Since $6W_{n-1} + W_{n+2} = 3W_{n+1} - 3W_{n-1} - W_{n-2} + 6W_{n-1}$, we get

$$W_{n+1} + W_{n-1} = 2W_{n-1} + \frac{1}{3}W_{n-2} + \frac{1}{3}W_{n+2}$$
$$= \frac{6}{3}\left(\frac{\gamma^{n-1} - \delta^{n-1}}{\gamma - \delta} - \frac{\alpha^{n-1} - \beta^{n-1}}{\alpha - \beta}\right)$$
$$+ \frac{1}{3}\left(\frac{\gamma^{n-2} - \delta^{n-2}}{\gamma - \delta} - \frac{\alpha^{n-2} - \beta^{n-2}}{\alpha - \beta}\right)$$

$$
\begin{aligned}
&+\frac{1}{3}\left(\frac{\gamma^{n+2}-\delta^{n+2}}{\gamma-\delta}-\frac{\alpha^{n+2}-\beta^{n+2}}{\alpha-\beta}\right)\\
&=\frac{1}{3(\gamma-\delta)}\left[\gamma^n\left(\frac{6}{\gamma}+\frac{1}{\gamma^2}+\gamma^2\right)+\delta^n\left(\frac{-6}{\delta}-\frac{1}{\delta^2}-\delta^2\right)\right]\\
&+\frac{1}{3(\alpha-\beta)}\left[\alpha^n\left(\frac{-6}{\alpha}-\frac{1}{\alpha^2}-\alpha^2\right)+\beta^n\left(\frac{6}{\beta}+\frac{1}{\beta^2}+\beta^2\right)\right]\\
&=(\gamma^n+\delta^n)-(\alpha^n+\beta^n),
\end{aligned}
$$

since $\frac{6}{\gamma}+\frac{1}{\gamma^2}+\gamma^2=\frac{-6}{\delta}-\frac{1}{\delta^2}-\delta^2=6\sqrt{2}$ and $\frac{-6}{\alpha}-\frac{1}{\alpha^2}-\alpha^2=\frac{6}{\beta}+\frac{1}{\beta^2}+\beta^2=-3\sqrt{5}$.

(3) Notice that $F_n=\frac{\alpha^n-\beta^n}{\alpha-\beta}$ and $P_n=\frac{\gamma^n-\delta^n}{\gamma-\delta}$. So we get $\sqrt{5}F_n=\alpha^n-\beta^n$ and $2\sqrt{2}P_n=\gamma^n-\delta^n$. Thus clearly, $\sqrt{5}F_n+2\sqrt{2}P_n=(\gamma^n-\delta^n)+(\alpha^n-\beta^n)$.

(4) It is easily seen that $P_{n+1}+P_{n-1}=\gamma^n+\delta^n$. Also $L_n=\alpha^n+\beta^n$. So $L_n+P_{n+1}+P_{n-1}=\alpha^n+\beta^n+\gamma^n+\delta^n$.

(5) Since $W_{n+1}=3W_n-3W_{n-2}-W_{n-3}$, we easily get

$$
\begin{aligned}
W_{n+1}-W_n&=2W_n-3W_{n-2}-W_{n-3}\\
&=2\left(\frac{\gamma^n-\delta^n}{\gamma-\delta}-\frac{\alpha^n-\beta^n}{\alpha-\beta}\right)-3\left(\frac{\gamma^{n-2}-\delta^{n-2}}{\gamma-\delta}-\frac{\alpha^{n-2}-\beta^{n-2}}{\alpha-\beta}\right)\\
&-\left(\frac{\gamma^{n-3}-\delta^{n-3}}{\gamma-\delta}-\frac{\alpha^{n-3}-\beta^{n-3}}{\alpha-\beta}\right)\\
&=\frac{1}{\gamma-\delta}\left[\gamma^n\left(2-\frac{3}{\gamma^2}-\frac{1}{\gamma^3}\right)+\delta^n\left(-2+\frac{3}{\delta^2}+\frac{1}{\delta^3}\right)\right]\\
&+\frac{1}{\alpha-\beta}\left[\alpha^{n-1}\left(2\alpha-\frac{3}{\alpha}-\frac{1}{\alpha^2}\right)-\beta^{n-1}\left(2\beta-\frac{3}{\beta}-\frac{1}{\beta^2}\right)\right]
\end{aligned}
$$

and hence

$$
\begin{aligned}
2W_{n+1}-2W_n&=\frac{2}{2\sqrt{2}}\left[\gamma^n\left(\frac{2\gamma^3-3\gamma-1}{\gamma^3}\right)+\delta^n\left(\frac{-2\delta^2+3\delta+1}{\delta^3}\right)\right]\\
&-\frac{2}{\alpha-\beta}\left[\alpha^{n-1}\left(\frac{2\alpha^3-3\alpha-1}{\alpha^2}\right)-\beta^{n-1}\left(\frac{2\beta^3-3\beta-1}{\beta^2}\right)\right]\\
\Leftrightarrow\quad 2W_{n+1}-2W_n&+\frac{2}{\alpha-\beta}\left[\alpha^{n-1}\left(\frac{2\alpha^3-3\alpha-1}{\alpha^2}\right)-\beta^{n-1}\left(\frac{2\beta^3-3\beta-1}{\beta^2}\right)\right]\\
&=\frac{1}{\sqrt{2}}\left[\gamma^n\left(\frac{2\gamma^3-3\gamma-1}{\gamma^3}\right)+\delta^n\left(\frac{-2\delta^3+3\delta+1}{\delta^3}\right)\right]\\
\Leftrightarrow\quad 2(W_{n+1}&-W_n+F_{n-1})=\gamma^n+\delta^n,
\end{aligned}
$$

since $\frac{2\gamma^3-3\gamma-1}{\gamma^3}=\frac{-2\delta^3+3\delta+1}{\delta^3}=\sqrt{2}$ and $\frac{2\alpha^3-3\alpha-1}{\alpha^2}=\frac{2\beta^3-3\beta-1}{\beta^2}=1$.

(6) It is just an algebraic computation, since $W_n=(\frac{\gamma^n-\delta^n}{\gamma-\delta})-(\frac{\alpha^n-\beta^n}{\alpha-\beta})$. □

Theorem 4 *The sum of the first n terms of W_n is*

$$\sum_{i=1}^{n}W_i=\frac{W_n+4W_{n-1}+4W_{n-2}+W_{n-3}+1}{2} \tag{8}$$

for $n\geq 3$.

Proof Recall that $W_n = 3W_{n-1} - 3W_{n-3} - W_{n-4}$. So

$$W_{n-3} + W_{n-4} = 3W_{n-1} - 2W_{n-3} - W_n. \tag{9}$$

Applying (9), we deduce that

$$\begin{aligned}
&W_1 + W_0 = 3W_3 - 2W_1 - W_4,\\
&W_2 + W_1 = 3W_4 - 2W_2 - W_5,\\
&W_3 + W_2 = 3W_5 - 2W_3 - W_6,\\
&\ldots,\\
&W_{n-4} + W_{n-5} = 3W_{n-2} - 2W_{n-4} - W_{n-1},\\
&W_{n-3} + W_{n-4} = 3W_{n-1} - 2W_{n-3} - W_n.
\end{aligned} \tag{10}$$

If we sum of both sides of (10), then we obtain $W_{n-3} + W_0 + 2(W_1 + \cdots + W_{n-4}) = 3(W_3 + W_4 + \cdots + W_{n-1}) - 2(W_1 + W_2 + \cdots + W_{n-3}) - (W_4 + W_5 + \cdots + W_n)$. So we get $W_{n-3} + 2(W_1 + W_2 + \cdots + W_{n-4}) = 1 - W_{n-2} - W_{n-1} - W_n + 3W_{n-2} + 3W_{n-1}$ and hence we get the desired result. □

Theorem 5 *The recurrence relations are*

$$\begin{aligned}
W_{2n} &= 9W_{2n-2} - 20W_{2n-4} + 9W_{2n-6} - W_{2n-8},\\
W_{2n+1} &= 9W_{2n-1} - 20W_{2n-3} + 9W_{2n-5} - W_{2n-7}
\end{aligned}$$

for $n \geq 4$.

Proof Recall that $W_n = 3W_{n-1} - 3W_{n-3} - W_{n-4}$. So $W_{2n} = 3W_{2n-1} - 3W_{2n-3} - W_{2n-4}$ and hence

$$\begin{aligned}
W_{2n} &= 3W_{2n-1} - 3W_{2n-3} - W_{2n-4}\\
&= 9W_{2n-2} - 9W_{2n-4} - 3W_{2n-5} - 9W_{2n-4} + 9W_{2n-6} + 3W_{2n-7}\\
&\quad + W_{2n-8} - W_{2n-8} - W_{2n-4}\\
&= -(3W_{2n-5} - 3W_{2n-7} - W_{2n-8}) + 9W_{2n-2} - 18W_{2n-4} + 9W_{2n-6}\\
&\quad - W_{2n-8} - W_{2n-4}\\
&= -W_{2n-4} + 9W_{2n-2} - 9W_{2n-4} - 9W_{2n-4} + 9W_{2n-6} - W_{2n-8} - W_{2n-4}\\
&= 9W_{2n-2} - 20W_{2n-4} + 9W_{2n-6} - W_{2n-8}.
\end{aligned}$$

The other assertion can be proved similarly. □

The rank of an integer N is defined to be

$$\rho(N) = \begin{cases} p & \text{if } p \text{ is the smallest prime with } p|N,\\ \infty & \text{if } N \text{ is prime.} \end{cases}$$

Thus we can give the following theorem.

Theorem 6 *The rank of* W_n *is*

$$\rho(W_n) = \begin{cases} 2 & \text{if } n = 5 + 6k, 6 + 6k, 7 + 6k, \\ 3 & \text{if } n = 8 + 12k, 9 + 12k, 15 + 12k, 16 + 12k, \\ 5 & \text{if } n = 14 + 60k, 46 + 60k \end{cases}$$

for an integer $k \geq 0$.

Proof Let $n = 5+6k$. We prove it by induction on k. Let $k = 0$. Then we get $W_5 = 24 = 2^3 \cdot 3$. So $\rho(W_5) = 2$. Let us assume that the rank of W_n is 2 for $n = k-1$, that is, $\rho(W_{6k-1}) = 2$, so $W_{5+6(k-1)} = W_{6k-1} = 2^a \cdot B$ for some integers $a \geq 1$ and $B > 0$. For $n = k$, we get

$$\begin{aligned} W_{6k+5} &= 3W_{6k+4} - 3W_{6k+2} - W_{6k+1} \\ &= 3(3W_{6k+3} - 3W_{6k+1} - W_{6k}) - 3W_{6k+2} - W_{6k+1} \\ &= 9W_{6k+3} - 9W_{6k+1} - 3W_{6k} - 3W_{6k+2} - W_{6k+1} \\ &= 9(3W_{6k+2} - 3W_{6k} - W_{6k-1}) - 9W_{6k+1} - 3W_{6k} - 3W_{6k+2} - W_{6k+1} \\ &= 27W_{6k+2} - 27W_{6k} - 9W_{6k-1} - 9W_{6k+1} - 3W_{6k} - 3W_{6k+2} - W_{6k+1} \\ &= 24W_{6k+2} - 30W_{6k} - 10W_{6k+1} - 9W_{6k-1} \\ &= 24W_{6k+2} - 30W_{6k} - 10W_{6k+1} - 9 \cdot 2^a B \\ &= 2\left[12W_{6k+2} - 15W_{6k} - 5W_{6k+1} - 9 \cdot 2^{a-1} B\right]. \end{aligned}$$

Therefore $\rho(W_{5+6k}) = 2$. Similarly it can be shown that $\rho(W_{6+6k}) = \rho(W_{7+6k}) = 2$.

Now let $n = 8 + 12k$. For $k = 0$, we get $W_8 = 387 = 3^2 \cdot 43$. So $\rho(W_8) = 3$. Let us assume that for $n = k-1$ the rank of W_n is 3, that is, $\rho(W_{8+12(k-1)}) = \rho(W_{12k-4}) = 3^b \cdot H$ for some integers $b \geq 1$ and $H > 0$ which is not even integer. For $n = k$, we get

$$\begin{aligned} W_{12k+8} &= 3W_{12k+7} - 3W_{12k+5} - W_{12k+4} \\ &= 3W_{12k+7} - 3W_{12k+5} - (3W_{12k+3} - 3W_{12k+1} - W_{12k}) \\ &= 3W_{12k+7} - 3W_{12k+5} - 3W_{12k+3} + 3W_{12k+1} + W_{12k} \\ &= 3W_{12k+7} - 3W_{12k+5} - 3W_{12k+3} + 3W_{12k+1} \\ &\quad + (3W_{12k-1} - 3W_{12k-3} - W_{12k-4}) \\ &= 3W_{12k+7} - 3W_{12k+5} - 3W_{12k+3} + 3W_{12k+1} + 3W_{12k-1} \\ &\quad - 3W_{12k-3} - W_{12k-4} \\ &= 3W_{12k+7} - 3W_{12k+5} - 3W_{12k+3} + 3W_{12k+1} + 3W_{12k-1} \\ &\quad - 3W_{12k-3} - 3^b \cdot H \\ &= 3\big(W_{12k+7} - W_{12k+5} - W_{12k+3} + W_{12k+1} + W_{12k-1} \\ &\quad - W_{12k-3} - 3^{b-1} \cdot H\big). \end{aligned}$$

So $\rho(W_{12k+8}) = 3$. The others can be proved similarly. □

Remark 1 Apart from the above theorem, we see that $\rho(W_{22}) = \rho(W_{26}) = \infty$, while $\rho(W_{70}) = \rho(W_{98}) = 13$ and $\rho(W_{10}) = \rho(W_{34}) = \rho(W_{50}) = 23$. But there is no general formula.

The companion matrix for W_n is

$$M = \begin{bmatrix} 3 & 0 & -3 & -1 \\ 1 & 0 & 0 & 0 \\ 0 & 1 & 0 & 0 \\ 0 & 0 & 1 & 0 \end{bmatrix}.$$

Set

$$N = \begin{bmatrix} 1 \\ 0 \\ 0 \\ 0 \end{bmatrix}$$

and

$$R = [3 \quad 1 \quad 0 \quad 0].$$

Then we can give the following theorem, which can be proved by induction on n.

Theorem 7 *For the sequence W_n, we have:*

(1) $RM^nN = W_{n+3} + P_n + 2(W_{n+1} - F_n)$ *for* $n \geq 1$.

(2) $R(M^T)^{n-3}N = W_n$ *for* $n \geq 3$.

(3) *If* $n \geq 7$ *is odd, then*

$$M^n = \begin{bmatrix} m_{11} & m_{12} & m_{13} & m_{14} \\ m_{21} & m_{22} & m_{23} & m_{24} \\ m_{31} & m_{32} & m_{33} & m_{34} \\ m_{41} & m_{42} & m_{43} & m_{44} \end{bmatrix},$$

where

$$m_{11} = W_{n+2}, \qquad m_{21} = W_{n+1}, \qquad m_{31} = W_n, \qquad m_{41} = W_{n-1},$$

$$m_{14} = -W_{n+1}, \qquad m_{24} = -W_n, \qquad m_{34} = -W_{n-1}, \qquad m_{44} = -W_{n-2},$$

$$m_{12} = -1 - W_{n+1} - 2\sum_{i=0}^{\frac{n-5}{2}} W_{n-1-2i}, \qquad m_{13} = -W_{n+2} - 2\sum_{i=0}^{\frac{n-3}{2}} W_{n-2i},$$

$$m_{22} = -W_n - 2\sum_{i=0}^{\frac{n-5}{2}} W_{n-2-2i}, \qquad m_{23} = -1 - W_{n+1} - 2\sum_{i=0}^{\frac{n-5}{2}} W_{n-1-2i},$$

$$m_{32} = -1 - W_{n-1} - 2\sum_{i=0}^{\frac{n-7}{2}} W_{n-3-2i}, \qquad m_{33} = -W_n - 2\sum_{i=0}^{\frac{n-5}{2}} W_{n-2-2i},$$

$$m_{42} = -W_{n-2} - 2\sum_{i=0}^{\frac{n-7}{2}} W_{n-4-2i}, \qquad m_{43} = -1 - W_{n-1} - 2\sum_{i=0}^{\frac{n-7}{2}} W_{n-3-2i},$$

and if $n \geq 8$ is even, then

$$M^n = \begin{bmatrix} m_{11} & m_{12} & m_{13} & m_{14} \\ m_{21} & m_{22} & m_{23} & m_{24} \\ m_{31} & m_{32} & m_{33} & m_{34} \\ m_{41} & m_{42} & m_{43} & m_{44} \end{bmatrix},$$

where

$$m_{11} = W_{n+2}, \quad m_{21} = W_{n+1}, \quad m_{31} = W_n, \quad m_{41} = W_{n-1},$$

$$m_{14} = -W_{n+1}, \quad m_{24} = -W_n, \quad m_{34} = -W_{n-1}, \quad m_{44} = -W_{n-2},$$

$$m_{12} = -W_{n+1} - 2\sum_{i=0}^{\frac{n-4}{2}} W_{n-1-2i}, \qquad m_{13} = -1 - W_{n+2} - 2\sum_{i=0}^{\frac{n-4}{2}} W_{n-2i},$$

$$m_{22} = -1 - W_n - 2\sum_{i=0}^{\frac{n-6}{2}} W_{n-2-2i}, \qquad m_{23} = -W_{n+1} - 2\sum_{i=0}^{\frac{n-4}{2}} W_{n-1-2i},$$

$$m_{32} = -W_{n-1} - 2\sum_{i=0}^{\frac{n-6}{2}} W_{n-3-2i}, \qquad m_{33} = -1 - W_n - 2\sum_{i=0}^{\frac{n-6}{2}} W_{n-2-2i},$$

$$m_{42} = -1 - W_{n-2} - 2\sum_{i=0}^{\frac{n-8}{2}} W_{n-4-2i}, \qquad m_{43} = -W_{n-1} - 2\sum_{i=0}^{\frac{n-6}{2}} W_{n-3-2i}.$$

A circulant matrix is a matrix $A = [a_{ij}]_{n\times n}$ defined to be

$$A = \begin{bmatrix} a_0 & a_1 & a_2 & \cdots & a_{n-1} \\ a_{n-1} & a_0 & a_1 & \cdots & a_{n-2} \\ a_{n-2} & a_{n-1} & a_0 & \cdots & a_{n-3} \\ . & . & . & \cdots & . \\ . & . & . & \cdots & . \\ a_1 & a_2 & a_3 & \cdots & a_0 \end{bmatrix},$$

where a_i are constants. The eigenvalues of A are

$$\lambda_j(A) = \sum_{k=0}^{n-1} a_k w^{-jk}, \tag{11}$$

where $w = e^{\frac{2\pi i}{n}}$, $i = \sqrt{-1}$, and $j = 0, 1, \ldots, n-1$. The spectral norm for a matrix $B = [b_{ij}]_{n\times m}$ is defined to be $\|B\|_{\text{spec}} = \max\{\sqrt{\lambda_i}\}$, where λ_i are the eigenvalues of $B^H B$ for $0 \leq j \leq n-1$ and B^H denotes the conjugate transpose of B.

For the circulant matrix

$$W = W(W_n) = \begin{bmatrix} W_0 & W_1 & W_2 & \cdots & W_{n-1} \\ W_{n-1} & W_0 & W_1 & \cdots & W_{n-2} \\ W_{n-2} & W_{n-1} & W_0 & \cdots & W_{n-3} \\ \cdot & \cdot & \cdot & \cdots & \cdot \\ \cdot & \cdot & \cdot & \cdots & \cdot \\ W_1 & W_2 & W_3 & \cdots & W_0 \end{bmatrix}$$

for W_n, we can give the following theorem.

Theorem 8 *The eigenvalues of W are*

$$\lambda_j(W) = \frac{\begin{Bmatrix} W_{n-1}w^{-3j} + (W_n + P_{n-1} - 2F_{n-1} + 1)w^{-2j} \\ + (P_n - 2F_n - W_{n-1})w^{-j} - W_n \end{Bmatrix}}{w^{-4j} + 3w^{-3j} - 3w^{-j} + 1}$$

for $j = 0, 1, 2, \ldots, n-1$.

Proof Applying (11) we easily get

$$\begin{aligned}
\lambda_j(W) &= \sum_{k=0}^{n-1} W_k w^{-jk} = \sum_{k=0}^{n-1} \left(\frac{\gamma^k - \delta^k}{\gamma - \delta} - \frac{\alpha^k - \beta^k}{\alpha - \beta} \right) w^{-jk} \\
&= \frac{1}{\gamma - \delta} \left[\frac{\gamma^n - 1}{\gamma w^{-j} - 1} - \frac{\delta^n - 1}{\delta w^{-j} - 1} \right] - \frac{1}{\alpha - \beta} \left[\frac{\alpha^n - 1}{\alpha w^{-j} - 1} - \frac{\beta^n - 1}{\beta w^{-j} - 1} \right] \\
&= \frac{1}{\gamma - \delta} \left[\frac{(\gamma^n - 1)(\delta w^{-j} - 1) - (\delta^n - 1)(\gamma w^{-j} - 1)}{(\gamma w^{-j} - 1)(\delta w^{-j} - 1)} \right] \\
&\quad - \frac{1}{\alpha - \beta} \left[\frac{(\alpha^n - 1)(\beta w^{-j} - 1) - (\beta^n - 1)(\alpha w^{-j} - 1)}{(\alpha w^{-j} - 1)(\beta w^{-j} - 1)} \right] \\
&= \frac{1}{\gamma - \delta} \left[\frac{w^{-j}(\gamma^n \delta - \delta^n \gamma + \gamma - \delta) + \delta^n - \gamma^n}{\delta \gamma w^{-2j} - w^{-j}(\delta + \gamma) + 1} \right] \\
&\quad - \frac{1}{\alpha - \beta} \left[\frac{w^{-j}(\alpha^n \beta - \beta^n \alpha + \alpha - \beta) + \beta^n - \alpha^n}{\beta \alpha w^{-2j} - w^{-j}(\beta + \alpha) + 1} \right] \\
&= \frac{\begin{Bmatrix} w^{-3j}[\sqrt{5}(\delta - \gamma + \gamma\delta^n - \delta\gamma^n) + 2\sqrt{2}(\alpha - \beta + \alpha^n\beta - \alpha\beta^n)] \\ + w^{-2j}[\sqrt{5}(\gamma^n - \delta^n + \delta - \gamma + \gamma\delta^n - \gamma^n\delta) + 2\sqrt{2}(\beta^n - \alpha^n) \\ + 4\sqrt{2}(\alpha - \beta + \alpha^n\beta - \alpha\beta^n)] + w^{-j}[\sqrt{5}(\gamma^n - \delta^n + \gamma - \delta \\ + \gamma^n\delta - \gamma\delta^n) + 2\sqrt{2}(\beta - \alpha + \beta^n\alpha - \alpha^n\beta) + 4\sqrt{2}(\beta^n - \alpha^n)] \\ + [\sqrt{5}(\delta^n - \gamma^n) + 2\sqrt{2}(\alpha^n - \beta^n)] \end{Bmatrix}}{2\sqrt{10}(w^{-4j} + 3w^{-3j} - 3w^{-j} + 1)} \\
&= \frac{\begin{Bmatrix} W_{n-1}w^{-3j} + (W_n + P_{n-1} - 2F_{n-1} + 1)w^{-2j} \\ + (P_n - 2F_n - W_{n-1})w^{-j} - W_n \end{Bmatrix}}{w^{-4j} + 3w^{-3j} - 3w^{-j} + 1},
\end{aligned}$$

since $\alpha\beta = -1$, $\gamma\delta = -1$, $\alpha + \beta = 1$, $\alpha - \beta = \sqrt{5}$, $\gamma + \delta = 2$, and $\gamma - \delta = 2\sqrt{2}$. □

After all, we consider the spectral norm of W. Let $n = 2$. Then $W_2 = [0]_{2\times 2}$. So $\| W_2 \|_{\text{spec}} = 0$. Similarly for $n = 3$, we get

$$W_3 = \begin{bmatrix} 0 & 0 & 1 \\ 1 & 0 & 0 \\ 0 & 1 & 0 \end{bmatrix}$$

and hence $W_3^H W_3 = I_3$. So $\| W_3 \|_{\text{spec}} = 1$. For $n \geq 4$, the spectral norm of W_n is given by the following theorem, which can be proved by induction on n.

Theorem 9 *The spectral norm of* W_n *is*

$$\| W_n \|_{\text{spec}} = \frac{W_{n-1} + 4W_{n-2} + 4W_{n-3} + W_{n-4} + 1}{2}$$

for $n \geq 4$.

For example, let $n = 6$. Then the eigenvalues of $W_6^H W_6$ are

$$\lambda_0 = 1{,}369, \qquad \lambda_1 = 289, \qquad \lambda_2 = \lambda_4 = 784 \quad \text{and} \quad \lambda_3 = \lambda_5 = 388.$$

So the spectral norm is $\| W_6 \|_{\text{spec}} = \sqrt{\lambda_0} = 37$. Also $\frac{W_5+4W_4+4W_3+W_2+1}{2} = 37$. Consequently,

$$\| W_6 \|_{\text{spec}} = \frac{W_5 + 4W_4 + 4W_3 + W_2 + 1}{2} = 37$$

as we claimed.

Competing interests

The author declares that they have no competing interests.

Acknowledgements

The author wishes to thank Professor Ahmet Tekcan of Uludag University for constructive suggestions.

References

1. Conway, JH, Guy, RK: Fibonacci numbers. In: The Book of Numbers. Springer, New York (1996)
2. Hilton, P, Holton, D, Pedersen, J: Fibonacci and Lucas numbers. In: Mathematical Reflections in a Room with Many Mirrors, Chapter 3. Springer, New York (1997)
3. Koshy, T: Fibonacci and Lucas Numbers with Applications. Wiley, New York (2001)
4. Niven, I, Zuckerman, HS, Montgomery, HL: An Introduction to the Theory of Numbers, 5th edn. Wiley, New York (1991)
5. Ogilvy, CS, Anderson, JT: Fibonacci numbers. In: Excursions in Number Theory, Chapter 11. Dover, New York (1988)
6. Ribenboim, P: My Numbers, My Friends: Popular Lectures on Number Theory. Springer, New York (2000)
7. Taşcı, D: On quadrapell numbers and quadrapell polynomials. Hacet. J. Math. Stat. **38**(3), 265-275 (2009)
8. Tekcan, A, Özkoç, A, Engür, M, Özbek, ME: On algebraic identities on a new integer sequence with four parameters. Ars Comb. (accepted)

14

Alternate control delayed systems

Yuming Feng[1,2], Dan Tu[3], Chuandong Li[1*] and Tingwen Huang[4]

*Correspondence: licd@cqu.edu.cn
[1]School of Electronic Information Engineering, Southwest University, Chongqing, 400715, P.R. China
Full list of author information is available at the end of the article

Abstract

In the previous paper (Feng *et al.* in Adv. Differ. Equ. 2014:305, 2014), we have already used sandwich control to control a system. But when we considered the influence of delay, can sandwich control also be applied in the delayed system? In order to answer this question, we first introduce the alternate delayed system, then we study the exponential stability of delayed chaotic neural networks by means of alternate control. Some sufficient conditions are given in terms of a set of linear matrix inequalities to ensure the exponential stability of the system. Numerical simulations are presented to verify the correction of the obtained results.

Keywords: alternate control delayed system; globally exponential stabilization; Lu oscillator

1 Introduction

Alternate control [1] is a special case of switching control [2] and is a generalization of intermittent control [3, 4]. In an alternate control system, two different controls are applied alternately. So there is not *rest time* for the control. This system is suitable for the case in which the time is precious.

In [1] Feng *et al.* studied the alternate control system without delay. They have obtained some conditions in terms of LIMs to ensure the stability of the non-delayed system. For delayed systems [5–8], we know that the methods used are different from the ones without delay. There are many papers about delayed system [9–11]. A delayed system is much more difficult to study than the non-delayed one, we are trying to get some conditions to ensure the stability of the delayed system in the theory of control [12–14].

In this paper, we consider the influence of the delay of the system by means of alternate control, that is to say, we study the delayed system by means of alternate control. First of all, we introduce an alternate delayed system. Then we investigate the stability of it by constructing a Lyapunov function, and we obtain stability conditions in terms of LMIs. Lastly we study the stability of Lu oscillator by using the results obtained in the paper.

2 Problem formulation and preliminaries

Consider a class of delayed nonlinear systems described by

$$\begin{cases} \dot{x}(t) = Ax(t) + f(x(t)) + g(x(t-\tau)) + u(t), & t > 0, \\ x(t) = \phi(t), & t \in [-\tau, 0], \end{cases} \tag{1}$$

where $x \in R^n$ presents state vector, f and g are continuous nonlinear functions of $R^n \to R^n$ with $f(0) = g(0) = 0$ and there exist two diagonal matrices $L_1 = \operatorname{diag}(l_1^{(1)}, l_2^{(1)}, \ldots, l_n^{(1)}) \geq 0$

and $L_2 = \mathrm{diag}(l_1^{(2)}, l_2^{(2)}, \ldots, l_n^{(2)}) \geq 0$ such that $\|f(x)\|^2 \leq x^T L_1 x$ and $\|g(x)\|^2 \leq x^T L_2 x$ for any $x \in R^n$, $A \in R^{n\times n}$ is a constant matrix, ϕ is a function of $R^n \to R^n$, $u(t)$ denotes the external input of system (1).

For stabilizing the origin of system (1) by means of periodically alternate control, we assume that the control imposed on the system is of the following form:

$$u(t) = \begin{cases} K_1 x(t), & mT < t \leq mT + \theta, \\ K_2 x(t), & mT + \theta < t \leq (m+1)T, \end{cases} \tag{2}$$

where $K_1, K_2 \in R^{n\times n}$ are constant matrices, $T > 0$ denotes the control period, $\theta \in (0, T)$ is a constant.

Our target is to design suitable T, θ, K_1 and K_2 such that system (1) can be stabilized at the origin.

By the control law (2), system (1) can be rewritten as follows with $m = 0, 1, 2, \ldots$:

$$\begin{cases} \dot{x}(t) = Ax(t) + f(x(t)) + g(x(t-\tau)) + K_1 x(t), & mT < t \leq mT + \theta, \\ \dot{x}(t) = Ax(t) + f(x(t)) + g(x(t-\tau)) + K_2 x(t), & mT + \theta < t \leq (m+1)T, \\ x(t) = \phi(t), \quad t \in [-\tau, 0]. \end{cases} \tag{3}$$

It is obvious that system (3) is a classical switched system where the switching rule only depends on the time. Specifically, the switching rule of system (3) depends on T and θ.

In the sequel, we will use the following definitions and lemmas.

Lemma 1 (Sanchez and Perez [15]) *Given any real matrices Σ_1, Σ_2, Σ_3 of appropriate dimensions and a scalar $\epsilon \geq 0$ such that $0 < \Sigma_3 = \Sigma_3^T$, the following inequality holds:*

$$\Sigma_1^T \Sigma_2 + \Sigma_2^T \Sigma_1 \leq \epsilon \Sigma_1^T \Sigma_3 \Sigma_1 + \epsilon^{-1} \Sigma_2^T \Sigma_3^{-1} \Sigma_2. \tag{4}$$

Lemma 2 (Boyd *et al.* [16], Horn and Johnson [17]) *The LMI*

$$\begin{bmatrix} Q(x) & S(x) \\ S^T(x) & R(x) \end{bmatrix} > 0,$$

where $Q(x) = Q^T(x)$, $R(x) = R^T(x)$ and $S(x)$ depend affinely on x, is equivalent to

$$R(x) > 0, \quad Q(x) - S(x)R^{-1}(x)S^T(x) > 0.$$

Definition 1 The zero solution of (1) is said to be globally exponentially stable if there are two constants $M(|\phi|) > 0$, $\gamma > 0$ such that

$$\|x(t)\| \leq M(|\phi|)\exp(-\gamma t), \quad t > 0,$$

where $|\phi| = \sup_{-\tau \leq t \leq 0} \|\phi(t)\|$.

Definition 2 Right-upper Dini's derivative of a function $V : R^+ \times R^n \to R^+$ is defined by

$$D^+ V(t, x(t)) = \lim\sup \frac{1}{h}\left[V(t+h, x(t+h)) - V(t, x(t))\right].$$

Note that $V(x(t))$ or $V(x)$ is short for $V(t, x(t))$.

Lemma 3 (Halany inequality [18]) *Assume that* $\tau > 0$ *and* $\omega : [\mu - \tau, \infty) \to [0, \infty)$ *is a continuous function such that*

$$\dot{\omega}(t) \leq -a\omega(t) + b \sup_{t-\tau\leq\theta\leq t} \omega(\theta)$$

is satisfied for all $t \geq \mu$. *If* $a > b > 0$, *then*

$$\omega(t) \leq \overline{\omega}(\mu) \exp\bigl(-\gamma(t-\mu)\bigr), \quad t \geq \mu,$$

where $\overline{\omega}(t) = \sup_{t-\tau\leq\theta\leq t} \omega(\theta)$ *and* $\gamma > 0$ *is the smallest real root of the equation*

$$a - b\exp(\gamma\tau) = \gamma.$$

Lemma 4 ([3]) *Assume that* $\tau > 0$ *and* $\omega : [\mu - \tau, \infty) \to [0, \infty)$ *is a continuous function such that*

$$\dot{\omega}(t) \leq a\omega(t) + b\omega(t-\tau)$$

is satisfied for all $t \geq \mu$. *If* $a > 0$ *and* $b > 0$, *then*

$$\omega(t) \leq \overline{\omega}(\mu) \exp\bigl(\eta(t-\mu+\tau)\bigr), \quad t \geq \mu,$$

where $\overline{\omega}(t) = \sup_{t-\tau\leq\theta\leq t} \omega(\theta)$ *and* $\eta > 0$ *is the unique root of the equation*

$$a + b\exp(-\eta\tau) = \eta.$$

Throughout this paper, we use P^T, $\lambda_M(P)$ and $\lambda_m(P)$ to denote the transpose, the maximum eigenvalue and the minimum eigenvalue of a square matrix P, respectively. $\|x\|$ is used to denote the Euclidean norm of the vector x. The matrix norm $\|\cdot\|$ is also referred to as the Euclidean norm. We use $P > 0$ ($< 0, \leq 0, \geq 0$) to denote a symmetrical positive (negative, semi-negative, semi-positive) definite matrix P. $f(x(t_1^-))$ is defined by $f(x(t_1^-)) = \lim_{t\to t_1^-} f(x(t))$.

3 Main results

Theorem 1 *If* $\theta > \tau$ *and there exist a symmetric and positive definite matrix* $P \in R^{n\times n}$, *positive scalar constants* $g_1 > 0$, $g_2 > 0$, $q_1 > 0$, $q_2 > 0$, $\epsilon_1 > 0$, $\epsilon_2 > 0$, $\eta_1 > 0$ *and* $\eta_2 > 0$ *such that the following hold*:

(1) $PA + A^TP + PK_1 + K_1^TP + (\epsilon_1 + \eta_1)P^2 + \epsilon_1^{-1}L_1 + g_1P \leq 0$,

(2) $PA + A^TP + PK_2 + K_2^TP + (\epsilon_2 + \eta_2)P^2 + \epsilon_2^{-1}L_1 - g_2P \leq 0$,

(3) $\eta_1^{-1}L_2 - q_1P \leq 0$,

(4) $\eta_2^{-1}L_2 - q_2P \leq 0$,

(5) $g_1 > q_1$ *and* $\gamma(\theta - \tau) - \eta(T - \theta + \tau) > 0$,

where $\gamma > 0$ *is the smallest real root of the equation* $g_1 - q_1\exp(\gamma\tau) = \gamma$ *and* $\eta > 0$ *is the unique root of the equation* $g_2 + q_2\exp(-\eta\tau) = \eta$, *then the origin of system* (3) *is globally exponentially stable, and*

$$\|x(t)\| < \sqrt{\frac{\lambda_M(P)}{\lambda_m(P)}}|\phi| \exp\left(-(\gamma(\theta-\tau)-\eta(T-\theta+\tau))\frac{t-\theta}{2T}\right), \quad t>0,$$

where $|\phi| = \sup_{-\tau\le t\le 0}\|\phi(t)\|$.

Proof Let us construct the following Lyapunov function:

$$V(x(t)) = x^T(x)Px(t), \tag{5}$$

from which we obtain that

$$\lambda_m(P)\|x(t)\|^2 \le V(x(t)) \le \lambda_M(P)\|x(t)\|. \tag{6}$$

If $mT < t \le mT+\theta$, then by (3), (4) and (5) we have that

$$\begin{aligned}
\dot V(x(t)) &= 2x^T(t)P\dot x(t)\\
&= 2x^T(t)P\big[Ax(t) + f(x(t)) + g(x(t-\tau)) + K_1x(t)\big]\\
&= 2x^T(t)PAx(t) + 2x^T(t)Pf(x) + 2x^TPg(x(t-\tau)) + 2x^TPK_1x(t)\\
&= x^T\big[PA + A^TP + PK_1 + K_1^TP\big]x + 2x^TPf(x) + 2x^TPg(x(t-\tau))\\
&\le x^T\big[PA + A^TP + PK_1 + K_1^TP\big]x\\
&\quad + \epsilon_1x^T(t)P^2x(t) + \epsilon_1^{-1}x^T(t)L_1x(t)\\
&\quad + \eta_1x^T(t)P^2x(t) + \eta_1^{-1}x^T(t-\tau)L_2x(t-\tau)\\
&= -g_1V(x(t)) + x^T\big[PA + A^TP + PK_1 + K_1^TP\\
&\quad + (\epsilon_1+\eta_1)P^2 + \epsilon_1^{-1}L_1 + g_1P\big]x + q_1V(x(t-\tau))\\
&\quad + x^T(t-\tau)(\eta_1^{-1}L_2 - q_1P)x(t-\tau)\\
&\le -g_1V(x(t)) + q_1V(x(t-\tau)),
\end{aligned}$$

which implies that

$$V(x(t)) \le \overline{V}(x(mT))\exp(-\gamma(t-mT)), \tag{7}$$

where $\gamma > 0$ is the smallest real root of the equation $g_1 - q_1\exp(\gamma\tau) = \gamma$.

Similarly, if $mT+\theta < t \le (m+1)T$, then we have that

$$\begin{aligned}
\dot V(x) &= 2x^TP\dot x\\
&\le g_2V(x(t)) + x^T\big[PA + A^TP + PK_2 + K_2^TP\\
&\quad + (\epsilon_2+\eta_2)P^2 + \epsilon_2^{-1}L_1 - g_2P\big]x + q_2V(x(t-\tau))\\
&\quad + x^T(t-\tau)(\eta_2^{-1}L_2 - q_2P)x(t-\tau)\\
&\le g_2V(x(t)) + q_2V(x(t-\tau)),
\end{aligned}$$

which implies that

$$V(x(t)) \le \overline{V}(x(mT+\theta))\exp(\eta(t-mT-\theta+\tau)), \tag{8}$$

where $\eta > 0$ is the unique root of the equation $g_2 + q_2\exp(-\eta\tau) = \eta$.

It follows from (7) and (8) that

(1) If $0 < t \le \theta$, then we have that

$$V(x(t)) \le \overline{V}(x(0))\exp(-\gamma t).$$

So

$$\overline{V}(x(\theta)) = \sup_{\theta-\tau\le t\le\theta} V(t) \le \sup_{\theta-\tau\le t\le\theta} (\overline{V}(x(0))\exp(-\gamma t)) = \overline{V}(x(0))\exp(-\gamma(\theta-\tau)).$$

(2) If $\theta < t \le T$, then we have that

$$\begin{aligned} V(x(t)) &\le \overline{V}(x(\theta))\exp(\eta(t-\theta+\tau)) \\ &\le \overline{V}(x(0))\exp(-\gamma(\theta-\tau)+\eta(t-\theta+\tau)). \end{aligned}$$

So

$$\overline{V}(x(T)) \le \overline{V}(x(0))\exp(-\gamma(\theta-\tau)+\eta(T-\theta+\tau)).$$

(3) If $T < t \le T+\theta$, then we have that

$$\begin{aligned} V(x(t)) &\le \overline{V}(x(T))\exp(-\gamma(t-T)) \\ &\le \overline{V}(x(0))\exp(-\gamma(t-T)-\gamma(\theta-\tau)+\eta(T-\theta+\tau)). \end{aligned}$$

So

$$\overline{V}(x(T+\theta)) \le \overline{V}(x(0))\exp(-\gamma(2\theta-2\tau)+\eta(T-\theta+\tau)).$$

(4) If $T+\theta < t \le 2T$, then we have that

$$\begin{aligned} V(x(t)) &\le \overline{V}(x(T+\theta))\exp(\eta(t-T-\theta+\tau)) \\ &\le \overline{V}(x(0))\exp(\eta(t-T-\theta+\tau)-\gamma(2\theta-2\tau)+\eta(T-\theta+\tau)). \end{aligned}$$

So

$$\overline{V}(x(2T)) \le \overline{V}(x(0))\exp(2\eta(T-\theta+\tau)-2\gamma(\theta-\tau)).$$

(5) If $2T < t \le 2T+\theta$, then we have that

$$\begin{aligned} V(x(t)) &\le \overline{V}(x(2T))\exp(-\gamma(t-2T)) \\ &\le \overline{V}(x(0))\exp(-\gamma(t-2T)+2\eta(T-\theta+\tau)-\gamma(2\theta-2\tau)). \end{aligned}$$

So

$$\overline{V}(x(2T+\theta) \le \overline{V}(x(0)) \exp(2\eta(T-\theta+\tau) - 3\gamma(\theta-\tau)).$$

(6) If $2T+\theta < t \le 3T$, then we have that

$$\begin{aligned} V(x(t)) &\le \overline{V}(x(2T+\theta)) \exp(\eta(t-2T-\theta+\tau)) \\ &\le \overline{V}(x(0)) \exp(\eta(t-2T-\theta+\tau) + 2\eta(T-\theta+\tau) - 3\gamma(\theta-\tau)). \end{aligned}$$

So

$$\overline{V}(x(3T)) \le \overline{V}(x(0)) \exp(3\eta(T-\theta+\tau) - 3\gamma(\theta-\tau)).$$

By induction, we have that

(7) If $mT < t \le mT+\theta$, *i.e.*, $\frac{t-\theta}{T} < m \le \frac{t}{T}$, then we have that

$$V(x(t)) \le \overline{V}(x(0)) \exp(-\gamma(t-m(T-\theta+\tau)) + m\eta(T-\theta+\tau)). \tag{9}$$

(8) If $mT+\theta < t \le (m+1)T$, *i.e.*, $\frac{t}{T} < m+1 \le \frac{t+T-\theta}{T}$, then we have that

$$\begin{aligned} V(x(t)) &\le \overline{V}(x(0)) \exp(\eta(t-mT-\theta+\tau) + m\eta(T-\theta+\tau) - (m+1)\gamma(\theta-\tau)) \\ &= \overline{V}(x(0)) \exp(-\gamma(m+1)(\theta-\tau) + \eta(t-(m+1)(\theta-\tau))). \end{aligned} \tag{10}$$

From (9) we know that

$$\begin{aligned} V(x(t)) &\le \overline{V}(x(0)) \exp(-\gamma(t-m(T-\theta+\tau)) + m\eta(T-\theta+\tau)) \\ &\le \overline{V}(x(0)) \exp(-\gamma(mT-m(T-\theta+\tau)) + m\eta(T-\theta+\tau)) \\ &= \overline{V}(x(0)) \exp(-(\gamma(\theta-\tau) - \eta(T-\theta+\tau))m) \\ &< \overline{V}(x(0)) \exp\left(-(\gamma(\theta-\tau) - \eta(T-\theta+\tau))\frac{t-\theta}{T}\right), \end{aligned} \tag{11}$$

where $mT < t \le mT+\tau$.

From (10) we know that

$$\begin{aligned} V(x(t)) &\le \overline{V}(x(0)) \exp(-\gamma(m+1)(\theta-\tau) + \eta(t-(m+1)(\theta-\tau))) \\ &\le \overline{V}(x(0)) \exp(-\gamma(m+1)(\theta-\tau) + \eta((m+1)T-(m+1)(\theta-\tau))) \\ &= \overline{V}(x(0)) \exp(-(\gamma(\theta-\tau) - \eta(T-\theta+\tau))(m+1)) \\ &< \overline{V}(x(0)) \exp\left(-(\gamma(\theta-\tau) - \eta(T-\theta+\tau))\frac{t}{T}\right) \\ &\le \overline{V}(x(0)) \exp\left(-(\gamma(\theta-\tau) - \eta(T-\theta+\tau))\frac{t-\theta}{T}\right), \end{aligned} \tag{12}$$

where $mT+\tau < t \le (m+1)T$.

It follows from (11) and (12) that, for any $t > 0$,

$$V(x(t)) < \overline{V}(x(0)) \exp\left(-(\gamma(\theta - \tau) - \eta(T - \theta + \tau))\frac{t - \theta}{T}\right). \tag{13}$$

By (5), (6) and (13), we conclude that

$$\|x(t)\| < \sqrt{\frac{\lambda_M(P)}{\lambda_m(P)}} |\phi| \exp\left(-(\gamma(\theta - \tau) - \eta(T - \theta + \tau))\frac{t - \theta}{2T}\right), \quad t > 0,$$

where $|\phi| = \sup_{-\tau \le t \le 0} \|\phi(t)\|$.

So we finish the proof. □

From Lemma 2, we know that the two conditions of Theorem 1 are equivalent to the following two LMIs, respectively:

$$\begin{bmatrix} PA + A^T P + PK_1 + K_1^T P + \epsilon_1^{-1} L_1 + g_1 P & -P \\ -P & -(\epsilon_1 + \eta_1)^{-1} I \end{bmatrix} \le 0, \tag{14}$$

$$\begin{bmatrix} PA + A^T P + PK_2 + K_2^T P + \epsilon_2^{-1} L_2 - g_2 P & -P \\ -P & -(\epsilon_2 + \eta_2)^{-1} I \end{bmatrix} \le 0. \tag{15}$$

Corollary 1 *If $\theta > \tau$ and there exist a symmetric and positive definite matrix $P \in R^{n \times n}$, positive scalar constants $\epsilon_1 > 0$, $\epsilon_2 > 0$, $\eta_1 > 0$, $\eta_2 > 0$, $q_1 > 0$, $q_2 > 0$ and $\eta > 0$ such that the following hold:*

(1) $PA + A^T P + PK_1 + K_1^T P + (\epsilon_1 + \eta_1)P^2 + \epsilon_1^{-1} L_1 + g_1 P \le 0$, *where* $g_1 = \gamma + q_1 \exp(\gamma\tau)$ *and* $\gamma = \frac{\eta(T - \theta + \tau)}{\theta - \tau} + q_1$,

(2) $PA + A^T P + PK_2 + K_2^T P + (\epsilon_2 + \eta_2)P^2 + \epsilon_2^{-1} L_1 - g_2 P \le 0$, *where* $g_2 = \eta - q_2 \exp(-\eta\tau) > 0$,

(3) $\eta_1^{-1} L_2 - q_1 P \le 0$,

(4) $\eta_2^{-1} L_2 - q_2 P \le 0$, *then the origin of system* (3) *is globally exponentially stable, and*

$$\|x(t)\| < \sqrt{\frac{\lambda_M(P)}{\lambda_m(P)}} |\phi| \exp\left(-(\gamma(\theta - \tau) - \eta(T - \theta + \tau))\frac{t - \theta}{2T}\right), \quad t > 0,$$

where $|\phi| = \sup_{-\tau \le t \le 0} \|\phi(t)\|$.

Proof In fact, the previous four conditions can imply

$$g_1 > q_1$$

and

$$\gamma(\theta - \tau) - \eta(T - \theta + \tau) > 0.$$

From condition (1) we know

$$g_1 = \gamma + q_1 \exp(\gamma\tau) = \frac{\eta(T - \theta + \tau)}{\theta - \tau} + q_1 + q_1 \exp(\gamma\tau) > q_1$$

and

$$\gamma = \frac{\eta(T-\theta+\tau)}{\theta-\tau} + q_1 > \frac{\eta(T-\theta+\tau)}{\theta-\tau},$$

which implies

$$\gamma(\theta-\tau) - \eta(T-\theta+\tau) > 0.$$

Thus, the fifth condition of Theorem 1 is valid. So the proof is completed. □

Remark 1 In order to judge the global exponential stability of system (3), Corollary 1 needs to determine the existence of a symmetric and positive definite matrix $P \in R^{n\times n}$ and seven positive scalar constants ϵ_1, ϵ_2, η_1, η_2, q_1, q_2 and η by the four linear matrix inequalities listed in it, while Theorem 1 has to determine the existence of a symmetric and positive definite matrix $P \in R^{n\times n}$ and eight positive scalar constants ϵ_1, ϵ_2, η_1, η_2, q_1, q_2, g_1 and g_2 by the five conditions of it. From this view of point, Corollary 1 is more applicative than Theorem 1.

4 Numerical example

Consider the neural oscillator model described by the following delayed differential equation:

$$\dot{x}(t) = Ax(t) + f(x(t)) + g(x(t-1)), \tag{16}$$

where

$$A = \begin{pmatrix} -1 & 0 \\ 0 & -1 \end{pmatrix}, \quad f(x(t)) = \begin{pmatrix} 2 & -0.1 \\ -5 & 3 \end{pmatrix} \begin{pmatrix} \tanh x_1(t) \\ \tanh x_2(t) \end{pmatrix}$$

and

$$g(x(t-1)) = \begin{pmatrix} -1.5 & -0.1 \\ -0.2 & -2.5 \end{pmatrix} \begin{pmatrix} \tanh x_1(t-1) \\ \tanh x_2(t-1) \end{pmatrix}.$$

This model was named *Lu oscillator* [19] and it is shown to be chaotic as in Figure 1. The time response curves are shown in Figure 2.

It is easy to obtain that

$$\begin{aligned} \|f(x(t))\|^2 &= f^T(x(t))f(x(t)) \\ &= \begin{pmatrix} \tanh x_1(t) \\ \tanh x_2(t) \end{pmatrix}^T \begin{pmatrix} 2 & -0.1 \\ -5 & 3 \end{pmatrix}^T \begin{pmatrix} 2 & -0.1 \\ -5 & 3 \end{pmatrix} \begin{pmatrix} \tanh x_1(t) \\ \tanh x_2(t) \end{pmatrix} \\ &\leq x^T L_1 x, \end{aligned}$$

where $L_1 = \begin{pmatrix} 29 & -15.2 \\ -15.2 & 9.01 \end{pmatrix}$.

Similarly, we can get that $\|g(x(t))\|^2 \leq x^T L_2 x$, where $L_2 = \begin{pmatrix} 2.29 & 0.65 \\ 0.65 & 6.26 \end{pmatrix}$.

Next, we will use Theorem 1 to judge the global exponential stability of system (16).

Figure 1 Chaotic behavior of Lu oscillator determined by system (16) with the initial values $x(\theta) = (2, -1)^T$, where $\theta \in [-1, 0]$.

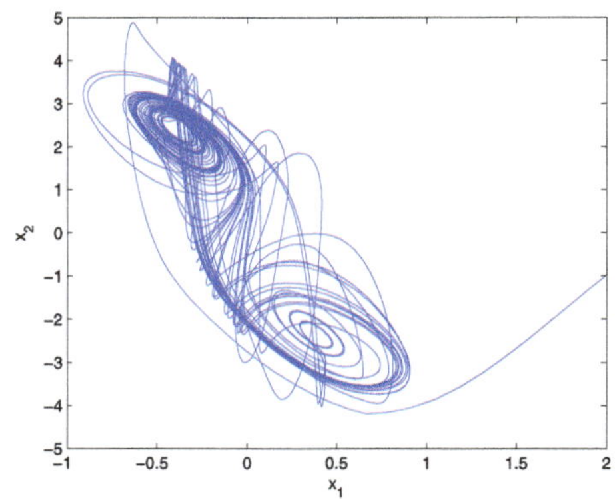

Figure 2 Time response curves of Lu oscillator without control when the initial values are $x(\theta) = (2, -1)^T$, where $\theta \in [-1, 0]$.

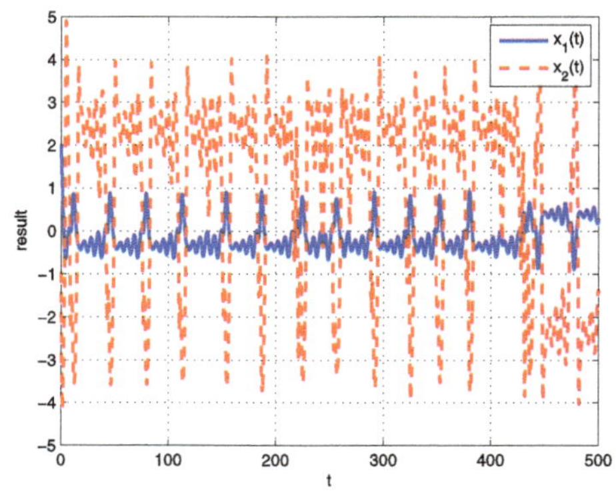

Choosing

$$K_1 = \text{diag}(-25, -25),$$
$$K_2 = \text{diag}(-10, -10),$$

with $T = 3$ and $\theta = 1.5$, solving LMIs (14), (15), $\eta_1^{-1} L_2 - q_1 P \leq 0$, $\eta_2^{-1} L_2 - q_2 P \leq 0$ and inequalities $g_1 > q_1$, $\gamma(\theta - 1) - \eta(T - \theta + 1) > 0$, where $\gamma > 0$ is the smallest real root of the equation $g_1 - q_1 \exp(\gamma) = \gamma$ and $\eta > 0$ is the unique root of the equation $g_2 + q_2 \exp(-\eta) = \eta$, we obtain a feasible solution:

$$\epsilon_1 = 8.6, \quad \epsilon_2 = 0.6, \quad \eta_1 = 9, \quad \eta_2 = 11, \quad g_1 = 28.2703, \quad g_2 = 0.3211,$$
$$q_1 = 0.72, \quad q_2 = 0.63,$$

and

$$P = \begin{bmatrix} 0.9498 & 0.0049 \\ 0.0049 & 1.0066 \end{bmatrix}.$$

Thus by Theorem 1 we obtain that the origin of system (3) is globally exponentially stable. The time response curves of Lu oscillator with alternate control are shown in Figure 3, while Figure 4 shows the corresponding control signal.

In the following, we will apply Corollary 1 to determine the global exponential stability of system (16).

Figure 3 Time response curves of Lu oscillator under alternate control with $T = 3$, $\theta = 1.5$, $K_1 = \mathrm{diag}(-10,-10)$, $K_2 = \mathrm{diag}(-6,-6)$ and the initial values $x(\theta) = (2,-1)^T$, where $\theta \in [-1,0]$.

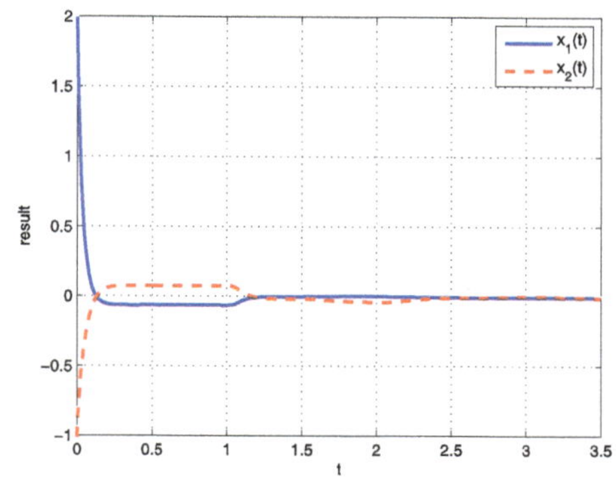

Figure 4 Control signal for Lu oscillator of $T = 3$, $\theta = 1.5$, $K_1 = \mathrm{diag}(-10,-10)$ and $K_2 = \mathrm{diag}(-6,-6)$.

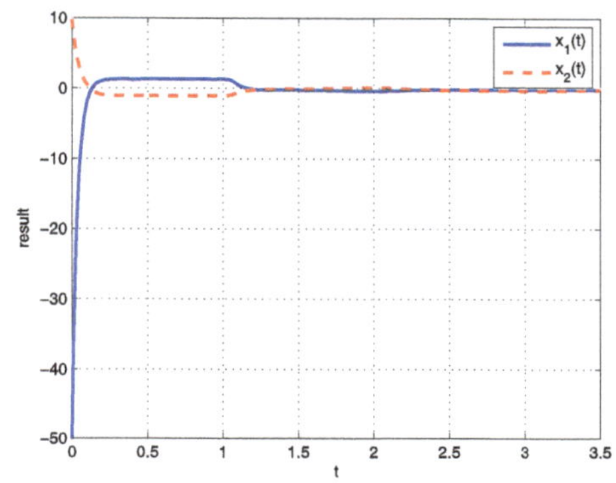

Choosing

$$K_1 = \mathrm{diag}(-15,-15),$$
$$K_2 = \mathrm{diag}(-9,-9),$$

with $T = 3$ and $\theta = 1.5$, solving LMIs (14), (15), $\eta_1^{-1}L_2 - q_1P \le 0$ and $\eta_2^{-1}L_2 - q_2P \le 0$, where $g_1 = \gamma + q_1 \exp(\gamma\tau)$ and $\gamma = \frac{\eta(T-\theta+\tau)}{\theta-\tau} + q_1$ and $g_2 = \eta - q_2 \exp(-\eta\tau) > 0$, we obtain a feasible solution:

$$\epsilon_1 = 4.6, \quad \epsilon_2 = 0.6, \quad \eta_1 = 9, \quad \eta_2 = 11,$$
$$q_1 = 0.52, \quad q_2 = 0.43, \quad \eta = 0.45$$

and

$$P = \begin{bmatrix} 1.2212 & 0.0532 \\ 0.0532 & 1.3590 \end{bmatrix}.$$

Thus by Corollary 1 we obtain that the origin of system (16) is globally exponentially stable. The time response curves of Lu oscillator with alternate control are shown in Figure 5. The control signal is shown in Figure 6.

Figure 5 Time response curves of Lu oscillator under alternate control with $T = 3$, $\theta = 1.5$, $K_1 = \text{diag}(-15,-15)$, $K_2 = \text{diag}(-11,-11)$ and the initial values $x(\theta) = (2,-1)^T$, where $\theta \in [-1,0]$.

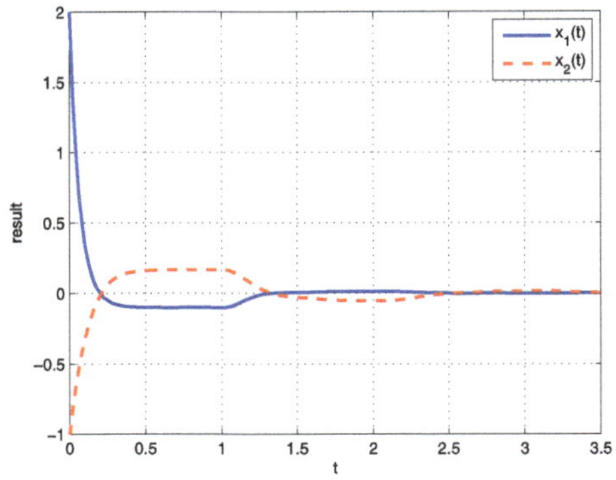

Figure 6 Control signal for Lu oscillator of $T = 3$, $\theta = 1.5$, $K_1 = \text{diag}(-15,-15)$ and $K_2 = \text{diag}(-11, -11)$.

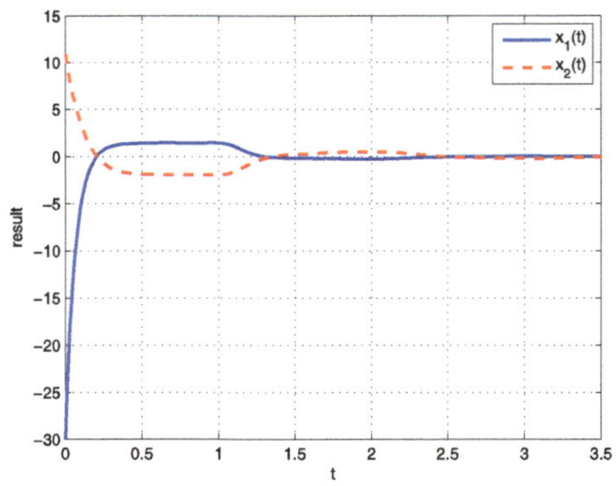

5 Conclusions

This paper studies the delayed system by using alternate control method. Some conditions to ensure the stability of the system are given in terms of linear matrix inequalities. By the results obtained, the Lu oscillate is controlled.

Competing interests

The authors declare that they have no competing interests.

Authors' contributions

The ideal of *alternate control delayed system* was proposed by CL and YF. The main theory was proved by YF and DT. The paper was typed by YF and TH and all the figures were provided by TH. All authors read and approved the final manuscript.

Author details

[1]School of Electronic Information Engineering, Southwest University, Chongqing, 400715, P.R. China. [2]School of Mathematics and Statistics, Chongqing Three Gorges University, Wanzhou, Chongqing, 404100, P.R. China. [3]Department of Mathematics, Texas A&M University at Qatar, P.O. Box 23874, Doha, Qatar. [4]School of Physical Education, Southwest University, Chongqing, 400715, P.R. China.

Acknowledgements

This research is supported by the Natural Science Foundation of China (Grant No: 61374078), NPRP Grant No. NPRP 4-1162-1-181 from the Qatar National Research Fund (a member of Qatar Foundation), Scientific & Technological Research Foundation of Chongqing Municipal Education Commission (Grant Nos. KJ1401006, KJ1401019) and the Fundamental Research Funds for the Central Universities (Grant No. XDJK2015D004).

References

1. Feng, Y, et al.: Alternate control systems. Adv. Differ. Equ. **2014**, 305 (2014)
2. Tanwani, A, Shim, H, Liberzon, D: Observability for switched linear systems: characterization and observer design. IEEE Trans. Autom. Control **58**(4), 891-904 (2013)
3. Huang, J, Li, C, Han, Q: Stabilization of delayed chaotic neural networks by periodically intermittent control. Circuits Syst. Signal Process. **28**, 567-579 (2009)

4. Huang, T, Li, C, Liu, X: Synchronization of chaotic systems with delay using intermittent linear state feedback. Chaos **18**, 033122 (2008)
5. Xia, W, Cao, J: Pinning synchronization of delayed dynamical networks via periodically intermittent control. Chaos **19**, 013120 (2009). doi:10.1063/1.3071933
6. Yang, X, Cao, J: Stochastic synchronization of coupled neural networks with intermittent control. Phys. Lett. A **373**(36), 3259-3272 (2009)
7. Yang, X: Can neural networks with arbitrary delays be finite-timely synchronized? Neurocomputing **143**, 275-281 (2014)
8. Zheng, G, Cao, J: Robust synchronization of coupled neural networks with mixed delays and uncertain parameters by intermittent pinning control. Neurocomputing **141**(2), 153-159 (2014)
9. El'sgol'ts, LE, Norkin, SB: Introduction to the Theory and Application of Differential Equations with Deviating Arguments. Academic Press, New York (1973)
10. Liao, X, Wang, L, Yu, P: Stability of Dynamical Systems. Monograph Series on Nonlinear Science and Complexity. Elsevier, Amsterdam (2007)
11. Liao, X, Yu, P: Absolute Stability of Nonlinear Control Systems. Springer, New York (2008)
12. Yu, J, Liu, L, Wang, L, Tan, M, Xu, D: Turning control of a multilink biomimetic robotic fish. IEEE Trans. Robot. **24**(1), 201-206 (2008)
13. Shen, Y, Xu, D, Tan, M, Yu, J: Mixed visual control method for robots with self-calibrated stereo rig. IEEE Trans. Instrum. Meas. **59**(2), 470-479 (2010)
14. Dong, D, Petersen, IR: Quantum control theory and applications: a survey. IET Control Theory Appl. **4**(12), 2651-2671 (2010)
15. Sanchez, EN, Perez, JP: Input-to-state stability (ISS) analysis for dynamic NN. IEEE Trans. Circuits Syst. I, Regul. Pap. **46**(11), 1395-1398 (1999)
16. Boyd, S, Ghaoui, L, Feron, EEI, Balakrishnan, V: Linear matrix inequalities in system and control theory. In: Linear Matrix Inequalities in System and Control Theory. SIAM, Philadephia (1994)
17. Horn, R, Johnson, C: Matrix Analysis. Cambridge University Press, Cambridge (1985)
18. Halanay, A: Differential Equations: Stability, Oscillations, Time Lags. Academic Press, New York (1966)
19. Lu, H: Chaotic attractors in delayed neural networks. Phys. Lett. A **298**, 109-116 (2002)

Existence results for multi-term fractional differential inclusions

Sotiris K Ntouyas[1,2], Sina Etemad[3] and Jessada Tariboon[4*]

*Correspondence:
jessadat@kmutnb.ac.th
[4]Nonlinear Dynamic Analysis Research Center, Department of Mathematics, Faculty of Applied Science, King Mongkut's University of Technology North Bangkok, Bangkok, 10800, Thailand
Full list of author information is available at the end of the article

Abstract

In this paper, we study the existence of solutions for a new class of boundary value problems for nonlinear multi-term fractional differential inclusions. Our main result relies on the multi-valued form of Krasnoselskii's fixed point theorem. An illustrative example is also presented.

MSC: 34A08; 34A60; 34B10; 34B15

Keywords: fractional differential inclusion; boundary value problem; existence; fixed point theorem

1 Introduction and preliminaries

In this paper we study the existence of solutions for the following multi-term fractional differential inclusions:

$$\begin{aligned} {}^cD^\alpha u(t) \in{} & F\big(t, u(t), u'(t), u''(t), {}^cD^{q_1}u(t), \dots, {}^cD^{q_k}u(t)\big) \\ & + G\big(t, u(t), u'(t), u''(t), {}^cD^{q_1}u(t), \dots, {}^cD^{q_k}u(t)\big) \end{aligned} \tag{1.1}$$

supplemented with boundary conditions

$$u(0) = 0, \qquad u'(0) = -u(1) - u'(1), \qquad u''(0) = -u''(1) - {}^cD^p u(1), \tag{1.2}$$

where ${}^cD^\alpha$, ${}^cD^{q_i}$ denote the Caputo fractional derivatives, $2 < \alpha \leq 3$, $1 < q_i \leq 2$, $i = 1, 2, \dots, k$, $t \in J := [0,1]$, $1 < p \leq 2$, $k \geq 1$, and $F, G : J \times \mathbb{R}^{k+3} \to \mathcal{P}(\mathbb{R})$ are multifunctions.

Many of published papers about fractional differential equations and inclusions apply the fixed point theory for proving the existence results. For instance, one can find a lot of papers in this field (see [1–25] and the references therein).

Let $\alpha > 0$, $n - 1 < \alpha < n$, $n = [\alpha] + 1$, and $u \in C([a,b], \mathbb{R})$. The Caputo derivative of fractional of order α for the function u is defined by ${}^cD^\alpha u(t) = \frac{1}{\Gamma(n-\alpha)} \int_0^t (t-\tau)^{n-\alpha-1} u^{(n)}(\tau)\, d\tau$ (see for more details [11, 23, 25–27]). Also, the Riemann-Liouville fractional order integral of the function u is defined by $I^\alpha u(t) = \frac{1}{\Gamma(\alpha)} \int_0^t \frac{u(\tau)}{(t-\tau)^{1-\alpha}}\, d\tau$ $(t > 0)$ whenever the integral exists [11, 23, 25–27]. In [28], it has been proved that the general solution of the fractional differential equation ${}^cD^\alpha u(t) = 0$ is given by $u(t) = c_0 + c_1 t + c_2 t^2 + \cdots + c_{n-1} t^{n-1}$, where $c_0, \dots, c_{n-1}$ are real constants and $n = [\alpha] + 1$. Also, for each $T > 0$ and $u \in C([0, T])$ we have

$$I^{\alpha c}D^{\alpha}u(t) = u(t) + c_0 + c_1 t + c_2 t^2 + \cdots + c_{n-1}t^{n-1},$$

where $c_0, \ldots, c_{n-1}$ are real constants and $n = [\alpha] + 1$ [28].

Now, we review some definitions and notations as regards multifunctions [29, 30].

For a normed space $(X, \|\cdot\|)$, let $\mathcal{P}_{cl}(X) = \{Y \in \mathcal{P}(X) : Y \text{ is closed}\}$, $\mathcal{P}_b(X) = \{Y \in \mathcal{P}(X) : Y \text{ is bounded}\}$, $\mathcal{P}_{cp}(X) = \{Y \in \mathcal{P}(X) : Y \text{ is compact}\}$, and $\mathcal{P}_{cp,cv}(X) = \{Y \in \mathcal{P}(X) : Y \text{ is compact and convex}\}$, $\mathcal{P}_{b,cl,cv}(X) = \{Y \in \mathcal{P}(X) : Y \text{ is bounded, closed, and convex}\}$. A multi-valued map $G : X \to \mathcal{P}(X)$ is convex (closed) valued if $G(x)$ is convex (closed) for all $x \in X$. The map G is bounded on bounded sets if $G(\mathbb{B}) = \bigcup_{x\in\mathbb{B}} G(x)$ is bounded in X for all $\mathbb{B} \in \mathcal{P}_b(X)$ (*i.e.*, $\sup_{x\in\mathbb{B}}\{\sup\{|y| : y \in G(x)\}\} < \infty$). G is called upper semi-continuous (u.s.c.) on X if for each $x_0 \in X$, the set $G(x_0)$ is a nonempty closed subset of X, and if for each open set N of X containing $G(x_0)$, there exists an open neighborhood $\mathcal{N}_0$ of x_0 such that $G(\mathcal{N}_0) \subseteq N$. G is said to be completely continuous if $G(\mathbb{B})$ is relatively compact for every $\mathbb{B} \in \mathcal{P}_b(X)$. If the multi-valued map G is completely continuous with nonempty compact values, then G is u.s.c. if and only if G has a closed graph, *i.e.*, $u_n \to u_*$, $y_n \to y_*$, $y_n \in G(u_n)$ imply $y_* \in G(u_*)$. G has a fixed point if there is $x \in X$ such that $x \in G(x)$. The fixed point set of the multi-valued operator G will be denoted by $\operatorname{Fix} G$. A multi-valued map $G : J \to \mathcal{P}_{cl}(\mathbb{R})$ is said to be measurable if for every $y \in \mathbb{R}$, the function $t \mapsto d(y, G(t)) = \inf\{|y - z| : z \in G(t)\}$ is measurable.

Consider the Pompeiu-Hausdorff metric $H_d : \mathcal{P}(X) \times \mathcal{P}(X) \to \mathbb{R} \cup \{\infty\}$ given by

$$H_d(A, B) = \max\Big\{\sup_{a\in A} d(a, B), \sup_{b\in B} d(A, b)\Big\},$$

where $d(A, b) = \inf_{a\in A} d(a; b)$ and $d(a, B) = \inf_{b\in B} d(a; b)$. A multi-valued operator $N : X \to \mathcal{P}_{cl}(X)$ is called contraction if there exists $\gamma \in (0, 1)$ such that $H_d(N(x), N(y)) \le \gamma d(x, y)$ for each $x, y \in X$.

We say that $F : J \times \mathbb{R}^{k+3} \to \mathcal{P}(\mathbb{R})$ is a Carathéodory multifunction if $t \mapsto F(t, u_1, \ldots, u_{k+3})$ is measurable for all $u_i \in \mathbb{R}$ and $(u_1, \ldots, u_{k+3}) \mapsto F(t, u_1, \ldots, u_{k+3})$ is upper semi-continuous for almost all $t \in J$ [29, 31]. Also, a Carathéodory multifunction $F : J \times \mathbb{R}^{k+3} \to \mathcal{P}(\mathbb{R})$ is called L^1-Carathéodory if for each $\rho > 0$ there exists $\phi_\rho \in L^1(J, \mathbb{R}^+)$ such that

$$\big\|F(t, u_1, \ldots, u_{k+3})\big\| = \sup_{t\in J}\big\{|s| : s \in F(t, u_1, \ldots, u_{k+3})\big\} \le \phi_\rho(t)$$

for all $|u_1|, \ldots, |u_{k+3}| \le \rho$ and for almost all $t \in J$ [29, 31].

Define the set of selections of F and G at $u \in C(J, \mathbb{R})$ by

$$S_{F,u} := \big\{v \in L^1(J, \mathbb{R}) : v(t) \in F\big(t, u(t), u'(t), u''(t), {}^cD^{q_1}u(t), \ldots, {}^cD^{q_k}u(t)\big)\big\}$$

and

$$S_{G,u} := \big\{v_1 \in L^1(J, \mathbb{R}) : v_1(t) \in G\big(t, u(t), u'(t), u''(t), {}^cD^{q_1}u(t), \ldots, {}^cD^{q_k}u(t)\big)\big\}$$

for almost all $t \in J$. If F is an arbitrary multifunction, then it has been proved that $S_F(u) \neq \emptyset$ for all $u \in C(J, X)$ if $\dim X < \infty$ [32].

The graph of a function F is the set $\mathrm{Gr}(F) = \{(x,y) \in X \times Y : y \in F(x)\}$ [29]. The graph $\mathrm{Gr}(F)$ of $F : X \to \mathcal{P}_{cl}(Y)$ is said to be a closed subset of $X \times Y$, if for every sequence $\{u_n\}_{n\in\mathbb{N}} \subset X$ and $\{y_n\}_{n\in\mathbb{N}} \subset Y$, when $n \to \infty$, $u_n \to u_0$, $y_n \to y_0$, and $y_n \in F(u_n)$, then $y_0 \in F(u_0)$ [29].

We will use the following lemmas and theorem in our main result.

Lemma 1.1 ([29], Proposition 1.2) *If $F : X \to \mathcal{P}_{cl}(Y)$ is u.s.c., then $\mathrm{Gr}(F)$ is a closed subset of $X \times Y$. Conversely, if F is completely continuous and has a closed graph, then it is upper semi-continuous.*

Lemma 1.2 ([32]) *Let X be a separable Banach space. Let $F : [0,1] \times X^{k+3} \to \mathcal{P}_{cp,cv}(X)$ be an L^1-Carathéodory function. Then the operator*

$$\Theta \circ S_F : C(J,X) \to \mathcal{P}_{cp,cv}\big(C(J,X)\big), \qquad x \mapsto (\Theta \circ S_F)(x) = \Theta(S_{F,x})$$

is a closed graph operator.

Theorem 1.3 ([33], Krasnoselskii's fixed point theorem) *Let X be a Banach space, $Y \in \mathcal{P}_{b,cl,cv}(X)$ and $A, B : Y \to \mathcal{P}_{cp,cv}(X)$ two multi-valued operators. If the following conditions are satisfied:*

(i) *$Ay + By \subset Y$ for all $y \in Y$;*
(ii) *A is a contraction;*
(iii) *B is u.s.c. and compact,*

then there exists $y \in Y$ such that $y \in Ay + By$.

2 Main results

Now, we are ready to prove our main result. Let $X = \{u : u, u', u'', {}^cD^{q_i}u \in C(J,\mathbb{R}), i = 1,2,\ldots,k\}$. Then $(X, \|\cdot\|)$ endowed with the norm

$$\|u\| = \sup_{t\in J}|u(t)| + \sup_{t\in J}|u'(t)| + \sup_{t\in J}|u''(t)| + \sup_{t\in J}|{}^cD^{q_i}u(t)| \quad (i = 1,\ldots,k)$$

is a Banach space [34].

We need the following auxiliary lemma. See also [35, 36].

Lemma 2.1 *Let $y \in C(J,\mathbb{R})$ and $u \in C^2([0,1],\mathbb{R})$ is a solution to the fractional boundary value problem*

$$\begin{cases} {}^cD^{\alpha}u(t) = y(t), \\ u(0) = 0, \qquad u'(0) = -u(1) - u'(1), \qquad u''(0) = -u''(1) - {}^cD^{p}u(1), \end{cases} \tag{2.1}$$

then

$$\begin{aligned} u(t) = {} & \int_0^t \frac{(t-s)^{\alpha-1}}{\Gamma(\alpha)} y(s)\,ds - \frac{t}{3}\int_0^1 \frac{(1-s)^{\alpha-1}}{\Gamma(\alpha)} y(s)\,ds - \frac{t}{3}\int_0^1 \frac{(1-s)^{\alpha-2}}{\Gamma(\alpha-1)} y(s)\,ds \\ & + \big(t-t^2\big)\Delta \int_0^1 \frac{(1-s)^{\alpha-3}}{\Gamma(\alpha-2)} y(s)\,ds + \big(t-t^2\big)\Delta \int_0^1 \frac{(1-s)^{\alpha-p-1}}{\Gamma(\alpha-p)} y(s)\,ds, \end{aligned} \tag{2.2}$$

and vice versa, where $\Delta = \frac{\Gamma(3-p)}{4\Gamma(3-p)+2} \neq 0$.

Proof It is well known that the solution of equation ${}^cD^\alpha u(t) = y(t)$ can be written as

$$u(t) = \int_0^t \frac{(t-s)^{\alpha-1}}{\Gamma(\alpha)} y(s)\, ds + c_0 + c_1 t + c_2 t^2, \tag{2.3}$$

where $c_0, c_1, c_2 \in \mathbb{R}$. Then we get

$$u'(t) = \int_0^t \frac{(t-s)^{\alpha-2}}{\Gamma(\alpha-1)} y(s)\, ds + c_1 + 2c_2 t,$$
$$u''(t) = \int_0^t \frac{(t-s)^{\alpha-3}}{\Gamma(\alpha-2)} y(s)\, ds + 2c_2$$

and

$${}^cD^p u(t) = \int_0^t \frac{(t-s)^{\alpha-p-1}}{\Gamma(\alpha-p)} y(s)\, ds + c_2 \frac{2t^{2-p}}{\Gamma(3-p)} \quad (1 < p \leq 2).$$

By using the boundary value conditions, we obtain $c_0 = 0$ and

$$c_1 = -\frac{1}{3}\int_0^1 \frac{(1-s)^{\alpha-1}}{\Gamma(\alpha)} y(s)\, ds - \frac{1}{3}\int_0^1 \frac{(1-s)^{\alpha-2}}{\Gamma(\alpha-1)} y(s)\, ds + \Delta \int_0^1 \frac{(1-s)^{\alpha-3}}{\Gamma(\alpha-2)} y(s)\, ds + \Delta \int_0^1 \frac{(1-s)^{\alpha-p-1}}{\Gamma(\alpha-p)} y(s)\, ds$$

and

$$c_2 = -\Delta \int_0^1 \frac{(1-s)^{\alpha-3}}{\Gamma(\alpha-2)} y(s)\, ds - \Delta \int_0^1 \frac{(1-s)^{\alpha-p-1}}{\Gamma(\alpha-p)} y(s)\, ds.$$

Substituting the values of c_0, c_1, and c_2 in (2.3) we get (2.2).

Conversely, applying the operator ${}^cD^\alpha$ on (2.2) and taking into account (2.1), it follows that ${}^cD^\alpha u(t) = y(t)$. From (2.2) it is easily to verify that the boundary conditions $u(0) = 0$, $u'(0) = -u(1) - u'(1)$, $u''(0) = -u''(1) - {}^cD^p u(1)$ are satisfied. This establishes the equivalence between (2.1) and (2.2). The proof is completed. □

Definition 2.2 A function $u \in C^2([0,1], \mathbb{R})$ is called a solution for the problem (1.1)-(1.2) if it satisfies the boundary value conditions $u(0) = 0$, $u'(0) = -u(1) - u'(1)$, and $u''(0) = -u''(1) - {}^cD^p u(1)$, there exist functions $v, v_1 \in L^1(J, \mathbb{R})$ such that $v(t) \in F(t, u(t), u'(t), u''(t), {}^cD^{q_1}u(t), \ldots, {}^cD^{q_k}u(t))$, $v_1(t) \in G(t, u(t), u'(t), u''(t), {}^cD^{q_1}u(t), \ldots, {}^cD^{q_k}u(t))$ for almost all $t \in J$ and

$$\begin{aligned} u(t) = {} & \int_0^t \frac{(t-s)^{\alpha-1}}{\Gamma(\alpha)} v(s)\, ds - \frac{t}{3}\int_0^1 \frac{(1-s)^{\alpha-1}}{\Gamma(\alpha)} v(s)\, ds \\ & - \frac{t}{3}\int_0^1 \frac{(1-s)^{\alpha-2}}{\Gamma(\alpha-1)} v(s)\, ds + (t-t^2)\Delta \int_0^1 \frac{(1-s)^{\alpha-3}}{\Gamma(\alpha-2)} v(s)\, ds \\ & + (t-t^2)\Delta \int_0^1 \frac{(1-s)^{\alpha-p-1}}{\Gamma(\alpha-p)} v(s)\, ds \\ & + \int_0^t \frac{(t-s)^{\alpha-1}}{\Gamma(\alpha)} v_1(s)\, ds - \frac{t}{3}\int_0^1 \frac{(1-s)^{\alpha-1}}{\Gamma(\alpha)} v_1(s)\, ds \end{aligned}$$

$$-\frac{t}{3}\int_0^1 \frac{(1-s)^{\alpha-2}}{\Gamma(\alpha-1)}v_1(s)\,ds + (t-t^2)\Delta\int_0^1 \frac{(1-s)^{\alpha-3}}{\Gamma(\alpha-2)}v_1(s)\,ds$$
$$+ (t-t^2)\Delta\int_0^1 \frac{(1-s)^{\alpha-p-1}}{\Gamma(\alpha-p)}v_1(s)\,ds. \tag{2.4}$$

Remark 2.3 For the sake of brevity, we set

$$\Lambda_1 = \frac{4}{3\Gamma(\alpha+1)} + \frac{1}{3\Gamma(\alpha)} + \frac{\Delta}{4\Gamma(\alpha-1)} + \frac{\Delta}{4\Gamma(\alpha-p+1)}, \tag{2.5}$$

$$\Lambda_2 = \frac{4}{3\Gamma(\alpha)} + \frac{1}{3\Gamma(\alpha+1)} + \frac{\Delta}{\Gamma(\alpha-1)} + \frac{\Delta}{\Gamma(\alpha-p+1)}, \tag{2.6}$$

$$\Lambda_3 = \frac{1+2\Delta}{\Gamma(\alpha-1)} + \frac{2\Delta}{\Gamma(\alpha-p+1)}, \tag{2.7}$$

and, for each $i = 1, \ldots, k$,

$$\Lambda_4^i = \frac{1}{\Gamma(\alpha-q_i+1)} + \frac{2\Delta}{\Gamma(3-q_i)\Gamma(\alpha-1)} + \frac{2\Delta}{\Gamma(3-q_i)\Gamma(\alpha-p+1)}. \tag{2.8}$$

Also in the following we use the notation $\|x\|_\infty = \sup\{|x(t)| : t \in J\}$.

Theorem 2.4 *Suppose that:*

(H_1) $F : J \times \mathbb{R}^{k+3} \to \mathcal{P}_{cp,c}(\mathbb{R})$ *is a multifunction and* $G : J \times \mathbb{R}^{k+3} \to \mathcal{P}_{cp,c}(\mathbb{R})$ *is a Carathéodory multifunction;*

(H_2) *there exist continuous functions* $p, m : J \to (0,\infty)$ *such that* $t \mapsto F(t, w_1, w_2, w_3, z_1, \ldots, z_k)$ *is measurable and*

$$\|F(t, w_1, w_2, w_3, z_1, \ldots, z_k)\| \le m(t), \qquad \|G(t, w_1, w_2, w_3, z_1, \ldots, z_k)\| \le p(t);$$

(H_3) *there exists a continuous function* $h : J \to (0,\infty)$ *such that*

$$H_d\big(F(t, w_1, w_2, w_3, z_1, \ldots, z_k), F(t, w_1', w_2', w_3', z_1', \ldots, z_k')\big)$$
$$\le h(t)\left[\sum_{i=1}^{3}|w_i - w_i'| + \sum_{i=1}^{k}|z_i - z_i'|\right]$$

for all $t \in J$ *and for each* $w_1, w_2, w_3, z_1, \ldots, z_k, w_1', w_2', w_3', z_1', \ldots, z_k' \in \mathbb{R}$.

If

$$L := \|h\|_\infty(\Lambda_1 + \Lambda_2 + \Lambda_3 + \Lambda_4^i) < 1$$

for $i = 1, 2, \ldots, k$, *where the* Λ_j $(j = 1, \ldots, 4)$ *are defined in* (2.5)-(2.8), *then the inclusion problem* (1.1)-(1.2) *has at least one solution.*

Proof We define the subset Y of X by $Y = \{u \in X : \|u\| \le M\}$, where

$$M = (\|p\|_\infty + \|m\|_\infty)(\Lambda_1 + \Lambda_2 + \Lambda_3 + \Lambda_4^i) \quad (i = 1, \ldots, k).$$

It is clear that Y is closed, bounded, and convex subset of Banach space X. We define the multi-valued operators $A, B : Y \to \mathcal{P}(X)$ such that for some $v \in S_{F,u}$,

$$A(u) = \Big\{u \in X : u(t) = \int_0^t \frac{(t-s)^{\alpha-1}}{\Gamma(\alpha)} v(s)\,ds - \frac{t}{3}\int_0^1 \frac{(1-s)^{\alpha-1}}{\Gamma(\alpha)} v(s)\,ds$$
$$- \frac{t}{3}\int_0^1 \frac{(1-s)^{\alpha-2}}{\Gamma(\alpha-1)} v(s)\,ds + \left(t-t^2\right)\Delta \int_0^1 \frac{(1-s)^{\alpha-3}}{\Gamma(\alpha-2)} v(s)\,ds$$
$$+ \left(t-t^2\right)\Delta \int_0^1 \frac{(1-s)^{\alpha-p-1}}{\Gamma(\alpha-p)} v(s)\,ds\Big\},$$

and for some $v_1 \in S_{G,u}$,

$$B(u) = \Big\{u \in X : u(t) = \int_0^t \frac{(t-s)^{\alpha-1}}{\Gamma(\alpha)} v_1(s)\,ds - \frac{t}{3}\int_0^1 \frac{(1-s)^{\alpha-1}}{\Gamma(\alpha)} v_1(s)\,ds$$
$$- \frac{t}{3}\int_0^1 \frac{(1-s)^{\alpha-2}}{\Gamma(\alpha-1)} v_1(s)\,ds + \left(t-t^2\right)\Delta \int_0^1 \frac{(1-s)^{\alpha-3}}{\Gamma(\alpha-2)} v_1(s)\,ds$$
$$+ \left(t-t^2\right)\Delta \int_0^1 \frac{(1-s)^{\alpha-p-1}}{\Gamma(\alpha-p)} v_1(s)\,ds\Big\}.$$

In this way, the fractional differential inclusion (1.1)-(1.2) is equivalent to the inclusion problem $u \in Au + Bu$. We show that the multi-valued operators A and B satisfy the conditions of Theorem 1.3 on Y.

First, we show that the operators A and B define the multi-valued operators $A, B : Y \to \mathcal{P}_{cp,cv}(X)$. First we prove that A is compact-valued on Y. Note that the operator A is equivalent to the composition $\mathcal{L} \circ S_F$, where $\mathcal{L}$ is the continuous linear operator on $L^1(J, \mathbb{R})$ into X, defined by

$$\mathcal{L}(v)(t) = \int_0^t \frac{(t-s)^{\alpha-1}}{\Gamma(\alpha)} v(s)\,ds - \frac{t}{3}\int_0^1 \frac{(1-s)^{\alpha-1}}{\Gamma(\alpha)} v(s)\,ds$$
$$- \frac{t}{3}\int_0^1 \frac{(1-s)^{\alpha-2}}{\Gamma(\alpha-1)} v(s)\,ds + \left(t-t^2\right)\Delta \int_0^1 \frac{(1-s)^{\alpha-3}}{\Gamma(\alpha-2)} v(s)\,ds$$
$$+ \left(t-t^2\right)\Delta \int_0^1 \frac{(1-s)^{\alpha-p-1}}{\Gamma(\alpha-p)} v(s)\,ds.$$

Suppose that $u \in Y$ is arbitrary and let $\{v_n\}$ be a sequence in $S_{F,u}$. Then, by definition of $S_{F,u}$, we have $v_n(t) \in F(t, u(t), u'(t), u''(t), {}^cD^{q_1}u(t), \ldots, {}^cD^{q_k}u(t))$ for almost all $t \in J$. Since $F(t, u(t), u'(t), u''(t), {}^cD^{q_1}u(t), \ldots, {}^cD^{q_k}u(t))$ is compact for all $t \in J$, there is a convergent subsequence of $\{v_n(t)\}$ (we denote it by $\{v_n(t)\}$ again) that converges in measure to some $v(t) \in S_{F,u}$ for almost all $t \in J$. On the other hand, $\mathcal{L}$ is continuous, so $\mathcal{L}(v_n)(t) \to \mathcal{L}(v)(t)$ pointwise on J.

In order to show that the convergence is uniform, we have to show that $\{\mathcal{L}(v_n)\}$ is an equi-continuous sequence. Let $t_1, t_2 \in J$ with $t_1 < t_2$. Then we have

$$\left|\mathcal{L}(v_n)(t_2) - \mathcal{L}(v_n)(t_1)\right|$$
$$\leq \left|\frac{1}{\Gamma(\alpha)}\int_0^{t_1}\left[(t_2-s)^{\alpha-1} - (t_1-s)^{\alpha-1}\right] v_n(s)\,ds + \frac{1}{\Gamma(\alpha)}\int_{t_1}^{t_2} (t_2-s)^{\alpha-1} v_n(s)\,ds\right|$$

$$+\frac{|t_2-t_1|}{3\Gamma(\alpha)}\int_0^1(1-s)^{\alpha-1}|v_n(s)|\,ds+\frac{|t_2-t_1|}{3\Gamma(\alpha-1)}\int_0^1(1-s)^{\alpha-2}|v_n(s)|\,ds$$
$$+\frac{\Delta|[(t_2-t_1)-(t_2^2-t_1^2)]|}{\Gamma(\alpha-2)}\int_0^1(1-s)^{\alpha-3}|v_n(s)|\,ds$$
$$+\frac{\Delta|[(t_2-t_1)-(t_2^2-t_1^2)]|}{\Gamma(\alpha-p)}\int_0^1(1-s)^{\alpha-p-1}|v_n(s)|\,ds$$
$$\le\|m\|_\infty\left\{\frac{|t_2^\alpha-t_1^\alpha|}{\Gamma(\alpha+1)}+\frac{|t_2-t_1|}{3\Gamma(\alpha+1)}+\frac{|t_2-t_1|}{3\Gamma(\alpha)}+\frac{\Delta|[(t_2-t_1)-(t_2^2-t_1^2)]|}{\Gamma(\alpha-1)}\right.$$
$$\left.+\frac{\Delta|[(t_2-t_1)-(t_2^2-t_1^2)]|}{\Gamma(\alpha-p+1)}\right\}.$$

Continuing this process, we have

$$\left|(\mathcal{L}'(v_n)(t_2))-(\mathcal{L}'(v_n)(t_1))\right|\le\|m\|_\infty\left\{\frac{|t_2^{\alpha-1}-t_1^{\alpha-1}|}{\Gamma(\alpha)}+\frac{\Delta|t_2-t_1|}{\Gamma(\alpha-1)}+\frac{\Delta|t_2-t_1|}{\Gamma(\alpha-p+1)}\right\}$$

and

$$\left|(\mathcal{L}''(v_n)(t_2))-(\mathcal{L}''(v_n)(t_1))\right|\le\left|\frac{1}{\Gamma(\alpha-2)}\int_0^{t_1}\left[(t_2-s)^{\alpha-3}-(t_1-s)^{\alpha-3}\right]v_n(s)\,ds\right.$$
$$\left.+\frac{1}{\Gamma(\alpha-2)}\int_{t_1}^{t_2}(t_2-s)^{\alpha-3}v_n(s)\,ds\right|$$
$$\le\|m\|_\infty\frac{|t_2^{\alpha-2}-t_1^{\alpha-2}|}{\Gamma(\alpha-1)},$$

and, finally, for every $i=1,\dots,k$,

$$\left|({}^cD^{q_i}\mathcal{L}(v_n)(t_2))-({}^cD^{q_i}\mathcal{L}(v_n)(t_1))\right|$$
$$\le\|m\|_\infty\left\{\frac{|t_2^{\alpha-q_i}-t_1^{\alpha-q_i}|}{\Gamma(\alpha-q_i+1)}+\frac{2\Delta|t_2^{2-q_i}-t_1^{2-q_i}|}{\Gamma(3-q_i)\Gamma(\alpha-1)}+\frac{2\Delta|t_2^{2-q_i}-t_1^{2-q_i}|}{\Gamma(3-q_i)\Gamma(\alpha-p+1)}\right\}.$$

We see that the right-hand sides of the above inequalities tend to zero as $t_2\to t_1$. Thus, the sequence $\{\mathcal{L}(v_n)\}$ is equi-continuous and by using the Arzelá-Ascoli theorem, we see that there is a uniformly convergent subsequence. So, there is a subsequence of $\{v_n\}$ (we denote it again by $\{v_n\}$) such that $\mathcal{L}(v_n)\to\mathcal{L}(v)$. Note that $\mathcal{L}(v)\in\mathcal{L}(S_{F,u})$. Hence, $A(u)=\mathcal{L}(S_{F,u})$ is compact for all $u\in Y$. So $A(u)$ is compact.

Now, we show that $A(u)$ is convex for all $u\in X$. Let $z_1,z_2\in A(u)$. We select $f_1,f_2\in S_{F,u}$ such that

$$z_i(t)=\int_0^t\frac{(t-s)^{\alpha-1}}{\Gamma(\alpha)}f_i(s)\,ds-\frac{t}{3}\int_0^1\frac{(1-s)^{\alpha-1}}{\Gamma(\alpha)}f_i(s)\,ds$$
$$-\frac{t}{3}\int_0^1\frac{(1-s)^{\alpha-2}}{\Gamma(\alpha-1)}f_i(s)\,ds+\left(t-t^2\right)\Delta\int_0^1\frac{(1-s)^{\alpha-3}}{\Gamma(\alpha-2)}f_i(s)\,ds$$
$$+\left(t-t^2\right)\Delta\int_0^1\frac{(1-s)^{\alpha-p-1}}{\Gamma(\alpha-p)}f_i(s)\,ds,\quad i=1,2$$

for almost all $t \in J$. Let $0 \leq \lambda \leq 1$. Then we have

$$\begin{aligned}
\left[\lambda z_1 + (1-\lambda)z_2\right](t) = & \int_0^t \frac{(t-s)^{\alpha-1}}{\Gamma(\alpha)}\left[\lambda f_1(s) + (1-\lambda)f_2(s)\right] ds \\
& - \frac{t}{3}\int_0^1 \frac{(1-s)^{\alpha-1}}{\Gamma(\alpha)}\left[\lambda f_1(s) + (1-\lambda)f_2(s)\right] ds \\
& - \frac{t}{3}\int_0^1 \frac{(1-s)^{\alpha-2}}{\Gamma(\alpha-1)}\left[\lambda f_1(s) + (1-\lambda)f_2(s)\right] ds \\
& + \left(t-t^2\right)\Delta \int_0^1 \frac{(1-s)^{\alpha-3}}{\Gamma(\alpha-2)}\left[\lambda f_1(s) + (1-\lambda)f_2(s)\right] ds \\
& + \left(t-t^2\right)\Delta \int_0^1 \frac{(1-s)^{\alpha-p-1}}{\Gamma(\alpha-p)}\left[\lambda f_1(s) + (1-\lambda)f_2(s)\right] ds.
\end{aligned}$$

Since F has convex values, $S_{F,u}$ is convex and $\lambda f_1(s) + (1-\lambda)f_2(s) \in S_{F,u}$. Thus

$$\lambda z_1 + (1-\lambda)z_2 \in A(u).$$

Consequently, A is convex-valued. Similarly, B is compact and convex-valued.

Here, we show that $A(u) + B(u) \subset Y$ for all $u \in Y$. Suppose that $u \in Y$ and $z_1 \in A(u)$, $z_2 \in B(u)$ are arbitrary elements. Choose $v_1 \in S_{F,u}$ and $v_2 \in S_{G,u}$ such that

$$\begin{aligned}
z_1(t) = & \int_0^t \frac{(t-s)^{\alpha-1}}{\Gamma(\alpha)} v_1(s)\, ds - \frac{t}{3}\int_0^1 \frac{(1-s)^{\alpha-1}}{\Gamma(\alpha)} v_1(s)\, ds \\
& - \frac{t}{3}\int_0^1 \frac{(1-s)^{\alpha-2}}{\Gamma(\alpha-1)} v_1(s)\, ds + \left(t-t^2\right)\Delta \int_0^1 \frac{(1-s)^{\alpha-3}}{\Gamma(\alpha-2)} v_1(s)\, ds \\
& + \left(t-t^2\right)\Delta \int_0^1 \frac{(1-s)^{\alpha-p-1}}{\Gamma(\alpha-p)} v_1(s)\, ds
\end{aligned}$$

and

$$\begin{aligned}
z_2(t) = & \int_0^t \frac{(t-s)^{\alpha-1}}{\Gamma(\alpha)} v_2(s)\, ds - \frac{t}{3}\int_0^1 \frac{(1-s)^{\alpha-1}}{\Gamma(\alpha)} v_2(s)\, ds \\
& - \frac{t}{3}\int_0^1 \frac{(1-s)^{\alpha-2}}{\Gamma(\alpha-1)} v_2(s)\, ds + \left(t-t^2\right)\Delta \int_0^1 \frac{(1-s)^{\alpha-3}}{\Gamma(\alpha-2)} v_2(s)\, ds \\
& + \left(t-t^2\right)\Delta \int_0^1 \frac{(1-s)^{\alpha-p-1}}{\Gamma(\alpha-p)} v_2(s)\, ds
\end{aligned}$$

for almost all $t \in J$. Hence, we get

$$\begin{aligned}
\left|z_1(t) + z_2(t)\right| \leq & \frac{1}{\Gamma(\alpha)}\int_0^t (t-s)^{\alpha-1}\left(\left|v_1(s)\right| + \left|v_2(s)\right|\right) ds \\
& + \frac{|t|}{3\Gamma(\alpha)}\int_0^1 (1-s)^{\alpha-1}\left(\left|v_1(s)\right| + \left|v_2(s)\right|\right) ds \\
& + \frac{|t|}{3\Gamma(\alpha-1)}\int_0^1 (1-s)^{\alpha-2}\left(\left|v_1(s)\right| + \left|v_2(s)\right|\right) ds \\
& + \frac{\Delta\left|t-t^2\right|}{\Gamma(\alpha-2)}\int_0^1 (1-s)^{\alpha-3}\left(\left|v_1(s)\right| + \left|v_2(s)\right|\right) ds
\end{aligned}$$

$$+\frac{\Delta|t-t^2|}{\Gamma(\alpha-p)}\int_0^1(1-s)^{\alpha-p-1}\left(|v_1(s)|+|v_2(s)|\right)ds$$
$$\le\left(\|p\|_\infty+\|m\|_\infty\right)\left\{\frac{4}{3\Gamma(\alpha+1)}+\frac{1}{3\Gamma(\alpha)}+\frac{\Delta}{4\Gamma(\alpha-1)}+\frac{\Delta}{4\Gamma(\alpha-p+1)}\right\}.$$

Hence, $\sup_{t\in J}|z_1(t)+z_2(t)|\le(\|p\|_\infty+\|m\|_\infty)\Lambda_1$. Also we have

$$\left|z_1'(t)+z_2'(t)\right|\le\left(\|p\|_\infty+\|m\|_\infty\right)\left\{\frac{4}{3\Gamma(\alpha)}+\frac{1}{3\Gamma(\alpha+1)}+\frac{\Delta}{\Gamma(\alpha-1)}+\frac{\Delta}{\Gamma(\alpha-p+1)}\right\},$$

which implies that $\sup_{t\in J}|z_1'(t)+z_2'(t)|\le(\|p\|_\infty+\|m\|_\infty)\Lambda_2$ and

$$\left|z_1''(t)+z_2''(t)\right|\le\left(\|p\|_\infty+\|m\|_\infty\right)\left\{\frac{1+2\Delta}{\Gamma(\alpha-1)}+\frac{2\Delta}{\Gamma(\alpha-p+1)}\right\}$$

from which $\sup_{t\in J}|z_1''(t)+z_2''(t)|\le(\|p\|_\infty+\|m\|_\infty)\Lambda_3$. Finally, for all $i=1,\dots,k$, we have

$$\left|{}^cD^{q_i}z_1(t)+{}^cD^{q_i}z_2(t)\right|$$
$$\le\left(\|p\|_\infty+\|m\|_\infty\right)\left\{\frac{1}{\Gamma(\alpha-q_i+1)}+\frac{2\Delta}{\Gamma(3-q_i)\Gamma(\alpha-1)}+\frac{2\Delta}{\Gamma(3-q_i)\Gamma(\alpha-p+1)}\right\},$$

and so $\sup_{t\in J}|{}^cD^{q_i}z_1(t)+{}^cD^{q_i}z_2(t)|\le(\|p\|_\infty+\|m\|_\infty)\Lambda_4^i$, $i=1,2,\dots,k$. Hence, it follows that

$$\|z_1+z_2\|\le\left(\|p\|_\infty+\|m\|_\infty\right)\left(\Lambda_1+\Lambda_2+\Lambda_3+\Lambda_4^i\right)=M,\quad i=1,2,\dots,k.$$

Now, we show that the operator B is compact on Y. To do this, it is enough to prove that $B(Y)$ is uniformly bounded and equi-continuous in X. Let $z\in B(Y)$ be arbitrary. For some $u\in Y$, choose $v_1\in S_{G,u}$ such that

$$\begin{aligned}z(t)=&\int_0^t\frac{(t-s)^{\alpha-1}}{\Gamma(\alpha)}v_1(s)\,ds-\frac{t}{3}\int_0^1\frac{(1-s)^{\alpha-1}}{\Gamma(\alpha)}v_1(s)\,ds\\&-\frac{t}{3}\int_0^1\frac{(1-s)^{\alpha-2}}{\Gamma(\alpha-1)}v_1(s)\,ds+\left(t-t^2\right)\Delta\int_0^1\frac{(1-s)^{\alpha-3}}{\Gamma(\alpha-2)}v_1(s)\,ds\\&+\left(t-t^2\right)\Delta\int_0^1\frac{(1-s)^{\alpha-p-1}}{\Gamma(\alpha-p)}v_1(s)\,ds,\quad t\in J.\end{aligned}\tag{2.9}$$

Hence,

$$\begin{aligned}|z(t)|&\le\|p\|_\infty\left\{\frac{4}{3\Gamma(\alpha+1)}+\frac{1}{3\Gamma(\alpha)}+\frac{\Delta}{4\Gamma(\alpha-1)}+\frac{\Delta}{4\Gamma(\alpha-p+1)}\right\},\\|z'(t)|&\le\|p\|_\infty\left\{\frac{4}{3\Gamma(\alpha)}+\frac{1}{3\Gamma(\alpha+1)}+\frac{\Delta}{\Gamma(\alpha-1)}+\frac{\Delta}{\Gamma(\alpha-p+1)}\right\},\\|z''(t)|&\le\|p\|_\infty\left\{\frac{1+2\Delta}{\Gamma(\alpha-1)}+\frac{2\Delta}{\Gamma(\alpha-p+1)}\right\},\\|{}^cD^{q_i}z(t)|&\le\|p\|_\infty\left\{\frac{1}{\Gamma(\alpha-q_i+1)}+\frac{2\Delta}{\Gamma(3-q_i)\Gamma(\alpha-1)}+\frac{2\Delta}{\Gamma(3-q_i)\Gamma(\alpha-p+1)}\right\}\end{aligned}$$

for $i=1,\dots,k$. Hence, $\|z\|\le\|p\|_\infty(\Lambda_1+\Lambda_2+\Lambda_3+\Lambda_4^i)$, $i=1,\dots,k$.

Now, we show that B maps Y to equi-continuous subsets of X. Let $t_1, t_2 \in J$ with $t_1 < t_2$, $u \in Y$, and $z \in B(u)$. Choose $v_1 \in S_{G,u}$ such that $z(t)$ is given by (2.9). Then we have

$$\begin{aligned}
|z(t_2) - z(t_1)| &\le \|p\|_\infty \left\{ \frac{|t_2^\alpha - t_1^\alpha|}{\Gamma(\alpha+1)} + \frac{|t_2 - t_1|}{3\Gamma(\alpha+1)} + \frac{|t_2 - t_1|}{3\Gamma(\alpha)} + \frac{\Delta|[(t_2 - t_1) - (t_2^2 - t_1^2)]|}{\Gamma(\alpha-1)} \right. \\
&\quad \left. + \frac{\Delta|[(t_2 - t_1) - (t_2^2 - t_1^2)]|}{\Gamma(\alpha - p + 1)} \right\}, \\
|z'(t_2) - z'(t_1)| &\le \|p\|_\infty \left\{ \frac{|t_2^{\alpha-1} - t_1^{\alpha-1}|}{\Gamma(\alpha)} + \frac{2\Delta|t_2 - t_1|}{\Gamma(\alpha-1)} + \frac{2\Delta|t_2 - t_1|}{\Gamma(\alpha - p + 1)} \right\}, \\
|z''(t_2) - z''(t_1)| &\le \|p\|_\infty \frac{|t_2^{\alpha-2} - t_1^{\alpha-2}|}{\Gamma(\alpha-1)}
\end{aligned}$$

and

$$\begin{aligned}
&|{}^cD^{q_i} z(t_2) - {}^cD^{q_i} z(t_1)| \\
&\le \|p\|_\infty \left\{ \frac{|t_2^{\alpha-q_i} - t_1^{\alpha-q_i}|}{\Gamma(\alpha - q_i + 1)} + \frac{2\Delta|t_2^{2-q_i} - t_1^{2-q_i}|}{\Gamma(3-q_i)\Gamma(\alpha-1)} + \frac{2\Delta|t_2^{2-q_i} - t_1^{2-q_i}|}{\Gamma(3-q_i)\Gamma(\alpha - p + 1)} \right\}
\end{aligned}$$

for each $i = 1, \dots, k$. It is seen that the right-hand sides of the above inequalities tend to zero as $t_2 \to t_1$. Hence, by using the Arzelá-Ascoli theorem, B is compact.

Next, we prove that B has a closed graph. Let $u_n \in Y$ and $z_n \in B(u_n)$ for all n such that $u_n \to u_0$ and $z_n \to z_0$. We show that $z_0 \in B(u_0)$. Associated with $z_n \in B(u_n)$ for each $n \in \mathbb{N}$, there exists $v_n \in S_{G,u_n}$ such that

$$\begin{aligned}
z_n(t) &= \int_0^t \frac{(t-s)^{\alpha-1}}{\Gamma(\alpha)} v_n(s)\,ds - \frac{t}{3}\int_0^1 \frac{(1-s)^{\alpha-1}}{\Gamma(\alpha)} v_n(s)\,ds \\
&\quad - \frac{t}{3}\int_0^1 \frac{(1-s)^{\alpha-2}}{\Gamma(\alpha-1)} v_n(s)\,ds + (t - t^2)\Delta \int_0^1 \frac{(1-s)^{\alpha-3}}{\Gamma(\alpha-2)} v_n(s)\,ds \\
&\quad + (t - t^2)\Delta \int_0^1 \frac{(1-s)^{\alpha-p-1}}{\Gamma(\alpha-p)} v_n(s)\,ds
\end{aligned}$$

for all $t \in J$. It suffices to show that there exists $v_0 \in S_{G,u_0}$ such that, for each $t \in J$,

$$\begin{aligned}
z_0(t) &= \int_0^t \frac{(t-s)^{\alpha-1}}{\Gamma(\alpha)} v_0(s)\,ds - \frac{t}{3}\int_0^1 \frac{(1-s)^{\alpha-1}}{\Gamma(\alpha)} v_0(s)\,ds \\
&\quad - \frac{t}{3}\int_0^1 \frac{(1-s)^{\alpha-2}}{\Gamma(\alpha-1)} v_0(s)\,ds + (t - t^2)\Delta \int_0^1 \frac{(1-s)^{\alpha-3}}{\Gamma(\alpha-2)} v_0(s)\,ds \\
&\quad + (t - t^2)\Delta \int_0^1 \frac{(1-s)^{\alpha-p-1}}{\Gamma(\alpha-p)} v_0(s)\,ds.
\end{aligned}$$

Consider the continuous linear operator $\Theta : L^1(J, \mathbb{R}) \to X$ by

$$\begin{aligned}
\Theta(v)(t) &= \int_0^t \frac{(t-s)^{\alpha-1}}{\Gamma(\alpha)} v(s)\,ds - \frac{t}{3}\int_0^1 \frac{(1-s)^{\alpha-1}}{\Gamma(\alpha)} v(s)\,ds \\
&\quad - \frac{t}{3}\int_0^1 \frac{(1-s)^{\alpha-2}}{\Gamma(\alpha-1)} v(s)\,ds + (t - t^2)\Delta \int_0^1 \frac{(1-s)^{\alpha-3}}{\Gamma(\alpha-2)} v(s)\,ds \\
&\quad + (t - t^2)\Delta \int_0^1 \frac{(1-s)^{\alpha-p-1}}{\Gamma(\alpha-p)} v(s)\,ds.
\end{aligned}$$

Notice that

$$\begin{aligned}\|z_n(t) - z_0(t)\| = \bigg\| &\int_0^t \frac{(t-s)^{\alpha-1}}{\Gamma(\alpha)}\big(v_n(s) - v_0(s)\big)\, ds - \frac{t}{3}\int_0^1 \frac{(1-s)^{\alpha-1}}{\Gamma(\alpha)}\big(v_n(s) - v_0(s)\big)\, ds \\ &- \frac{t}{3}\int_0^1 \frac{(1-s)^{\alpha-2}}{\Gamma(\alpha-1)}\big(v_n(s) - v_0(s)\big)\, ds \\ &+ \big(t - t^2\big)\Delta \int_0^1 \frac{(1-s)^{\alpha-3}}{\Gamma(\alpha-2)}\big(v_n(s) - v_0(s)\big)\, ds \\ &+ \big(t - t^2\big)\Delta \int_0^1 \frac{(1-s)^{\alpha-p-1}}{\Gamma(\alpha-p)}\big(v_n(s) - v_0(s)\big)\, ds\bigg\| \to 0 \quad \text{as } n \to \infty.\end{aligned}$$

By using Lemma 1.2, $\Theta \circ S_G$ is a closed graph operator. Since $z_n(t) \in \Theta(S_{G,u_n})$ for all n, and $u_n \to u_0$, there is $v_0 \in S_{G,u_0}$ such that

$$\begin{aligned}z_0(t) = &\int_0^t \frac{(t-s)^{\alpha-1}}{\Gamma(\alpha)} v_0(s)\, ds - \frac{t}{3}\int_0^1 \frac{(1-s)^{\alpha-1}}{\Gamma(\alpha)} v_0(s)\, ds \\ &- \frac{t}{3}\int_0^1 \frac{(1-s)^{\alpha-2}}{\Gamma(\alpha-1)} v_0(s)\, ds + \big(t - t^2\big)\Delta \int_0^1 \frac{(1-s)^{\alpha-3}}{\Gamma(\alpha-2)} v_0(s)\, ds \\ &+ \big(t - t^2\big)\Delta \int_0^1 \frac{(1-s)^{\alpha-p-1}}{\Gamma(\alpha-p)} v_0(s)\, ds.\end{aligned}$$

Hence, $z_0 \in B(u_0)$. So, it follows that B has a closed graph and this implies that the operator B is upper semi-continuous.

Finally, we show that A is a contraction multifunction. Let $u, w \in X$ and $z_1 \in A(w)$ is given. Then we can select $v_1 \in S_{F,w}$ such that

$$\begin{aligned}z_1(t) = &\int_0^t \frac{(t-s)^{\alpha-1}}{\Gamma(\alpha)} v_1(s)\, ds - \frac{t}{3}\int_0^1 \frac{(1-s)^{\alpha-1}}{\Gamma(\alpha)} v_1(s)\, ds \\ &- \frac{t}{3}\int_0^1 \frac{(1-s)^{\alpha-2}}{\Gamma(\alpha-1)} v_1(s)\, ds + \big(t - t^2\big)\Delta \int_0^1 \frac{(1-s)^{\alpha-3}}{\Gamma(\alpha-2)} v_1(s)\, ds \\ &+ \big(t - t^2\big)\Delta \int_0^1 \frac{(1-s)^{\alpha-p-1}}{\Gamma(\alpha-p)} v_1(s)\, ds\end{aligned}$$

for all $t \in J$. Since

$$\begin{aligned}&H_d\Big(F\big(t, u(t), u'(t), u''(t), {}^cD^{q_1}u(t), \ldots, {}^cD^{q_k}u(t)\big) \\ &\quad - F\big(t, w(t), w'(t), w''(t), {}^cD^{q_1}w(t), \ldots, {}^cD^{q_k}w(t)\big)\Big) \\ &\quad \le h(t)\left[\big|u(t) - w(t)\big| + \big|u'(t) - w'(t)\big| + \big|u''(t) - w''(t)\big| + \sum_{i=1}^{k}\big|{}^cD^{q_i}u(t) - {}^cD^{q_i}w(t)\big|\right]\end{aligned}$$

for almost all $t \in J$, there exists $y \in F(t, u(t), u'(t), u''(t), {}^cD^{q_1}u(t), \ldots, {}^cD^{q_k}u(t))$ such that

$$\begin{aligned}\big|v_1(t) - y\big| \le m(t)\Bigg[&\big|u(t) - w(t)\big| + \big|u'(t) - w'(t)\big| + \big|u''(t) - w''(t)\big| \\ &+ \sum_{i=1}^{k}\big|{}^cD^{q_i}u(t) - {}^cD^{q_i}w(t)\big|\Bigg]\end{aligned}$$

for almost all $t \in J$. Consider the multifunction $\mathcal{U} : J \to \mathcal{P}(\mathbb{R})$ by

$$\mathcal{U}(t) = \{s \in \mathbb{R} : |v_1(t) - s| \leq m(t)g(t) \text{ for almost all } t \in J\},$$

where

$$g(t) = \Big[|u(t) - w(t)| + |u'(t) - w'(t)| + |u''(t) - w''(t)| + \sum_{i=1}^{k} |{}^cD^{q_i}u(t) - {}^cD^{q_i}w(t)| \Big].$$

Since v_1 and $\varphi = mg$ are measurable, $\mathcal{U}(\cdot) \cap F(\cdot, u(\cdot), u'(\cdot), u''(\cdot), {}^cD^{q_1}u(\cdot), \ldots, {}^cD^{q_k}u(\cdot))$ is a measurable multifunction. Thus, we can choose

$$v_2(t) \in F\big(t, u(t), u'(t), u''(t), {}^cD^{q_1}u(t), \ldots, {}^cD^{q_k}u(t)\big)$$

such that

$$|v_1(t) - v_2(t)| \leq m(t) \Big[|u(t) - w(t)| + |u'(t) - w'(t)| + |u''(t) - w''(t)| + \sum_{i=1}^{k} |{}^cD^{q_i}u(t) - {}^cD^{q_i}w(t)| \Big]$$

and

$$\begin{aligned} z_2(t) = {} & \int_0^t \frac{(t-s)^{\alpha-1}}{\Gamma(\alpha)} v_2(s)\, ds - \frac{t}{3} \int_0^1 \frac{(1-s)^{\alpha-1}}{\Gamma(\alpha)} v_2(s)\, ds \\ & - \frac{t}{3} \int_0^1 \frac{(1-s)^{\alpha-2}}{\Gamma(\alpha-1)} v_2(s)\, ds + \big(t - t^2\big)\Delta \int_0^1 \frac{(1-s)^{\alpha-3}}{\Gamma(\alpha-2)} v_2(s)\, ds \\ & + \big(t - t^2\big)\Delta \int_0^1 \frac{(1-s)^{\alpha-p-1}}{\Gamma(\alpha-p)} v_2(s)\, ds \end{aligned}$$

for all $t \in J$. Now, we have

$$\begin{aligned} |z_1(t) - z_2(t)| \leq {} & \frac{1}{\Gamma(\alpha)} \int_0^t (t-s)^{\alpha-1} |v_1(s) - v_2(s)|\, ds \\ & + \frac{|t|}{3\Gamma(\alpha)} \int_0^1 (1-s)^{\alpha-1} |v_1(s) - v_2(s)|\, ds \\ & + \frac{|t|}{3\Gamma(\alpha-1)} \int_0^1 (1-s)^{\alpha-2} |v_1(s) - v_2(s)|\, ds \\ & + \frac{\Delta|t - t^2|}{\Gamma(\alpha-2)} \int_0^1 (1-s)^{\alpha-3} |v_1(s) - v_2(s)|\, ds \\ & + \frac{\Delta|t - t^2|}{\Gamma(\alpha-p)} \int_0^1 (1-s)^{\alpha-p-1} |v_1(s) - v_2(s)|\, ds \\ \leq {} & \|h\|_\infty \left\{ \frac{4}{3\Gamma(\alpha+1)} + \frac{1}{3\Gamma(\alpha)} + \frac{\Delta}{4\Gamma(\alpha-1)} + \frac{\Delta}{4\Gamma(\alpha-p+1)} \right\} \|u - w\|. \end{aligned}$$

Similarly,

$$|z_1'(t) - z_2'(t)| \le \|h\|_\infty \left\{ \frac{4}{3\Gamma(\alpha)} + \frac{1}{3\Gamma(\alpha+1)} + \frac{\Delta}{\Gamma(\alpha-1)} + \frac{\Delta}{\Gamma(\alpha-p+1)} \right\} \|u-w\|,$$

$$|z_1''(t) - z_2''(t)| \le \|h\|_\infty \left\{ \frac{1+2\Delta}{\Gamma(\alpha-1)} + \frac{2\Delta}{\Gamma(\alpha-p+1)} \right\} \|u-w\|,$$

$$|{}^cD^{q_i}z_1(t) - {}^cD^{q_i}z_2(t)|$$

$$\le \|h\|_\infty \left\{ \frac{1}{\Gamma(\alpha-q_i+1)} + \frac{2\Delta}{\Gamma(3-q_i)\Gamma(\alpha-1)} + \frac{2\Delta}{\Gamma(3-q_i)\Gamma(\alpha-p+1)} \right\} \|u-w\|.$$

Hence,

$$\sup_{t\in J} |z_1(t) - z_2(t)| \le \|h\|_\infty \Lambda_1 \|u-w\|,$$

$$\sup_{t\in J} |z_1'(t) - z_2'(t)| \le \|h\|_\infty \Lambda_2 \|u-w\|,$$

$$\sup_{t\in J} |z_1''(t) - z_2''(t)| \le \|h\|_\infty \Lambda_3 \|u-w\|,$$

$$\sup_{t\in J} |{}^cD^{q_i}z_1(t) - {}^cD^{q_i}z_2(t)| \le \|h\|_\infty \Lambda_4^i \|u-w\|$$

for each $1 \le i \le k$. So

$$\|z_1 - z_2\| \le \|h\|_\infty (\Lambda_1 + \Lambda_2 + \Lambda_3 + \Lambda_4^i) \|u-w\|, \quad i = 1, 2, \ldots, k.$$

This implies that $H_d(A(u), A(w)) \le L\|u - w\|$. Thus A and B satisfy all the conditions of Theorem 1.3 and so the inclusion $u \in A(u) + B(u)$ has a solution in Y. Therefore the inclusion problem (1.1)-(1.2) has a solution in Y and the proof is completed. □

Finally, we give an example to illustrate the validity of our main result.

Example 2.5 Consider the following fractional differential inclusion:

$$\begin{aligned} {}^cD^{\frac{5}{2}}u(t) \in & \left[0, \frac{t|u(t)|^3}{100(1+|u(t)|^3)} + \frac{t|2\sin(u'(t))|}{200(|\sin(u'(t))|+1)} + \frac{0.01t|u''(t)|}{|u''(t)|+1} \right. \\ & \left. + \frac{t|\cos({}^cD^{\frac{3}{2}}u(t))|}{100(1+|\cos({}^cD^{\frac{3}{2}}u(t))|)} + \frac{t^2|\sin\frac{\pi}{2}t||{}^cD^{\frac{3}{2}}u(t)|^2}{100t(|{}^cD^{\frac{3}{2}}u(t)|^2+1)} \right] \\ & + \left[0, \frac{e^{-t}|u(t)|}{(1+e^t)(1+|u(t)|)} + \frac{|\cos\pi t||u'(t)|e^{-t}}{(1+e^t)(1+|u'(t)|)} + \frac{e^{-t}|u''(t)|^2}{(1+|u''(t)|^2)(1+e^t)} \right. \\ & \left. + \frac{e^{-2t}|\sin({}^cD^{\frac{3}{2}}u(t))|}{(1+|\sin({}^cD^{\frac{3}{2}}u(t))|)e^t(1+e^t)} + \frac{e^{-3t}|{}^cD^{\frac{3}{2}}u(t)|^3}{(e^{2t}+e^{3t})(1+|{}^cD^{\frac{3}{2}}u(t)|^3)} \right], \end{aligned} \tag{2.10}$$

with the following boundary conditions:

$$u(0) = 0, \qquad u'(0) = -u(1) - u'(1), \qquad u''(0) = -u''(1) - {}^cD^{\frac{3}{2}}u(1), \tag{2.11}$$

where $t \in [0,1]$. In the above inclusion problem, we have $\alpha = 5/2$, $p = 3/2$, $k = 2$, and $q_1 = q_2 = 3/2$. Also, we have $\Delta = 0.1597$.

Now, we define $F: [0,1] \times \mathbb{R} \times \mathbb{R} \times \mathbb{R} \times \mathbb{R} \times \mathbb{R} \to \mathcal{P}(\mathbb{R})$ by

$$F(t,x,y,z,v,w) = \left[0, \frac{t|x|^3}{100(1+|x|^3)} + \frac{t|2\sin y|}{200(|\sin y|+1)} + \frac{0.01t|z|}{|z|+1} + \frac{t|\cos v|}{100(1+|\cos v|)} + \frac{t^2|\sin\frac{\pi}{2}t|w^2}{100t(w^2+1)}\right],$$

and $G: [0,1] \times \mathbb{R} \times \mathbb{R} \times \mathbb{R} \times \mathbb{R} \times \mathbb{R} \to \mathcal{P}(\mathbb{R})$ by

$$G(t,x,y,z,v,w) = \left[0, \frac{e^{-t}|x|}{(1+e^t)(1+|x|)} + \frac{|\cos\pi t||y|e^{-t}}{(1+e^t)(1+|y|)} + \frac{e^{-t}z^2}{(1+z^2)(1+e^t)} + \frac{e^{-2t}|\sin v|}{(1+|\sin v|)e^t(1+e^t)} + \frac{e^{-3t}|w|^3}{(e^{2t}+e^{3t})(1+|w|^3)}\right].$$

Then there exist continuous functions $m, p: [0,1] \to (0,\infty)$ given by

$$m(t) = 5 + \frac{t}{100}, \qquad p(t) = \frac{e^{-t}}{1+e^t}.$$

On the other hand, we can easily check that, for every $x_i, y_i, z_i, v_i, w_i \in \mathbb{R}$ $(i = 1,2)$,

$$H_d\big(F(t,x_1,y_1,z_1,v_1,w_1) - F(t,x_2,y_2,z_2,v_2,w_2)\big) \le h(t)\big(|x_1-x_2| + |y_1-y_2| + |z_1-z_2| + |v_1-v_2| + |w_1-w_2|\big),$$

where $h: [0,1] \to (0,\infty)$ is defined by $h(t) = \frac{t}{100}$. It can easily be found that $\Lambda_1 = 0.7369$, $\Lambda_2 = 1.4434$, $\Lambda_3 = 1.8102$, $\Lambda_4^1 = 1.7687$, and $\Lambda_4^2 = 1.7687$. Since $\|h\|_\infty = \frac{1}{100}$, we have $L := \|h\|_\infty(\Lambda_1 + \Lambda_2 + \Lambda_3 + \Lambda_4^1 + \Lambda_4^2) = 0.01 \times 7.5279 = 0.075279 < 1$. Consequently all assumptions and conditions of Theorem 2.4 are satisfied. Hence, Theorem 2.4 implies that the fractional differential inclusion problem (2.10)-(2.11) has a solution.

Competing interests
The authors declare that they have no competing interests.

Authors' contributions
All authors contributed equally in this article. They read and approved the final manuscript.

Author details
[1]Department of Mathematics, University of Ioannina, Ioannina, 451 10, Greece. [2]Nonlinear Analysis and Applied Mathematics (NAAM) Research Group, Department of Mathematics, Faculty of Science, King Abdulaziz University, P.O. Box 80203, Jeddah, 21589, Saudi Arabia. [3]Young Researchers and Elite Club, Tabriz Branch, Islamic Azad University, Tabriz, Iran. [4]Nonlinear Dynamic Analysis Research Center, Department of Mathematics, Faculty of Applied Science, King Mongkut's University of Technology North Bangkok, Bangkok, 10800, Thailand.

Acknowledgements
We would like to thank the reviewers for their valuable comments and suggestions on the manuscript.

References
1. Agarwal, RP, Ahmad, B: Existence theory for anti-periodic boundary value problems of fractional differential equations and inclusions. J. Appl. Math. Comput. **62**, 1200-1214 (2011)
2. Agarwal, RP, Belmekki, M, Benchohra, M: A survey on semilinear differential equations and inclusions involving Riemann-Liouville fractional derivative. Adv. Differ. Equ. **2009**, Article ID 981728 (2009)
3. Agarwal, RP, Benchohra, M, Hamani, S: A survey on existence results for boundary value problems of nonlinear fractional differential equations and inclusions. Acta Appl. Math. **109**, 973-1033 (2010)

4. Agarwal, RP, Ntouyas, SK, Ahmad, B, Alhothuali, M: Existence of solutions for integro-differential equations of fractional order with nonlocal three-point fractional boundary conditions. Adv. Differ. Equ. **2013**, 128 (2013)
5. Ahmad, B, Nieto, JJ: Existence of solutions for anti-periodic boundary value problems involving fractional differential equations via Leray-Schauder degree theory. Topol. Methods Nonlinear Anal. **35**, 295-304 (2010)
6. Ahmad, B, Ntouyas, SK: Boundary value problem for fractional differential inclusions with four-point integral boundary conditions. Surv. Math. Appl. **6**, 175-193 (2011)
7. Ahmad, B, Ntouyas, SK, Alsedi, A: On fractional differential inclusions with anti-periodic type integral boundary conditions. Bound. Value Probl. **2013**, 82 (2013)
8. Alsaedi, A, Ntouyas, SK, Ahmad, B: Existence of solutions for fractional differential inclusions with separated boundary conditions in Banach space. Abstr. Appl. Anal. **2013**, Article ID 869837 (2013)
9. Bai, Z, Sun, W: Existence and multiplicity of positive solutions for singular fractional boundary value problems. Comput. Math. Appl. **63**, 1369-1381 (2012)
10. Baleanu, D, Agarwal, RP, Mohammadi, H, Rezapour, S: Some existence results for a nonlinear fractional differential equation on partially ordered Banach spaces. Bound. Value Probl. **2013**, 112 (2013)
11. Baleanu, D, Diethelm, K, Scalas, E, Trujillo, JJ: Fractional Calculus Models and Numerical Methods. World Scientific, Singapore (2012)
12. Baleanu, D, Mohammadi, H, Rezapour, S: The existence of solutions for a nonlinear mixed problem of singular fractional differential equations. Adv. Differ. Equ. **2013**, 359 (2013)
13. Baleanu, D, Mohammadi, H, Rezapour, S: Positive solutions of a boundary value problem for nonlinear fractional differential equations. Abstr. Appl. Anal. **2012**, Article ID 837437 (2012)
14. Baleanu, D, Mohammadi, H, Rezapour, S: Some existence results on nonlinear fractional differential equations. Philos. Trans. R. Soc. Lond. A **371**, 20120144 (2013)
15. Baleanu, D, Mohammadi, H, Rezapour, S: On a nonlinear fractional differential equation on partially ordered metric spaces. Adv. Differ. Equ. **2013**, 83 (2013)
16. Baleanu, D, Nazemi, Z, Rezapour, S: The existence of positive solutions for a new coupled system of multi-term singular fractional integro-differential boundary value problems. Abstr. Appl. Anal. **2013**, Article ID 368659 (2013)
17. Baleanu, D, Nazemi, Z, Rezapour, S: Existence and uniqueness of solutions for multi-term nonlinear fractional integro-differential equations. Adv. Differ. Equ. **2013**, 368 (2013)
18. Baleanu, D, Nazemi, Z, Rezapour, S: Attractivity for a k-dimensional system of fractional functional differential equations and global attractivity for a k-dimensional system of nonlinear fractional differential equations. J. Inequal. Appl. **2014**, 31 (2014)
19. Bragdi, M, Debbouche, A, Baleanu, D: Existence of solutions for fractional differential inclusions with separated boundary conditions in Banach space. Adv. Math. Phys. **2013**, Article ID 426061 (2013)
20. Campos, LM: On the solution of some simple fractional differential equations. Int. J. Math. Sci. **13**(3), 481-496 (1990)
21. Luchko, Y, Srivastava, H: The exact solution of certain differential equations of fractional order by using operational calculus. Comput. Math. Appl. **29**, 73-85 (1995)
22. Ouahab, A: Some results for fractional boundary value problem of differential inclusions. Nonlinear Anal. **69**, 3877-3896 (2008)
23. Podlubny, I: Fractional Differential Equations. Academic Press, San Diego (1999)
24. Phung, PD, Truong, LX: On a fractional differential inclusion with integral boundary conditions in Banach space. Fract. Calc. Appl. Anal. **16**, 538-558 (2013)
25. Samko, G, Kilbas, A, Marichev, O: Fractional Integrals and Derivatives: Theory and Applications. Gordon & Breach, New York (1993)
26. Diethelm, K: Analysis of Fractional Differential Equations. Springer, Berlin (2010)
27. Oldham, KB, Spanier, J: The Fractional Calculus. Academic Press, San Diego (1974)
28. Miller, S, Ross, B: An Introduction to the Fractional Calculus and Fractional Differential Equations. Wiley, New York (1993)
29. Deimling, K: Multi-Valued Differential Equations. de Gruyter, Berlin (1992)
30. Hu, S, Papageorgiou, N: Handbook of Multivalued Analysis. Volume I: Theory. Kluwer Academic, Dordrecht (1997)
31. Aubin, J, Ceuina, A: Differential Inclusions: Set-Valued Maps and Viability Theory. Springer, Berlin (1984)
32. Lasota, A, Opial, Z: An application of the Kakutani-Ky Fan theorem in the theory of ordinary differential equations. Bull. Acad. Pol. Sci., Sér. Sci. Math. Astron. Phys. **13**, 781-786 (1965)
33. Petrusel, A: Fixed points and selections for multi-valued operators. Semin. Fixed Point Theory Cluj-Napoca **2**, 3-22 (2001)
34. Su, X: Boundary value problem for a coupled system of nonlinear fractional differential equations. Appl. Math. Lett. **22**, 64-69 (2009)
35. Ibrahim, AG: Fractional differential inclusions with anti-periodic boundary conditions in Banach spaces. Electron. J. Qual. Theory Differ. Equ. **2014**, 65 (2014)
36. Lan, Q, Lin, W: Positive solutions of systems of Caputo fractional differential equations. Commun. Appl. Anal. **17**, 61-86 (2013)

Existence of positive periodic solutions for abstract evolution equations

Qiang Li* and Yongxiang Li

*Correspondence:
lznwnuliqiang@126.com
Department of Mathematics, Northwest Normal University, Lanzhou, 730070, People's Republic of China

Abstract

In this paper, we discuss the existence of the positive time periodic mild solutions for the evolution equation in an ordered Banach space E, $u'(t) + Au(t) = f(t, u(t))$, $t \in \mathbb{R}$, where $A : D(A) \subset E \rightarrow E$ is a closed linear operator and $-A$ generates a positive compact semigroup $T(t)$ $(t \geq 0)$ in E, the nonlinear function $f : \mathbb{R} \times E \rightarrow E$ is continuous and $f(t, x)$ is ω-periodic in t. We apply the operator semigroup theory and the Leray-Schauder fixed point theorem to obtain the existence of a positive ω-periodic mild solution under the condition that the nonlinear function satisfies a linear growth condition concerning the growth exponent of the semigroup $T(t)$ $(t \geq 0)$. In the end, an example is given to illustrate the applicability of our abstract results.

MSC: 34K30; 47H07; 47H08

Keywords: abstract evolution equation; positive periodic mild solutions; positive compact semigroup; the growth exponent of the semigroup; fixed point theorem

1 Introduction

Let E be an ordered Banach space, whose positive cone K is normal cone with normal constant N. In this paper, we discuss the existence of positive time periodic mild solutions for the evolution equation in the ordered Banach space E

$$u'(t) + Au(t) = f\big(t, u(t)\big), \quad t \in \mathbb{R}, \tag{1.1}$$

where $A : D(A) \subset E \rightarrow E$ is a closed linear operator and $-A$ generates a positive compact semigroup $T(t)$ $(t \geq 0)$ in E, the nonlinear mapping $f : \mathbb{R} \times E \rightarrow E$ is continuous, and, for every $x \in E$, $f(t, x)$ is ω-periodic in t.

It is well known that a variety of partial differential equations with time t, such as the reaction-diffusion equation, the heat equation, the wave equation, the telegraph equation and so on, can be classified as a nonlinear evolution equation (1.1) in some Banach space E. In the nonlinear evolution equation (1.1), A corresponds to the linear partial differential operator with certain boundary conditions, f corresponds to the nonlinear term.

It is noteworthy at this point that the problem concerning periodic solutions of partial differential equations has become an important area of investigation in recent years. Specially, the periodic problems of abstract evolution equations in the form of (1.1) have been considered by several authors; see [1–11].

In [9], the existence and uniqueness for the periodic boundary value problem of the semilinear evolution equation in a Hilbert space H,

$$\begin{cases} u'(t) + Au(t) = f(t, u(t)), & 0 < t < \omega, \\ u(0) = u(\omega), \end{cases}$$

was studied, where $A : D(A) \subset H \to H$ is a positive definite self-adjoint operator and A has compact resolvent. The author presented some spectral conditions for the nonlinearity $f(t,x)$ to guarantee the existence and uniqueness by applying the Schauder fixed point theorem and the contraction mapping principle. Specially, in [10], the author established the existence and uniqueness results of periodic solutions for the linear evolution equation corresponding to (1.1) and accurately estimated the spectral radius of periodic resolvent operator. With the aid of the estimation, under the assumption that the nonlinearity $f(t,x)$ satisfies some ordered conditions concerning the growth exponent of the semigroup $T(t)$ $(t \geq 0)$, the existence and uniqueness results of periodic mild solutions were obtained by applying the operator semigroup theorem and monotone iterative method.

Since in many practice models, only positive periodic solutions are significant, motivated by the papers mentioned above, we research the existence of positive ω-periodic mild solutions for (1.1) in the positive cone K. In this paper, we will use a completely different method to prove the existence of positive ω-periodic mild solutions for (1.1) under some new conditions by applying the Leray-Schauder fixed point theorem in an ordered Banach space E. More precisely, the nonlinear term satisfies a linear growth condition concerning the growth exponent of the semigroup $T(t)$ $(t \geq 0)$ or the first eigenvalue of the operator A.

For the C_0-semigroup $T(t)$ $(t \geq 0)$, there exist $M > 0$ and $\gamma \in \mathbb{R}$ such that (see [12])

$$\|T(t)\| \leq Me^{\gamma t}, \quad t \geq 0. \tag{1.2}$$

Let

$$\nu_0 = \inf\{\gamma \in \mathbb{R} \mid \text{there exists } M > 0 \text{ such that } \|T(t)\| \leq Me^{\gamma t}, \forall t \geq 0\},$$

then ν_0 is called the growth exponent of the semigroup $T(t)$ $(t \geq 0)$. Furthermore, ν_0 can also be obtained by the following formula:

$$\nu_0 = \limsup_{t \to +\infty} \frac{\ln \|T(t)\|}{t}. \tag{1.3}$$

If C_0-semigroup $T(t)$ is continuous in the uniform operator topology for every $t > 0$ in E, it is well known that ν_0 can also be determined by $\sigma(A)$ (see [13])

$$\nu_0 = -\inf\{\operatorname{Re}\lambda \mid \lambda \in \sigma(A)\}, \tag{1.4}$$

where $-A$ is the infinitesimal generator of the C_0-semigroup $T(t)$ $(t \geq 0)$. We know that $T(t)$ $(t \geq 0)$ is continuous in the uniform operator topology for $t > 0$ if $T(t)$ $(t \geq 0)$ is a compact semigroup (see [1]).

Our main results are as follows:

Theorem 1.1 *Let E be an ordered Banach space, whose positive cone K is a normal cone, $A : D(A) \subset E \to E$ be a closed linear operator, and $-A$ generate an exponentially stable positive compact semigroup $T(t)$ $(t \geq 0)$ in E, that is, $\nu_0 < 0$. Assume that $f : \mathbb{R} \times K \to K$ is continuous and $f(t,x)$ is ω-periodic in t. If the following condition is satisfied:*

(H1) *there is a constant $c \in (0, |\nu_0|)$ and a function $h_0 \in C_\omega(\mathbb{R}, K)$ such that*

$$f(t,x) \leq cx + h_0(t), \quad t \in \mathbb{R}, x \in K,$$

then (1.1) *has at least one positive ω-periodic mild solution u.*

Theorem 1.2 *Let E be an ordered Banach space, whose positive cone K is a normal cone, $A : D(A) \subset E \to E$ be a closed linear operator, and $-A$ generate an exponentially stable positive compact semigroup $T(t)$ $(t \geq 0)$ in E. Assume that $f : \mathbb{R} \times K \to K$ is continuous and $f(t,x)$ is ω-periodic in t. If the following condition is satisfied:*

(H2) *there is a constant $c \in (0, |\nu_0|)$ such that for $x, y \in K$ with $y \leq x$,*

$$f(t,x) - f(t,y) \leq c(x - y), \quad t \in \mathbb{R},$$

then (1.1) *has a unique positive ω-periodic mild solution u.*

Furthermore, we assume that the positive cone K is a regeneration cone. By the characteristic of positive semigroups (see [14]), for sufficiently large $\lambda_0 > -\inf\{\mathrm{Re}\,\lambda \mid \lambda \in \sigma(A)\}$, we see that $\lambda_0 I + A$ has a positive bounded inverse operator $(\lambda_0 I + A)^{-1}$. Since $\sigma(A) \neq \emptyset$, the spectral radius $r((\lambda_0 I + A)^{-1}) = \frac{1}{\mathrm{dist}(-\lambda_0, \sigma(A))} > 0$. By the famous Krein-Rutmann theorem, A has the first eigenvalue λ_1, which has a positive eigenfunction e_1, and

$$\lambda_1 = \inf\{\mathrm{Re}\,\lambda \mid \lambda \in \sigma(A)\}, \tag{1.5}$$

that is, $\nu_0 = -\lambda_1$. Hence, by Theorem 1.1 and Theorem 1.2, we have the following.

Corollary 1.3 *Let E be an ordered Banach space, whose positive cone K is a normal regeneration cone. $-A$ generates an exponentially stable positive compact semigroup $T(t)$ $(t \geq 0)$ in E. Assume that $f : \mathbb{R} \times K \to K$ is continuous and $f(t,x)$ is ω-periodic in t. If the following condition is satisfied:*

(H1)′ *there is a constant $c \in (0, \lambda_1)$ and a function $h_0 \in C_\omega(\mathbb{R}, K)$ such that*

$$f(t,x) \leq cx + h_0(t), \quad t \in \mathbb{R}, x \in K,$$

then (1.1) *has at least one positive ω-periodic mild solution u.*

Corollary 1.4 *Let E be an ordered Banach space, whose positive cone K is a normal regeneration cone. $-A$ generates an exponentially stable positive compact semigroup $T(t)$ $(t \geq 0)$ in E. Assume that $f : \mathbb{R} \times K \to K$ is continuous and $f(t,x)$ is ω-periodic in t. If the following condition is satisfied:*

(H2)′ *there is a constant* $c \in (0, \lambda_1)$ *such that for* $x, y \in K$ *with* $y \leq x$,

$$f(t,x) - f(t,y) \leq c(x - y), \quad t \in \mathbb{R},$$

then (1.1) *has a unique positive* ω*-periodic mild solution* u.

Remark 1.5 In Corollary 1.3 and Corollary 1.4, since λ_1 is the first eigenvalue of A, the condition $c < \lambda_1$ in (H1)′ and (H2)′ cannot be extended to $c \leq \lambda_1$. Otherwise, (1.1) does not always have a mild solution. For example, $f(t,x) = \lambda_1 x$.

The paper is organized as follows. Section 2 provides the definitions and preliminary results to be used in theorems stated and proved in the paper. In Section 3, we apply the operator semigroup theory and the Leray-Schauder fixed point theorem to prove Theorem 1.1 and Theorem 1.2. In the last section, we give an example to illustrate the applicability of the abstract results.

2 Preliminaries

In this section, we introduce some notions, definitions, and preliminary facts which are used through this paper.

Let J denote the infinite interval $[0, +\infty)$ and $h : J \to E$, consider the initial value problem of the linear evolution equation

$$\begin{cases} u'(t) + Au(t) = h(t), & t \in J, \\ u(0) = x_0. \end{cases} \tag{2.1}$$

It is well known [12], Chapter 4, Theorem 2.9, when $x_0 \in D(A)$ and $h \in C^1(J, E)$, the initial value problem (2.1) has a unique classical solution $u \in C^1(J, E) \cap C(J, E_1)$ expressed by

$$u(t) = T(t)x_0 + \int_0^t T(t-s)h(s)\, ds, \tag{2.2}$$

where $E_1 = D(A)$ is Banach space with the graph norm $\|\cdot\|_1 = \|\cdot\| + \|A\cdot\|$. Generally, for $x_0 \in E$ and $h \in C(J, E)$, the function u given by (2.2) belongs to $C(J, E)$ and it is called a mild solution of the linear evolution equation (2.1).

Let $C_\omega(\mathbb{R}, E)$ denote the Banach space $\{u \in C(\mathbb{R}, E) \mid u(t+\omega) = u(t), t \in \mathbb{R}\}$ endowed with the maximum norm $\|u\|_C = \max_{t \in [0,\omega]} \|u(t)\|$. Evidently, $C_\omega(\mathbb{R}, E)$ is also an ordered Banach space with the partial order '$\leq$' induced by the positive cone $K_C = \{u \in C_\omega(\mathbb{R}, E) \mid u(t) \geq \theta, t \in \mathbb{R}\}$ and K_C is also normal with the normal constant N.

Given $h \in C_\omega(\mathbb{R}, E)$, for the following linear evolution equation corresponding to (1.1):

$$u'(t) + Au(t) = h(t), \quad t \in \mathbb{R}, \tag{2.3}$$

we have the following result.

Lemma 2.1 ([10]) *If* $-A$ *generates an exponentially stable positive* C_0*-semigroup* $T(t)$ $(t \geq 0)$ *in* E, *then for* $h \in C_\omega(\mathbb{R}, E)$, *for the linear evolution equation* (2.3) *there exists a unique* *ω-periodic mild solution u, which can be expressed by*

$$u(t) = \big(I - T(\omega)\big)^{-1} \int_{t-\omega}^{t} T(t-s)h(s)\,ds := (Ph)(t), \tag{2.4}$$

and the solution operator $P : C_\omega(\mathbb{R}, E) \to C_\omega(\mathbb{R}, E)$ is a positive bounded linear operator with the spectral radius $r(P) \le \frac{1}{|\nu_0|}$.

Proof For any $\nu \in (0, |\nu_0|)$, there exists $M > 0$ such that

$$\|T(t)\| \le Me^{-\nu t} \le M, \quad t \ge 0. \tag{2.5}$$

In E, define the equivalent norm $|\cdot|$ by

$$|x| = \sup_{t\ge 0} \|e^{\nu t}T(t)x\|,$$

then $\|x\| \le |x| \le M\|x\|$. By $|T(t)|$ we denote the norm of $T(t)$ in $(E, |\cdot|)$, then for $t \ge 0$, it is easy to obtain $|T(t)| < e^{-\nu t}$. Hence, $(I - T(\omega))$ has a bounded inverse operator

$$\big(I - T(\omega)\big)^{-1} = \sum_{n=0}^{\infty} T(n\omega), \tag{2.6}$$

and its norm satisfies

$$|\big(I - T(\omega)\big)^{-1}| \le \frac{1}{1 - |T(\omega)|} \le \frac{1}{1 - e^{-\nu\omega}}. \tag{2.7}$$

Set

$$x_0 = \big(I - T(\omega)\big)^{-1} \int_0^{\omega} T(t-s)h(s)\,ds := Bh, \tag{2.8}$$

then the mild solution $u(t)$ of the linear initial value problem (2.1) given by (2.2) satisfies the periodic boundary condition $u(0) = u(\omega) = x_0$. For $t \in \mathbb{R}^+$, by (2.2) and the properties of the semigroup $T(t)$ $(t \ge 0)$, we have

$$\begin{aligned} u(t+\omega) &= T(t+\omega)u(0) + \int_0^{t+\omega} T(t+\omega-s)h(s)\,ds \\ &= T(t)\Big(T(\omega)u(0) + \int_0^{\omega} T(\omega-s)h(s)\,ds\Big) + \int_0^{t} T(t-s)h(s-\omega)\,ds \\ &= T(t)u(0) + \int_0^{t} T(t-s)h(s)\,ds = u(t). \end{aligned}$$

Therefore, the ω-periodic extension of u on $\mathbb{R}$ is a unique ω-periodic mild solution of (2.3). By (2.2) and (2.8), the ω-periodic mild solution can be expressed by

$$\begin{aligned} u(t) &= T(t)B(h) + \int_0^{t} T(t-s)h(s)\,ds \\ &= \big(I - T(\omega)\big)^{-1} \int_{t-\omega}^{t} T(t-s)h(s)\,ds := (Ph)(t). \end{aligned} \tag{2.9}$$

Evidently, by the positivity of the semigroup $T(t)$ $(t \geq 0)$, we can see that $P : C_\omega(\mathbb{R}, E) \to C_\omega(\mathbb{R}, E)$ is a positive bounded linear operator. By (2.7) and (2.9), we have

$$\begin{aligned} |(Ph)(t)| &\leq |(I - T(\omega))^{-1}| \int_{t-\omega}^{t} |T(t-s)h(s)|\, ds \\ &\leq \frac{1}{1 - e^{-\nu\omega}} \int_{t-\omega}^{t} e^{-\nu(t-s)} |h|_C \, ds \leq \frac{1}{\nu} |h|_C, \end{aligned}$$

which implies that $|P| \leq \frac{1}{\nu}$. Therefore, $r(P) \leq |P| \leq \frac{1}{\nu}$. Hence, by the arbitrariness of $\nu \in (0, |\nu_0|)$, we have $r(P) \leq \frac{1}{|\nu_0|}$. This completes the proof of Lemma 2.1. □

In the proof of our main results, we need the following fixed point theorem.

Lemma 2.2 (Leray-Schauder fixed point theorem [15]) *Let Ω be a convex subset of Banach space E with $\theta \in \Omega$, and let $Q : \Omega \to \Omega$ be a compact operator. If the set $\{u \in \Omega \mid u = \eta Q u, 0 < \eta < 1\}$ is bounded, then Q has a fixed point in Ω.*

3 Proof of the main results

Proof of Theorem 1.1 Evidently, the normal cone K_C is a convex subset of the Banach space $C_\omega(\mathbb{R}, E)$ and $\theta \in K_C$. Consider the operator Q defined by

$$Qu = (P \circ F)(u), \tag{3.1}$$

where

$$F(u) := f(\cdot, u(\cdot)), \quad u \in K_C. \tag{3.2}$$

From the positivity of semigroup of $T(t)$ $(t \geq 0)$ and the conditions of Theorem 1.1, it is easy to see that $Q : K_C \to K_C$ is well defined. It is clear that

$$(Qu)(t) = (I - T(\omega))^{-1} \int_{t-\omega}^{t} T(t-s) f(s, u(s))\, ds, \quad t \in \mathbb{R}. \tag{3.3}$$

By the definition of P, the positive ω-periodic mild solution of (1.1) is equivalent to the fixed point of the operator Q. In the following, we will prove Q has a fixed point by applying the famous Leray-Schauder fixed point theorem.

At first, we prove that Q is continuous on K_C. To this end, let $\{u_m\} \subset K_C$ be a sequence such that $u_m \to u \in K_C$ as $m \to \infty$, so for every $t \in \mathbb{R}$, $\lim_{m\to\infty} u_m(t) = u(t)$. Since $f : \mathbb{R} \times K \to K$ is continuous, for every $t \in \mathbb{R}$, we get

$$f(t, u_m(t)) \to f(t, u(t)), \quad m \to \infty. \tag{3.4}$$

By (3.3) and the Lebesgue dominated convergence theorem, for every $t \in \mathbb{R}$, we have

$$\begin{aligned} &\|(Qu_m)(t) - (Qu)(t)\| \\ &= \left\| (I - T(\omega))^{-1} \left(\int_{t-\omega}^{t} T(t-s) f(s, u_m(s))\, ds - \int_{t-\omega}^{t} T(t-s) f(s, u(s))\, ds \right) \right\| \end{aligned}$$

$$\leq \|(I - T(\omega))^{-1}\| \cdot \int_{t-\omega}^{t} \|T(t-s)\| \cdot \|f(s, u_m(s)) - f(s, u(s))\| \, ds$$

$$\leq CM \cdot \int_{t-\omega}^{t} \|f(s, u_m(s)) - f(s, u(s))\| \, ds, \tag{3.5}$$

where $C = \|(I - T(\omega))^{-1}\|$. Therefore, we can conclude that

$$\|Qu_m - Qu\| \to 0, \quad m \to \infty. \tag{3.6}$$

Thus, $Q : K_C \to K_C$ is continuous.

Subsequently, we show that Q maps every bounded set in K_C into a bounded set. For any $R > 0$, let

$$\Omega_R := \{u \in K_C \mid \|u\|_C \leq R\}. \tag{3.7}$$

For each $u \in \Omega_R$, from the continuity of f, we know that there exists $M_1 > 0$ such that

$$\|f(t, u(t))\| \leq M_1, \quad t \in \mathbb{R}, \tag{3.8}$$

hence, we get

$$\begin{aligned}
\|(Qu)(t)\| &= \left\|(I - T(\omega))^{-1} \int_{t-\omega}^{t} T(t-s)f(s, u(s)) \, ds\right\| \\
&\leq \|(I - T(\omega))^{-1}\| \int_{t-\omega}^{t} \|T(t-s)\| \cdot \|f(s, u(s))\| \, ds \\
&\leq CM \int_{t-\omega}^{t} M_1 \, ds \\
&\leq CMM_1\omega := \overline{R}.
\end{aligned}$$

Therefore, $Q(\Omega_R)$ is bounded.

Next, we demonstrate that $Q(\Omega_R)$ is equicontinuous. For every $u \in \Omega_R$, by the periodicity of u, we only consider it on $[0, \omega]$. Set $0 \leq t_1 < t_2 \leq \omega$, we get

$$\begin{aligned}
&Qu(t_2) - Qu(t_1) \\
&\quad = (I - T(\omega))^{-1} \int_{t_2-\omega}^{t_2} T(t_2 - s)f(s, u(s)) \, ds \\
&\qquad - (I - T(\omega))^{-1} \int_{t_1-\omega}^{t_1} T(t_1 - s)f(s, u(s)) \, ds \\
&\quad = (I - T(\omega))^{-1} \int_{t_2-\omega}^{t_1} (T(t_2 - s) - T(t_1 - s))f(s, u(s)) \, ds \\
&\qquad - (I - T(\omega))^{-1} \int_{t_1-\omega}^{t_2-\omega} T(t_1 - s)f(s, u(s)) \, ds \\
&\qquad + (I - T(\omega))^{-1} \int_{t_1}^{t_2} T(t_2 - s)f(s, u(s)) \, ds \\
&\quad := I_1 + I_2 + I_3,
\end{aligned}$$

where

$$I_1 = (I - T(\omega))^{-1} \int_{t_2-\omega}^{t_1} (T(t_2 - s) - T(t_1 - s))f(s, u(s))\, ds,$$

$$I_2 = -(I - T(\omega))^{-1} \int_{t_1-\omega}^{t_2-\omega} T(t_1 - s)f(s, u(s))\, ds,$$

$$I_3 = (I - T(\omega))^{-1} \int_{t_1}^{t_2} T(t_2 - s)f(s, u(s))\, ds.$$

It is clear that

$$\|Qu(t_2) - Qu(t_1)\| \leq \|I_1\| + \|I_2\| + \|I_3\|. \tag{3.9}$$

Now, we only need to check that the $\|I_i\|$ tend to 0 independently of $u \in \Omega_R$ when $t_2 - t_1 \to 0$, $i = 1, 2, 3$. From the definition of the I_i, we can easily see

$$\begin{aligned}
\|I_1\| &\leq C \cdot \int_{t_2-\omega}^{t_1} \|(T(t_2 - s) - T(t_1 - s))\| \cdot \|f(s, u(s))\|\, ds \\
&\leq CM_1 \int_{t_2-\omega}^{t_1} \|(T(t_2 - s) - T(t_1 - s))\|\, ds \\
&\to 0, \quad \text{as } t_2 - t_1 \to 0, \\
\|I_2\| &\leq C \cdot \int_{t_1-\omega}^{t_2-\omega} \|(T(t_1 - s))\| \cdot \|f(s, u(s))\|\, ds \\
&\leq CMM_1(t_2 - t_1) \\
&\to 0, \quad \text{as } t_2 - t_1 \to 0, \\
\|I_3\| &\leq C \cdot \int_{t_1}^{t_2} \|(T(t_2 - s))\| \cdot \|f(s, u(s))\|\, ds \\
&\leq CMM_1(t_2 - t_1)\, ds \\
&\to 0, \quad \text{as } t_2 - t_1 \to 0.
\end{aligned}$$

As a result, $\|Qu(t_2) - Qu(t_1)\|$ tends to 0 independently of $u \in \Omega_R$ as $t_2 - t_1 \to 0$, which means that $Q(\Omega_R)$ is equicontinuous.

Now, we prove that $(Q\Omega_R)(t)$ is relatively compact in K for all $t \in \mathbb{R}$. To this end, we define a set $(Q_\varepsilon \Omega_R)(t)$ by

$$(Q_\varepsilon \Omega_R)(t) := \{(Q_\varepsilon u)(t) \mid u \in \Omega_R, 0 < \varepsilon < \omega, t \in \mathbb{R}\}, \tag{3.10}$$

where

$$\begin{aligned}
(Q_\varepsilon u)(t) &= (I - T(\omega))^{-1} \int_{t-\omega}^{t-\varepsilon} T(t - s)f(s, u(s))\, ds \\
&= T(\varepsilon)(I - T(\omega))^{-1} \int_{t-\omega}^{t-\varepsilon} T(t - s - \varepsilon)f(s, u(s))\, ds.
\end{aligned}$$

Then the set $(Q_\varepsilon \Omega_R)(t)$ is relatively compact in K since the operator $T(\varepsilon)$ is compact in K. For any $u \in \Omega_R$ and $t \in \mathbb{R}$, from the following inequality:

$$\begin{aligned}
&\|Qu(t) - Q_\varepsilon u(t)\| \\
&= \left\| (I - T(\omega))^{-1} \left(\int_{t-\omega}^{t} T(t-s)f(s,u(s))\,ds - \int_{t-\omega}^{t-\varepsilon} T(t-s)f(s,u(s)) \right) ds \right\| \\
&\le C \int_{t-\varepsilon}^{t} \|T(t-s)f(s,u(s))\|\,ds \\
&\le CMM_1\varepsilon,
\end{aligned} \tag{3.11}$$

one see that the set $(Q\Omega_R)(t)$ is relatively compact in K for all $t \in \mathbb{R}$.

Thus, the Arzela-Ascoli theorem guarantees that $Q : K_C \to K_C$ is a compact operator.

Finally, we prove the set $\Lambda(Q) := \{u \in K_C \mid u = \eta Qu, \forall 0 < \eta < 1\}$ is bounded. For every $u \in K_C$, by (3.2) and the condition (H1), we have

$$\theta \le F(u)(t) = f(t,u(t)) \le cu(t) + h_0(t), \quad t \in \mathbb{R}. \tag{3.12}$$

Let $u \in \Lambda(Q)$, then there is a constant $\eta \in (0,1)$ such that $u = \eta Qu$. Therefore, by the definition of Q, Lemma 2.1, and (3.12), we have

$$\begin{aligned}
\theta \le u(t) &= \eta(Qu)(t) < (Qu)(t) \\
&= P \circ F(u)(t) \le P(cu(t) + h_0(t)) \\
&= cPu(t) + Ph_0(t) < c(P \circ Qu)(t) + Ph_0(t) \\
&\le cP(cPu(t) + Ph_0(t)) + Ph_0(t) \\
&= c^2P^2u(t) + cP^2h_0(t) + Ph_0(t),
\end{aligned}$$

inductively, we can see

$$u(t) \le c^n P^n u(t) + \mathcal{P}h_0(t), \quad n = 1, 2, \ldots, \tag{3.13}$$

where $\mathcal{P} = c^{n-1}P^n + c^{n-2}P^{n-1} + \cdots + cP^2 + P$ is a bounded linear operator, and there exists a constant $M_2 > 0$ such that $\|\mathcal{P}\| \le M_2$. Hence, by the normality of the cone K_C, we can see

$$\|u\|_C < c^n N \|P^n\| \cdot \|u\|_C + M_2 \|h_0\|_C. \tag{3.14}$$

From the spectral radius of the Gelfand formula $\lim_{n\to\infty} \sqrt[n]{\|P^n\|} = r(P) = \frac{1}{|\nu_0|}$, and the condition (H1), when n is large enough, we get $c^n\|P^n\| < \frac{1}{N}$, then

$$\|u\|_C < \frac{M_2\|h_0\|_C}{1 - Nc^n\|P^n\|}, \tag{3.15}$$

which implies that $\Lambda(Q)$ is bounded. By the Leray-Schauder fixed point theorem of a compact operator, the operator Q has at least one fixed point u in K_C, which is a positive ω-periodic mild solution of (1.1). This completes the proof of Theorem 1.1. □

Proof of Theorem 1.2 From the condition (H2), it is easy to see that the condition (H1) holds. Hence by Theorem 1.1, (1.1) has positive ω-periodic mild solutions. Let $u_1, u_2 \in K_C$ be the positive ω-periodic solutions of (1.1), then they are the fixed points of the operator $Q = P \circ F$. Let us assume $u_1 \leq u_2$, by the definition of F and the condition (H2), we see that $F(u_2)(t) - F(u_1)(t) \leq c(u_2(t) - u_1(t))$. So by Lemma 2.1, we obtain

$$\begin{aligned}\theta &\leq u_2(t) - u_1(t) = (Qu_2)(t) - (Qu_1)(t) = P\big((Fu_2)(t) - (Fu_1)(t)\big)\\ &\leq cP\big(u_2(t) - u_1(t)\big) \leq \cdots \leq c^n P^n\big(u_2(t) - u_1(t)\big).\end{aligned}$$

By the normality of the cone K_C, we can see that

$$\|u_2 - u_1\|_C \leq Nc^n \big\|P^n\big\| \cdot \|u_2 - u_1\|_C. \tag{3.16}$$

From the proof of Theorem 1.1, when n is large enough, $Nc^n\|P^n\| < 1$, so $\|u_2 - u_1\|_C = 0$, it follows that $u_2 \equiv u_1$. Thus, (1.1) has only one positive ω-periodic mild solution. □

4 Application

In this section, we present one example, which indicates how our abstract results can be applied to concrete problems.

Let $\Omega \in \mathbb{R}^N$ be a bounded domain with a sufficiently smooth boundary $\partial\Omega$. Let

$$A(x,D)u = -\sum_{i,j=1}^{N} a_{ij}(x)D_iD_ju + \sum_{j=1}^{N} a_j(x)D_ju + a_0(x)u \tag{4.1}$$

be a uniformly elliptic differential operator in $\overline{\Omega}$, whose coefficients $a_{ij}(x)$, $a_j(x)$ $(i,j = 1,\ldots,N)$ and $a_0(x)$ are Hölder-continuous on $\overline{\Omega}$ and $a_0(x) \geq 0$. We let $B = B(x,D)$ be a boundary operator on $\partial\Omega$ of the form

$$Bu := b_0(x)u + \delta\frac{\partial u}{\partial \beta}, \tag{4.2}$$

where either $\delta = 0$ and $b_0(x) \equiv 1$ (Dirichlet boundary operator), or $\delta = 1$ and $b_0(x) \geq 0$ (regular oblique derivative boundary operator; at this point, we further assume that $a_0(x) \not\equiv 0$ or $b_0(x) \not\equiv 0$), β is an outward pointing, nowhere tangent vector field on $\partial\Omega$.

Let λ_1 be the first eigenvalue of elliptic operator $A(x,D)$ under the boundary condition $Bu = 0$. It is well known ([16], Theorem 1.16) that $\lambda_1 > 0$.

Under the above assumptions, we discuss the existence and uniqueness of positive time ω-periodic solutions of the semilinear parabolic boundary value problem

$$\begin{cases}\frac{\partial}{\partial t}u(x,t) + A(x,D)u(x,t) = g(x,t,u(x,t)), & (x,t) \in \Omega \times \mathbb{R},\\ Bu = 0, \quad x \in \partial\Omega,\end{cases} \tag{4.3}$$

where $g : \overline{\Omega} \times \mathbb{R} \times \mathbb{R} \to \mathbb{R}$ is a local Hölder-continuous function which is ω-periodic in t.

Let $E = L^p(\Omega)$ $(p > 1)$, $K = \{u \in E \mid u(x) \geq 0 \text{ a.e. } x \in \Omega\}$, then E is an ordered Banach space, whose positive cone K is a normal regeneration cone. Define an operator $A : D(A) \subset E \to E$ by

$$D(A) = \big\{u \in W^{2,p}(\Omega) \mid B(x,D)u = 0, x \in \partial\Omega\big\}, \qquad Au = A(x,D)u. \tag{4.4}$$

If $a_0(x) \geq 0$, then $-A$ generates an exponentially stable analytic semigroup $T_p(t)$ $(t \geq 0)$ in E (see [17]). By the maximum principle of elliptic operators, we know that $(\lambda I + A)$ has a positive bounded inverse operator $(\lambda I + A)^{-1}$ for $\lambda > 0$, hence $T_p(t)$ $(t \geq 0)$ is a positive semigroup (see [14]). From the operator A has compact resolvent in $L^p(\Omega)$, we obtain $T_p(t)$ $(t \geq 0)$ is also a compact semigroup (see [12]).

Therefore, by Corollary 1.3, we have the following result.

Theorem 4.1 *Assume that* $g : \overline{\Omega} \times \mathbb{R} \times \mathbb{R} \to \mathbb{R}$ *is a local Hölder-continuous function which is* ω*-periodic in* t *and satisfies* $g(x, t, u) \geq 0$ *for* $(x, t, u) \in (\overline{\Omega} \times \mathbb{R} \times \mathbb{R}^+)$. *If the following condition holds*:

(H3) *there is a constant* $c \in (0, \lambda_1)$ *and a function* $h_0 \in C_\omega(\Omega \times \mathbb{R})$ *satisfying* $h_0(x, t) \geq 0$ *such that*

$$g(x, t, u) \leq cu + h_0(x, t), \quad (x, t) \in \Omega \times \mathbb{R}, u \geq 0,$$

then the parabolic boundary value problem (4.3) *has at least one positive* ω*-periodic solution* $u \in C^{2,1}(\overline{\Omega} \times \mathbb{R})$.

Proof Let $u(t) = u(\cdot, t), f(t, u(t)) = g(\cdot, t, u(\cdot, t))$, then the parabolic boundary value problem (4.3) can be reformulated as the abstract evolution (1.1) in E. It is easy to see that the conditions of Corollary 1.3 are satisfied. By Corollary 1.3, the parabolic boundary value problem (4.3) has a time positive ω-periodic mild solution $u \in C_\omega(\mathbb{R}, E)$. By the analyticity of the semigroup $T_p(t)$ $(t \geq 0)$ and the regularization method used in [17], we can see that $u \in C^{2,1}(\overline{\Omega} \times \mathbb{R})$ is a classical time ω-periodic solution of the problem (4.3). This completes the proof of the theorem. □

From Corollary 1.4 and Theorem 4.1, we obtain the uniqueness result.

Theorem 4.2 *Assume that* $g : \overline{\Omega} \times \mathbb{R} \times \mathbb{R} \to \mathbb{R}$ *is a local Hölder-continuous function which is* ω*-periodic in* t *and satisfies* $g(x, t, u) \geq 0$ *for* $(x, t, u) \in (\overline{\Omega} \times \mathbb{R} \times \mathbb{R}^+)$. *If the following condition holds*:

(H4) *there is a constant* $c \in (0, \lambda_1)$ *such that for* $0 < v < w$

$$g(x, t, w) - g(x, t, v) \leq c(w - v), \quad (x, t) \in \Omega \times \mathbb{R},$$

then the parabolic boundary value problem (4.3) *has a unique positive* ω*-periodic solution* $u \in C^{2,1}(\overline{\Omega} \times \mathbb{R})$.

Competing interests
The authors declare that they have no competing interests.

Authors' contributions
All authors contributed equally to the writing of this paper. All authors read and approved the final manuscript.

Acknowledgements
Research supported by NNSF of China (11261053) and NSF of Gansu Province (1208RJZA129).

References

1. Hino, Y, Murakami, S: Periodic solutions of linear Voltera systems. In: Differential Equations. Lect. Notes Pure Appl. Math., vol. 118, pp. 319-326. Dekker, New York (1987)
2. Vrabie, I: Periodic solutions for nonlinear evolution equations in a Banach space. Proc. Am. Math. Soc. **109**(3), 653-661 (1990)
3. Liu, J: Bounded and periodic solutions of differential equations in Banach space. Appl. Math. Comput. **65**, 141-150 (1994)
4. Liu, J: Bounded and periodic solutions of semi-linear evolution equations. Dyn. Syst. Appl. **4**, 341-350 (1995)
5. Li, Y: Periodic solutions of semilinear evolution equations in Banach spaces. Acta Math. Sin. **41**, 629-636 (1998) (in Chinese)
6. Li, Y, Cong, F, Lin, Z, Lin, W: Periodic solutions for evolution equations. Nonlinear Anal. **36**, 275-293 (1999)
7. Akrid, T, Maniar, L, Ouhinou, A: Periodic solutions for some nonautonomous semilinear boundary evolution equations. Nonlinear Anal. **71**, 1000-1011 (2009)
8. Zitane, M, Bensoida, C: Massera problem for non-autonomous retarded differential equations. J. Math. Anal. Appl. **402**, 453-462 (2013)
9. Li, Y: Existence and uniqueness of periodic solution for a class of semilinear evolution equations. J. Math. Anal. Appl. **349**, 226-234 (2009)
10. Li, Y: Existence and uniqueness of positive periodic solution for abstract semilinear evolution equations. J. Syst. Sci. Math. Sci. **25**, 720-728 (2005) (in Chinese)
11. Li, Y: Existence and asymptotic stability of periodic solution for evolution equations with delays. J. Funct. Anal. **261**, 1309-1324 (2011)
12. Pazy, A: Semigroup of Linear Operators and Applications to Partial Differential Equations. Applied Mathematical Sciences, vol. 44. Springer, New York (1983)
13. Triggiani, R: On the stabilizability problem in Banach space. J. Math. Anal. Appl. **52**, 383-403 (1975)
14. Li, Y: The positive solutions of abstract semilinear evolution equations and their applications. Acta Math. Sin. **39**, 666-672 (1996) (in Chinese)
15. Deimling, K: Nonlinear Functional Analysis. Springer, New York (1985)
16. Amann, H: Nonlinear operators in ordered Banach spaces and some applications to nonlinear boundary value problem. In: Nonlinear Operators and the Calculus of Variations. Lecture Notes in Mathematics, pp. 1-55. Springer, Berlin (1976)
17. Amann, H: Periodic solutions of semilinear parabolic equations. In: Cesari, L, Kannan, R, Weinberger, R (eds.) Nonlinear Analysis: A Collection of Papers in Honor of Erich H Rothe, pp. 1-29. Academic Press, New York (1978)

17

On Malmquist type theorem of systems of complex difference equations

Jianjun Zhang[*]

[*]Correspondence:
zhangjianjun1982@163.com
Mathematics and Information Technology School, Jiangsu Second Normal University, Nanjing, Jiangsu 210013, P.R. China

Abstract

The main purpose of this paper is to give the Malmquist type result of the meromorphic solutions of a system of complex difference equations of the following form:

$$\begin{cases} \sum_{\lambda_1\in I_1,\mu_1\in J_1} \alpha_{\lambda_1,\mu_1}(z)(\prod_{\nu=1}^{n} f(z+c_\nu)^{l_{\lambda_1,\nu}} \prod_{\nu=1}^{n} g(z+c_\nu)^{m_{\mu_1,\nu}}) = \frac{\sum_{i=0}^{p} a_i(z)g(z)^i}{\sum_{j=0}^{q} b_j(z)g(z)^j}, \\ \sum_{\lambda_2\in I_2,\mu_2\in J_2} \beta_{\lambda_2,\mu_2}(z)(\prod_{\nu=1}^{n} f(z+c_\nu)^{l_{\lambda_2,\nu}} \prod_{\nu=1}^{n} g(z+c_\nu)^{m_{\mu_2,\nu}}) = \frac{\sum_{k=0}^{s} d_k(z)f(z)^k}{\sum_{l=0}^{t} e_l(z)f(z)^l}, \end{cases}$$

where $c_1, c_2, \ldots, c_n$ are distinct, nonzero complex numbers, the coefficients $\alpha_{\lambda_1,\mu_1}(z)$ ($\lambda_1 \in I_1$, $\mu_1 \in J_1$), $\beta_{\lambda_2,\mu_2}(z)$ ($\lambda_2 \in I_2$, $\mu_2 \in J_2$), $a_i(z)$ ($i = 0, 1, \ldots, p$), $b_j(z)$ ($j = 0, 1, \ldots, q$), $d_k(z)$ ($k = 0, 1, \ldots, s$), and $e_l(z)$ ($l = 0, 1, \ldots, t$) are small functions relative to $f(z)$ and $g(z)$, $I_i = \{\lambda_i = (l_{\lambda_i,1}, l_{\lambda_i,2}, \ldots, l_{\lambda_i,n}) | l_{\lambda_i,\nu} \in N \cup \{0\}, \nu = 1, 2, \ldots, n\}$ ($i = 1, 2$) and $J_j = \{\mu_j = (m_{\mu_j,1}, m_{\mu_j,2}, \ldots, m_{\mu_j,n}) | m_{\mu_j,\nu} \in N \cup \{0\}, \nu = 1, 2, \ldots, n\}$ ($j = 1, 2$) are finite index sets. The growth of meromorphic solutions of a related system of complex functional equations is also investigated.

Keywords: systems of complex difference equations; meromorphic functions; Malmquist type theorem; functional equation

1 Introduction and main results

Let $f(z)$ be a meromorphic function in the complex plane C. We assume that the reader is familiar with the standard notations and results in Nevanlinna's value distribution theory of meromorphic functions (see *e.g.* [1–3]). We use $\rho(f)$ to denote the growth order of a meromorphic function $f(z)$. The notation $S(r,f)$ denotes any quantity that satisfies the condition $S(r,f) = o(T(r,f))$ as $r \to \infty$ possibly outside an exceptional set of r of finite logarithmic measure. A meromorphic function $a(z)$ is called a small function of $f(z)$ if and only if $T(r, a(z)) = S(r,f)$.

In the last ten years, there has been a great deal of interest in studying the properties of complex difference equations (see *e.g.* [4–20]). Especially, a number of papers (see *e.g.* [4, 6, 10, 11, 15, 17–19]) focusing on a Malmquist type theorem of the complex difference equations emerged. In 2000, Ablowitz *et al.* [4] proved some results on the Malmquist theorem of the complex difference equations by utilizing Nevanlinna theory. They obtained the following two results.

Theorem A *If the second-order difference equation*

$$f(z+1)+f(z-1)=\frac{a_0(z)+a_1(z)f+\cdots+a_p(z)f^p}{b_0(z)+b_1(z)f+\cdots+b_q(z)f^q},$$

with polynomial coefficients a_i $(i=1,2,\ldots,p)$ *and* b_j $(j=1,2,\ldots,q)$, *admits a transcendental meromorphic solution of finite order, then* $d=\max\{p,q\}\leq 2$.

Theorem B *If the second-order difference equation*

$$f(z+1)f(z-1)=\frac{a_0(z)+a_1(z)f+\cdots+a_p(z)f^p}{b_0(z)+b_1(z)f+\cdots+b_q(z)f^q},$$

with polynomial coefficients a_i $(i=1,2,\ldots,p)$ *and* b_j $(j=1,2,\ldots,q)$, *admits a transcendental meromorphic solution of finite order, then* $d=\max\{p,q\}\leq 2$.

Subsequently, Heittokangas *et al.* [10], Laine *et al.* [15] and Huang *et al.* [11], respectively, gave some generalizations of the above two results. In 2010, the first author in this paper and Liao [18] obtained the following more general result.

Theorem C *Let* $c_1,c_2,\ldots,c_n$ *be distinct, nonzero complex numbers, and suppose that* $f(z)$ *is a transcendental meromorphic solution of the difference equation*

$$\sum_{\lambda\in I}\alpha_\lambda(z)\left(\prod_{\nu=1}^{n}f(z+c_\nu)^{l_{\lambda,\nu}}\right)=\frac{a_0(z)+a_1(z)f(z)+\cdots+a_p(z)f(z)^p}{b_0(z)+b_1(z)f(z)+\cdots+b_q(z)f(z)^q}, \tag{1}$$

with coefficients $\alpha_\lambda(z)$ $(\lambda\in I)$, $a_i(z)$ $(i=0,1,\ldots,p)$, *and* $b_j(z)$ $(j=0,1,\ldots,q)$, *which are small functions relative to* $f(z)$, *where* $I=\{\lambda=(l_{\lambda,1},l_{\lambda,2},\ldots,l_{\lambda,n})|l_{\lambda,\nu}\in N\cup\{0\},\nu=1,2,\ldots,n\}$ *is a finite index set, and denote*

$$\sigma_\nu=\max_{\lambda}\{l_{\lambda,\nu}\}\quad(\nu=1,2,\ldots,n),\qquad \sigma=\sum_{\nu=1}^{n}\sigma_\nu.$$

If the order $\rho(f)$ *is finite, then* $d=\max\{p,q\}\leq\sigma$.

If all the coefficients in the complex difference equation (1) are rational functions, then in [19], we have the following Malmquist type result, which is reminiscent of the classical Malmquist theorem in complex differential equations.

Theorem D *Let* $c_1,c_2,\ldots,c_n$ *be distinct, nonzero complex numbers and suppose that* $f(z)$ *is a transcendental meromorphic solution of the equation*

$$P[z,f]:=\sum_{\lambda\in I}\alpha_\lambda(z)\left(\prod_{\nu=1}^{n}f(z+c_\nu)^{l_{\lambda,\nu}}\right)=R(z,f(z))=\frac{P(z,f(z))}{Q(z,f(z))},$$

where $I=\{\lambda=(l_{\lambda,1},l_{\lambda,2},\ldots,l_{\lambda,n})|l_{\lambda,\nu}\in N\cup\{0\},\nu=1,2,\ldots,n\}$ *is a finite index set, P and Q are relatively prime polynomials in f over the field of rational functions, the coefficients* α_λ $(\lambda\in I)$ *are rational functions. Denoting the degree of* $P[z,f]$ *by*

$$\gamma_P:=\max_{\lambda\in I}\left\{l_{\lambda,1}+l_{\lambda,2}+\cdots+l_{\lambda,n}|\lambda=(l_{\lambda,1},l_{\lambda,2},\ldots,l_{\lambda,n})\right\}.$$

If $f(z)$ is finite order and has at most finitely many poles, then $R(z,f)$ reduces to a polynomial in f of degree $d \le \gamma_P$.

More recently, people began to study the properties of meromorphic solutions of systems of complex difference equations. In [21], Gao discussed the proximity function and counting function of meromorphic solutions of some classes of systems of complex difference equations. In 2013, Wang *et al.* [16] investigated the growth of meromorphic solutions of systems of complex difference equations.

Now, we give the Malmquist type result of a system of complex difference equations as follows.

Theorem 1 *Let $c_1, c_2, \ldots, c_n$ be distinct, nonzero complex numbers and suppose that $(f(z), g(z))$ is a transcendental meromorphic solution of a system of complex difference equations of the form*

$$\begin{cases} \sum_{\lambda_1\in I_1,\mu_1\in J_1} \alpha_{\lambda_1,\mu_1}(z)(\prod_{\nu=1}^{n} f(z+c_\nu)^{l_{\lambda_1,\nu}} \prod_{\nu=1}^{n} g(z+c_\nu)^{m_{\mu_1,\nu}}) = \frac{\sum_{i=0}^{p} a_i(z)g(z)^i}{\sum_{j=0}^{q} b_j(z)g(z)^j}, \\ \sum_{\lambda_2\in I_2,\mu_2\in J_2} \beta_{\lambda_2,\mu_2}(z)(\prod_{\nu=1}^{n} f(z+c_\nu)^{l_{\lambda_2,\nu}} \prod_{\nu=1}^{n} g(z+c_\nu)^{m_{\mu_2,\nu}}) = \frac{\sum_{k=0}^{s} d_k(z)f(z)^k}{\sum_{l=0}^{t} e_l(z)f(z)^l}, \end{cases} \tag{2}$$

with coefficients $\alpha_{\lambda_1,\mu_1}(z)$ ($\lambda_1 \in I_1$, $\mu_1 \in J_1$), $\beta_{\lambda_2,\mu_2}(z)$ ($\lambda_2 \in I_2$, $\mu_2 \in J_2$), $a_i(z)$ ($i = 0,1,\ldots,p$), $b_j(z)$ ($j = 0,1,\ldots,q$), $d_k(z)$ ($k = 0,1,\ldots,s$), and $e_l(z)$ ($l = 0,1,\ldots,t$) are small functions relative to $f(z)$ and $g(z)$, $a_p(z), b_q(z), d_s(z), e_t(z) \not\equiv 0$, where $I_i = \{\lambda_i = (l_{\lambda_i,1}, l_{\lambda_i,2}, \ldots, l_{\lambda_i,n}) | l_{\lambda_i,\nu} \in N \cup \{0\}, \nu = 1,2,\ldots,n\}$ ($i = 1,2$), and $J_j = \{\mu_j = (m_{\mu_j,1}, m_{\mu_j,2}, \ldots, m_{\mu_j,n}) | m_{\mu_j,\nu} \in N \cup \{0\}, \nu = 1,2,\ldots,n\}$ ($j = 1,2$) are finite index sets, and denote

$$\xi_{1,\nu} = \max_{\lambda_1\in I_1}\{l_{\lambda_1,\nu}\}, \qquad \eta_{1,\nu} = \max_{\mu_1\in J_1}\{m_{\mu_1,\nu}\},$$

$$\xi_{2,\nu} = \max_{\lambda_2\in I_2}\{l_{\lambda_2,\nu}\}, \qquad \eta_{2,\nu} = \max_{\mu_2\in J_2}\{m_{\mu_2,\nu}\}$$

($\nu = 1,2,\ldots,n$), and

$$\sigma_{11} = \sum_{\nu=1}^{n} \xi_{1,\nu}, \qquad \sigma_{12} = \sum_{\nu=1}^{n} \eta_{1,\nu}, \qquad \sigma_{21} = \sum_{\nu=1}^{n} \xi_{2,\nu}, \qquad \sigma_{22} = \sum_{\nu=1}^{n} \eta_{2,\nu}.$$

If $\max\{p,q\} > \sigma_{12}$, $\max\{s,t\} > \sigma_{21}$, and $\max\{\rho(f),\rho(g)\} < +\infty$, then $\rho(f) = \rho(g)$ and $(\max\{p,q\} - \sigma_{12}) \cdot (\max\{s,t\} - \sigma_{21}) \le \sigma_{11}\sigma_{22}$.

Example 1 It is easy to check that $(f(z), g(z)) = (\tan z, \cot z)$ satisfies the following system of difference equations:

$$\begin{cases} f(z+\frac{\pi}{4})g(z+\frac{\pi}{3})^2 + zg(z+\frac{\pi}{4}) \\ \quad = \frac{(3z+1)g^4+[(2\sqrt{3}-6)z+2-2\sqrt{3}]g^3+(4-4\sqrt{3})(z+1)g^2+[(2\sqrt{3}-2)z+6-2\sqrt{3}]g+z+3}{3g^4+2\sqrt{3}g^3-2g^2-2\sqrt{3}g-1}, \\ f(z+\frac{\pi}{3})^2 g(z+\frac{\pi}{4}) + f(z+\frac{\pi}{3})g(z+\frac{\pi}{4})^2 = \frac{-(\sqrt{3}+1)f^4-2f^3+2f^2-2f+3+\sqrt{3}}{3f^4+(6-2\sqrt{3})f^3+(4-4\sqrt{3})f^2+(2-2\sqrt{3})f+1}. \end{cases}$$

In Example 1, we have $\max\{p,q\} = 4$, $\max\{s,t\} = 4$, $\sigma_{11} = 1$, $\sigma_{12} = 3$, $\sigma_{21} = 2$, $\sigma_{22} = 2$, $\rho(f) = \rho(g) = 1 < +\infty$, and $(\max\{p,q\} - \sigma_{12}) \cdot (\max\{s,t\} - \sigma_{21}) = (4-3)(4-2) = 2 = 1 \times 2 = \sigma_{11}\sigma_{22}$. Therefore, the estimation in Theorem 1 is sharp.

Remark 1 Obviously, if the condition $\max\{p,q\} > \sigma_{12}$, $\max\{s,t\} > \sigma_{21}$ in Theorem 1 is replaced by $(\max\{p,q\} - \sigma_{12})(\max\{s,t\} - \sigma_{21}) = 0$ or $(\max\{p,q\} - \sigma_{12})(\max\{s,t\} - \sigma_{21}) < 0$, the estimation $(\max\{p,q\} - \sigma_{12}) \cdot (\max\{s,t\} - \sigma_{21}) \leq \sigma_{11}\sigma_{22}$ is still correct. If $\sigma_{11} = 0$ or $\sigma_{22} = 0$, then the first or second equation in (2) gets the form of (1). For some results as regards (1), the reader may refer to the paper [18].

Remark 2 If the condition $\max\{p,q\} > \sigma_{12}$, $\max\{s,t\} > \sigma_{21}$ in Theorem 1 is replaced by $\max\{p,q\} < \sigma_{12}$, $\max\{s,t\} < \sigma_{21}$, then the estimation $(\max\{p,q\} - \sigma_{12}) \cdot (\max\{s,t\} - \sigma_{21}) \leq \sigma_{11}\sigma_{22}$ is not true generally. For example, $(f(z), g(z)) = (\tan z, \cot z)$ satisfies the following system of complex difference equations:

$$\begin{cases} f(z+\frac{\pi}{4})^2 g(z-\frac{\pi}{4}) g(z+\frac{\pi}{4})^5 + 2g(z+\frac{\pi}{4})^2 = \frac{g^2-2g+1}{g^2+2g+1}, \\ f(z+\frac{\pi}{4})^4 f(z-\frac{\pi}{4}) g(z+\frac{\pi}{4}) + zf(z+\frac{\pi}{4}) = \frac{-(z+1)f^2-2f+z+1}{f^2-2f+1}, \end{cases} \tag{3}$$

where $\max\{p,q\} = 2$, $\max\{s,t\} = 2$, $\sigma_{11} = 2$, $\sigma_{12} = 6$, $\sigma_{21} = 5$, $\sigma_{22} = 1$, $\max\{p,q\} < \sigma_{12}$, $\max\{s,t\} < \sigma_{21}$. However, $(\max\{p,q\} - \sigma_{12}) \cdot (\max\{s,t\} - \sigma_{21}) = (-4) \times (-3) = 12 > 2 = \sigma_{11}\sigma_{22}$.

If $\sigma_{12} = \sigma_{21} = 0$, then we have the following simpler result.

Corollary 1 *Let $c_1, c_2, \ldots, c_n$ be distinct, nonzero complex numbers, and suppose that $(f(z), g(z))$ is a transcendental meromorphic solution of a system of complex difference equations of the form*

$$\begin{cases} \sum_{\lambda\in I} \alpha_\lambda(z)(\prod_{\nu=1}^{n} f(z+c_\nu)^{l_{\lambda,\nu}}) = \frac{\sum_{i=0}^{p} a_i(z) g(z)^i}{\sum_{j=0}^{q} b_j(z) g(z)^j}, \\ \sum_{\mu\in J} \beta_\mu(z)(\prod_{\nu=1}^{n} g(z+c_\nu)^{m_{\mu,\nu}}) = \frac{\sum_{k=0}^{s} d_k(z) f(z)^k}{\sum_{l=0}^{t} e_l(z) f(z)^l}, \end{cases} \tag{4}$$

with coefficients $\alpha_\lambda(z)$ $(\lambda \in I)$, $\beta_\mu(z)$ $(\mu \in J)$, $a_i(z)$ $(i = 0,1,\ldots,p)$, $b_j(z)$ $(j = 0,1,\ldots,q)$, $d_k(z)$ $(k = 0,1,\ldots,s)$, and $e_l(z)$ $(l = 0,1,\ldots,t)$ are small functions relative to $f(z)$ and $g(z)$, $a_p(z), b_q(z), d_s(z), e_t(z) \not\equiv 0$, where $I = \{\lambda = (l_{\lambda,1}, l_{\lambda,2}, \ldots, l_{\lambda,n}) | l_{\lambda,\nu} \in N \cup \{0\}, \nu = 1,2,\ldots,n\}$ and $J = \{\mu = (m_{\mu,1}, m_{\mu,2}, \ldots, m_{\mu,n}) | m_{\mu,\nu} \in N \cup \{0\}, \nu = 1,2,\ldots,n\}$ are two finite index sets, and denote

$$\xi_\nu = \max_\lambda\{l_{\lambda,\nu}\} \quad (\nu = 1,2,\ldots,n), \qquad \sigma_1 = \sum_{\nu=1}^{n} \xi_\nu$$

and

$$\eta_\nu = \max_\mu\{m_{\mu,\nu}\} \quad (\nu = 1,2,\ldots,n), \qquad \sigma_2 = \sum_{\nu=1}^{n} \eta_\nu.$$

If $\rho(f) < +\infty$ or $\rho(g) < +\infty$, then $\rho(f) = \rho(g)$ and $\max\{p,q\} \cdot \max\{s,t\} \leq \sigma_1\sigma_2$.

Example 2 Let $c_1 = \arctan 2$, $c_2 = \arctan(-2)$. It is easy to check that $(f(z), g(z)) = (\tan z, \cot z)$ satisfies the following system of difference equations:

$$\begin{cases} f(z+c_1)^2 f(z+c_2) + f(z+c_1) f(z+c_2)^2 = \frac{-40g^3+10g}{g^4-8g^2+16}, \\ g(z+c_1) g(z+c_2) + g(z+c_1)^2 = \frac{-20f^2+10f}{f^3+2f^2-4f-8}. \end{cases}$$

In Example 2, we have $\max\{p,q\} = 4$, $\max\{s,t\} = 3$, $\sigma_1 = 4$, $\sigma_2 = 3$, $\rho(f) = \rho(g) = 1 < +\infty$, and $\max\{p,q\} \cdot \max\{s,t\} = \sigma_1 \cdot \sigma_2 = 12$. Therefore, the estimation in Corollary 1 is sharp.

In [15], Laine *et al.* also considered the growth of meromorphic solutions of some classes of complex difference functional equations and obtained the following result.

Theorem E *Suppose that f is a transcendental meromorphic solution of the equation*

$$\sum_{\{J\}} \alpha_J(z)\Big(\prod_{j\in J} f(z+c_j)\Big) = f\big(p(z)\big),$$

where $p(z)$ is a polynomial of degree $k \geq 2$, $\{J\}$ is the collection of all subsets of $\{1,2,\ldots,n\}$. Moreover, we assume that the coefficients $\alpha_J(z)$ are small functions relative to f and that $n \geq k$. Then

$$T(r,f) = O\big((\log r)^{\alpha+\varepsilon}\big),$$

where $\alpha = \frac{\log n}{\log k}$.

In 2010, Zhang *et al.* [18] got a more generalized result than Theorem E. Next we give the growth of meromorphic solutions of a system of complex functional equations as follows.

Theorem 2 *Let $c_1, c_2, \ldots, c_n$ be distinct, nonzero complex numbers and suppose that $(f(z), g(z))$ is a transcendental meromorphic solution of a system of complex functional equations of the form*

$$\begin{cases} \sum_{\lambda_1\in I_1,\mu_1\in J_1} \alpha_{\lambda_1,\mu_1}(z)\big(\prod_{\nu=1}^n f(z+c_\nu)^{l_{\lambda_1,\nu}} \prod_{\nu=1}^n g(z+c_\nu)^{m_{\mu_1,\nu}}\big) = f(p(z)), \\ \sum_{\lambda_2\in I_2,\mu_2\in J_2} \beta_{\lambda_2,\mu_2}(z)\big(\prod_{\nu=1}^n f(z+c_\nu)^{l_{\lambda_2,\nu}} \prod_{\nu=1}^n g(z+c_\nu)^{m_{\mu_2,\nu}}\big) = g(p(z)), \end{cases} \tag{5}$$

where $p(z)$ is a polynomial of degree $k \geq 2$, $I_i = \{\lambda_i = (l_{\lambda_i,1}, l_{\lambda_i,2}, \ldots, l_{\lambda_i,n}) | l_{\lambda_i,\nu} \in N \cup \{0\}, \nu = 1,2,\ldots,n\}$ $(i = 1,2)$ and $J_j = \{\mu_j = (m_{\mu_j,1}, m_{\mu_j,2}, \ldots, m_{\mu_j,n}) | m_{\mu_j,\nu} \in N \cup \{0\}, \nu = 1,2,\ldots,n\}$ $(j = 1,2)$ are finite index sets, and denote

$$\xi_{1,\nu} = \max_{\lambda_1\in I_1}\{l_{\lambda_1,\nu}\}, \qquad \eta_{1,\nu} = \max_{\mu_1\in J_1}\{m_{\mu_1,\nu}\},$$

$$\xi_{2,\nu} = \max_{\lambda_2\in I_2}\{l_{\lambda_2,\nu}\}, \qquad \eta_{2,\nu} = \max_{\mu_2\in J_2}\{m_{\mu_2,\nu}\}$$

$(\nu = 1,2,\ldots,n)$,

$$\sigma_{11} = \sum_{\nu=1}^n \xi_{1,\nu}, \qquad \sigma_{12} = \sum_{\nu=1}^n \eta_{1,\nu}, \qquad \sigma_{21} = \sum_{\nu=1}^n \xi_{2,\nu}, \qquad \sigma_{22} = \sum_{\nu=1}^n \eta_{2,\nu}$$

and

$$\sigma = \max\{\sigma_{11}, \sigma_{12}, \sigma_{21}, \sigma_{22}\}.$$

Moreover, we assume that the coefficients $\alpha_{\lambda_1,\mu_1}(z)$ $(\lambda_1 \in I_1, \mu_1 \in J_1)$, $\beta_{\lambda_2,\mu_2}(z)$ $(\lambda_2 \in I_2, \mu_2 \in J_2)$ *are small functions relative to* $f(z)$ *and* $g(z)$, *and that* $2\sigma \geq k$. *Then*

$$T(r,f) = O\big((\log r)^{\alpha+\varepsilon}\big), \qquad T(r,g) = O\big((\log r)^{\alpha+\varepsilon}\big),$$

where $\alpha = \frac{\log 2\sigma}{\log k}$.

If $\sigma_{12} = \sigma_{21} = 0$, then we can obtain the following result easily.

Corollary 2 *Let* $c_1, c_2, \ldots, c_n$ *be distinct, nonzero complex numbers and suppose that* $(f(z), g(z))$ *is a transcendental meromorphic solution of a system of complex functional equations of the form*

$$\begin{cases} \sum_{\lambda\in I} \alpha_\lambda(z)(\prod_{\nu=1}^{n} f(z+c_\nu)^{l_{\lambda,\nu}}) = g(p(z)), \\ \sum_{\mu\in J} \beta_\mu(z)(\prod_{\nu=1}^{n} g(z+c_\nu)^{m_{\mu,\nu}}) = f(p(z)), \end{cases} \tag{6}$$

where $p(z)$ *is a polynomial of degree* $k \geq 2$, $I = \{\lambda = (l_{\lambda,1}, l_{\lambda,2}, \ldots, l_{\lambda,n}) | l_{\lambda,\nu} \in N \cup \{0\}, \nu = 1, 2, \ldots, n\}$ *and* $J = \{\mu = (m_{\mu,1}, m_{\mu,2}, \ldots, m_{\mu,n}) | m_{\mu,\nu} \in N \cup \{0\}, \nu = 1, 2, \ldots, n\}$ *are two finite index sets, and denote*

$$\xi_\nu = \max_\lambda \{l_{\lambda,\nu}\} \quad (\nu = 1, 2, \ldots, n), \qquad \sigma_1 = \sum_{\nu=1}^{n} \xi_\nu,$$

$$\eta_\nu = \max_\mu \{m_{\mu,\nu}\} \quad (\nu = 1, 2, \ldots, n), \qquad \sigma_2 = \sum_{\nu=1}^{n} \eta_\nu$$

and

$$\sigma = \max\{\sigma_1, \sigma_2\}.$$

Moreover, we assume that the coefficients $\alpha_\lambda(z)$ $(\lambda \in I)$, $\beta_\mu(z)$ $(\mu \in J)$ *are small functions relative to* $f(z)$ *and* $g(z)$, *and that* $2\sigma \geq k$. *Then*

$$T(r,f) = O\big((\log r)^{\alpha+\varepsilon}\big), \qquad T(r,g) = O\big((\log r)^{\alpha+\varepsilon}\big),$$

where $\alpha = \frac{\log 2\sigma}{\log k}$.

2 Some lemmas

In order to prove our results, we need the following lemmas.

Lemma 1 (see [3]) *Let* $f(z)$ *be a meromorphic function. Then for all irreducible rational functions in* f,

$$R(z,f) = \frac{P(z,f)}{Q(z,f)} = \frac{\sum_{i=0}^{p} a_i(z) f^i}{\sum_{j=0}^{q} b_j(z) f^j},$$

such that the meromorphic coefficients $a_i(z)$, $b_j(z)$ *satisfy*

$$\begin{cases} T(r, a_i) = S(r,f), & i = 0, 1, \ldots, p, \\ T(r, b_j) = S(r,f), & j = 0, 1, \ldots, q, \end{cases}$$

we have

$$T\big(r,R(z,f)\big)=\max\{p,q\}\cdot T(r,f)+S(r,f).$$

In [22], AZ Mokhon'ko and VD Mokhon'ko gave an estimation of Nevanlinna's characteristic function of

$$F(z)=\frac{P(z)}{Q(z)}=\frac{\sum_{\lambda\in I}f_1^{l_{\lambda,1}}f_2^{l_{\lambda,2}}\cdots f_n^{l_{\lambda,n}}}{\sum_{\mu\in J}f_1^{m_{\mu,1}}f_2^{m_{\mu,2}}\cdots f_n^{m_{\mu,n}}},$$

where $f_1,f_2,\ldots,f_n$ are distinct meromorphic functions, $I=\{\lambda=(l_{\lambda,1},l_{\lambda,2},\ldots,l_{\lambda,n})|l_{\lambda,\nu}\in N\cup\{0\},\nu=1,2,\ldots,n\}$ and $J=\{\mu=(m_{\mu,1},m_{\mu,2},\ldots,m_{\mu,n})|m_{\mu,\nu}\in N\cup\{0\},\nu=1,2,\ldots,n\}$ are two finite index sets. However, the method of the proof was too complex. For $F(z)$ of the form $\sum_{\lambda\in I}f_1^{l_{\lambda,1}}f_2^{l_{\lambda,2}}\cdots f_n^{l_{\lambda,n}}$, Zheng *et al.* [20] gave a simpler proof, but the estimation of $T(r,F)$ was not sharp. For completeness, we give the proof of the following lemma.

Lemma 2 *Let $f_1,f_2,\ldots,f_n$ be distinct meromorphic functions. Then*

$$T\left(r,\sum_{\lambda\in I}f_1^{l_{\lambda,1}}f_2^{l_{\lambda,2}}\cdots f_n^{l_{\lambda,n}}\right)\le\sum_{j=1}^{n}\sigma_jT(r,f_j)+\log t,$$

where $I=\{(l_{\lambda,1},l_{\lambda,2},\ldots,l_{\lambda,n})|l_{\lambda,j}\in N\cup\{0\},j=1,2,\ldots,n\}$ is an finite index set consisting of t elements and $\sigma_j=\max_{\lambda\in I}\{l_{\lambda,j}\}$ $(j=1,2,\ldots,n)$.

Proof All the poles of the function $\sum_{\lambda\in I}f_1^{l_{\lambda,1}}f_2^{l_{\lambda,2}}\cdots f_n^{l_{\lambda,n}}$ are generated by the poles of the functions f_j $(j=1,2,\ldots,n)$, and every pole of multiplicity k of f_j $(j=1,2,\ldots,n)$ has order at most $k\sigma_j$. This implies that

$$n\left(r,\sum_{\lambda\in I}f_1^{l_{\lambda,1}}f_2^{l_{\lambda,2}}\cdots f_n^{l_{\lambda,n}}\right)\le\sum_{j=1}^{n}\sigma_jn(r,f_j).$$

Thus we obtain

$$N\left(r,\sum_{\lambda\in I}f_1^{l_{\lambda,1}}f_2^{l_{\lambda,2}}\cdots f_n^{l_{\lambda,n}}\right)\le\sum_{j=1}^{n}\sigma_jN(r,f_j).\tag{7}$$

We next prove that

$$m\left(r,\sum_{\lambda\in I}f_1^{l_{\lambda,1}}f_2^{l_{\lambda,2}}\cdots f_n^{l_{\lambda,n}}\right)\le\sum_{j=1}^{n}\sigma_jm(r,f_j)+\log t,\tag{8}$$

and we define

$$\begin{cases}f_j^*(z)=f_j(z), & |f_j(z)|>1,\\ f_j^*(z)=1, & |f_j(z)|\le1,\end{cases}$$

for $j = 1, 2, \ldots, n$. Thus we have

$$\begin{aligned}
\left|\sum_{\lambda\in I} f_1^{l_{\lambda,1}} f_2^{l_{\lambda,2}} \cdots f_n^{l_{\lambda,n}}\right| &\leq \sum_{\lambda\in I} |f_1^{l_{\lambda,1}} f_2^{l_{\lambda,2}} \cdots f_n^{l_{\lambda,n}}| \\
&\leq \sum_{\lambda\in I} |f_1^{*l_{\lambda,1}} f_2^{*l_{\lambda,2}} \cdots f_n^{*l_{\lambda,n}}| \\
&= |f_1^{*\sigma_1} f_2^{*\sigma_2} \cdots f_n^{*\sigma_n}| \left(\sum_{\lambda\in I} \frac{|f_1^{*l_{\lambda,1}} f_2^{*l_{\lambda,2}} \cdots f_n^{*l_{\lambda,n}}|}{|f_1^{*\sigma_1} f_2^{*\sigma_2} \cdots f_n^{*\sigma_n}|}\right) \\
&\leq t |f_1^{*\sigma_1} f_2^{*\sigma_2} \cdots f_n^{*\sigma_n}|.
\end{aligned}$$

By the definition of $m(r,f)$, we immediately conclude that

$$\begin{aligned}
m\left(r, \sum_{\lambda\in I} f_1^{l_{\lambda,1}} f_2^{l_{\lambda,2}} \cdots f_n^{l_{\lambda,n}}\right) &= \frac{1}{2\pi}\int_0^{2\pi} \log^+\left|\sum_{\lambda\in I} f_1^{l_{\lambda,1}} f_2^{l_{\lambda,2}} \cdots f_n^{l_{\lambda,n}}\right| d\theta \\
&\leq \frac{1}{2\pi}\int_0^{2\pi} \log^+ |f_1^{*\sigma_1} f_2^{*\sigma_2} \cdots f_n^{*\sigma_n}|\, d\theta + \log t \\
&= \sum_{j=1}^{n} \sigma_j m(r,f_j) + \log t.
\end{aligned}$$

By (7) and (8), the assertion follows. □

Remark 3 If we suppose that $\alpha_\lambda(z) = o(T(r,f_j))$ $(\lambda \in I)$ hold for all $j \in \{1, 2, \ldots, n\}$, and denote $T(r, a_\lambda) = S(r,f)$ $(\lambda \in I)$, then we have the following estimation:

$$T\left(r, \sum_{\lambda\in I} \alpha_\lambda(z) f_1^{l_{\lambda,1}} f_2^{l_{\lambda,2}} \cdots f_n^{l_{\lambda,n}}\right) \leq \sum_{j=1}^{n} \sigma_j T(r,f_j) + S(r,f).$$

Lemma 3 (see [6]) *Let $f(z)$ be a meromorphic function with order $\rho = \rho(f)$, $\rho < +\infty$, and c be a fixed non zero complex number, then for each $\varepsilon > 0$, we have*

$$T\big(r, f(z+c)\big) = T(r,f) + O\big(r^{\rho-1+\varepsilon}\big) + O(\log r).$$

Lemma 4 (see [3]) *Let $g : (0, +\infty) \to R$, $h : (0, +\infty) \to R$ be monotone increasing functions such that $g(r) \leq h(r)$ outside of an exceptional set E of finite linear measure. Then, for any $\alpha > 1$, there exists $r_0 > 0$ such that $g(r) \leq h(\alpha r)$ for all r_0.*

Lemma 5 (see [23]) *Let f be a transcendental meromorphic function, and $p(z) = a_k z^k + a_{k-1} z^{k-1} + \cdots + a_1 z + a_0$, $a_k \neq 0$, be a nonconstant polynomial of degree k. Given $0 < \delta < |a_k|$, denote $\lambda = |a_k| + \delta$ and $\mu = |a_k| - \delta$. Then given $\varepsilon > 0$ and $a \in C \cup \{\infty\}$, we have*

$$\begin{aligned}
&kn\big(\mu r^k, a, f\big) \leq n\big(r, a, f(p(z))\big) \leq kn\big(\lambda r^k, a, f\big), \\
&N\big(\mu r^k, a, f\big) + O(\log r) \leq N\big(r, a, f(p(z))\big) \leq N\big(\lambda r^k, a, f\big) + O(\log r), \\
&(1-\varepsilon) T\big(\mu r^k, f\big) \leq T\big(r, f(p(z))\big) \leq (1+\varepsilon) T\big(\lambda r^k, f\big),
\end{aligned}$$

for all r large enough.

3 Proofs of theorems

Proof of Theorem 1 We assume that $(f(z), g(z))$ is a transcendental meromorphic solution of the system of complex difference equations (2). By the first equation in (2), Lemma 1, Lemma 2, and Lemma 3, we have, for each $\varepsilon > 0$,

$$\begin{aligned}
&\max\{p,q\}T(r,g)\\
&= T\left(r, \sum_{\lambda_1\in I_1,\mu_1\in J_1} \alpha_{\lambda_1,\mu_1}(z)\left(\prod_{\nu=1}^{n} f(z+c_\nu)^{l_{\lambda_1,\nu}} \prod_{\nu=1}^{n} g(z+c_\nu)^{m_{\mu_1,\nu}}\right)\right) + S(r,g)\\
&\le \sum_{\nu=1}^{n} \xi_{1,\nu} T\big(r,f(z+c_\nu)\big) + \sum_{\nu=1}^{n} \eta_{1,\nu} T\big(r,g(z+c_\nu)\big) + S(r,f) + S(r,g)\\
&= \sum_{\nu=1}^{n} \xi_{1,\nu} T\big(r,f(z)\big) + O\big(r^{\rho(f)-1+\varepsilon}\big) + \sum_{\nu=1}^{n} \eta_{1,\nu} T\big(r,g(z)\big) + O\big(r^{\rho(g)-1+\varepsilon}\big)\\
&\quad + O(\log r) + S(r,f) + S(r,g)\\
&= \left(\sum_{\nu=1}^{n} \xi_{1,\nu}\right) T\big(r,f(z)\big) + \left(\sum_{\nu=1}^{n} \eta_{1,\nu}\right) T\big(r,g(z)\big)\\
&\quad + O\big(r^{\rho(g)-1+\varepsilon}\big) + O(\log r) + S(r,f) + S(r,g)\\
&= \sigma_{11} T\big(r,f(z)\big) + \sigma_{12} T\big(r,g(z)\big) + O\big(r^{\rho(f)-1+\varepsilon}\big) + O\big(r^{\rho(g)-1+\varepsilon}\big)\\
&\quad + O(\log r) + S(r,f) + S(r,g).
\end{aligned} \tag{9}$$

By the above inequality, we get, for each $\varepsilon > 0$,

$$\begin{aligned}
&\big(\max\{p,q\} - \sigma_{12}\big) T(r,g)\\
&\le \sigma_{11} T\big(r,f(z)\big) + O\big(r^{\rho(f)-1+\varepsilon}\big) + O\big(r^{\rho(g)-1+\varepsilon}\big)\\
&\quad + O(\log r) + S(r,f) + S(r,g).
\end{aligned} \tag{10}$$

Since $\max\{p,q\} > \sigma_{12}$ by the assumption, we have, for each $\varepsilon > 0$,

$$\begin{aligned}
T(r,g) \le{}& \frac{\sigma_{11}}{\max\{p,q\} - \sigma_{12}} T\big(r,f(z)\big) + O\big(r^{\rho(f)-1+\varepsilon}\big) + O\big(r^{\rho(g)-1+\varepsilon}\big)\\
&+ O(\log r) + S(r,f) + S(r,g).
\end{aligned} \tag{11}$$

Similarly, by the second equation in (2), we obtain, for each $\varepsilon > 0$,

$$\begin{aligned}
&\max\{s,t\}T(r,f)\\
&= T\left(r, \sum_{\lambda_2\in I_2,\mu_2\in J_2} \beta_{\lambda_2,\mu_2}(z)\left(\prod_{\nu=1}^{n} f(z+c_\nu)^{l_{\lambda_2,\nu}} \prod_{\nu=1}^{n} g(z+c_\nu)^{m_{\mu_2,\nu}}\right)\right) + S(r,f)\\
&\le \sum_{\nu=1}^{n} \xi_{2,\nu} T\big(r,f(z+c_\nu)\big) + \sum_{\nu=1}^{n} \eta_{2,\nu} T\big(r,g(z+c_\nu)\big) + S(r,f) + S(r,g)\\
&= \sum_{\nu=1}^{n} \xi_{2,\nu} T\big(r,f(z)\big) + O\big(r^{\rho(f)-1+\varepsilon}\big) + \sum_{\nu=1}^{n} \eta_{2,\nu} T\big(r,g(z)\big)
\end{aligned}$$

$$
\begin{aligned}
&+ O\left(r^{\rho(g)-1+\varepsilon}\right) + O(\log r) + S(r,f) + S(r,g) \\
&= \left(\sum_{\nu=1}^{n} \xi_{2,\nu}\right) T\left(r,f(z)\right) + \left(\sum_{\nu=1}^{n} \eta_{2,\nu}\right) T\left(r,g(z)\right) \\
&\quad + O(\log r) + S(r,f) + S(r,g) \\
&= \sigma_{21} T\left(r,f(z)\right) + \sigma_{22} T\left(r,g(z)\right) + O\left(r^{\rho(f)-1+\varepsilon}\right) + O\left(r^{\rho(g)-1+\varepsilon}\right) \\
&\quad + O(\log r) + S(r,f) + S(r,g). \qquad (12)
\end{aligned}
$$

By (12) and $\max\{s,t\} > \sigma_{21}$, we have, for each $\varepsilon > 0$,

$$
\begin{aligned}
&\left(\max\{s,t\} - \sigma_{21}\right) T(r,f) \\
&\quad \le \sigma_{22} T\left(r,g(z)\right) + O\left(r^{\rho(f)-1+\varepsilon}\right) + O\left(r^{\rho(g)-1+\varepsilon}\right) \\
&\qquad + O(\log r) + S(r,f) + S(r,g) \qquad (13)
\end{aligned}
$$

and

$$
\begin{aligned}
T(r,f) \le &\frac{\sigma_{22}}{\max\{s,t\} - \sigma_{21}} T\left(r,g(z)\right) + O\left(r^{\rho(f)-1+\varepsilon}\right) + O\left(r^{\rho(g)-1+\varepsilon}\right) \\
&+ O(\log r) + S(r,f) + S(r,g). \qquad (14)
\end{aligned}
$$

Using (11), we can obtain $\rho(g) \le \rho(f)$. Similarly, we can get $\rho(f) \le \rho(g)$ from (14). Therefore, we have $\rho(f) = \rho(g)$.

It follows from (10) and (13) that

$$
\begin{aligned}
&\left(\max\{p,q\} - \sigma_{12}\right)\left(\max\{s,t\} - \sigma_{21}\right) T(r,f) T(r,g) \\
&\quad \le \sigma_{11}\sigma_{22} T\left(r,f(z)\right) T\left(r,g(z)\right) + o\left(T(r,f)T(r,g)\right). \qquad (15)
\end{aligned}
$$

From (15), we conclude that

$$
\left(\max\{p,q\} - \sigma_{12}\right) \cdot \left(\max\{s,t\} - \sigma_{21}\right) \le \sigma_{11}\sigma_{22}.
$$

This yields the asserted result. □

Proof of Theorem 2 We assume that $(f(z), g(z))$ is a transcendental meromorphic solution of a system of complex functional equations (5). Let $C = \max\{|c_1|, |c_2|, \ldots, |c_n|\}$. According to the first equation in (5), Lemma 2, Lemma 3, and the last assertion of Lemma 5, we get

$$
\begin{aligned}
&(1-\varepsilon) T\left(\mu r^k, f\right) \\
&\quad \le T\left(r, f(p(z))\right) \\
&\quad = T\left(r, \sum_{\lambda_1 \in I_1, \mu_1 \in J_1} \alpha_{\lambda_1,\mu_1}(z) \left(\prod_{\nu=1}^{n} f(z+c_\nu)^{l_{\lambda_1,\nu}} \prod_{\nu=1}^{n} g(z+c_\nu)^{m_{\mu_1,\nu}}\right)\right) \\
&\quad \le \sum_{\nu=1}^{n} \xi_{1,\nu} T\left(r, f(z+c_\nu)\right) + \sum_{\nu=1}^{n} \eta_{1,\nu} T\left(r, g(z+c_\nu)\right) + S(r,f) + S(r,g)
\end{aligned}
$$

$$\leq \sum_{\nu=1}^{n} \xi_{1,\nu} T\big(r + C, f(z)\big) + \sum_{\nu=1}^{n} \eta_{1,\nu} T\big(r + C, g(z)\big) + S(r,f) + S(r,g)$$

$$= \left(\sum_{\nu=1}^{n} \xi_{1,\nu}\right) T\big(r + C, f(z)\big) + \left(\sum_{\nu=1}^{n} \eta_{1,\nu}\right) T\big(r + C, g(z)\big) + S(r,f) + S(r,g)$$

$$= \sigma_{11} T\big(r + C, f(z)\big) + \sigma_{12} T\big(r + C, g(z)\big) + S(r,f) + S(r,g).$$

Since $T(r + C, f) \leq T(\beta r, f)$ and $T(r + C, g) \leq T(\beta r, g)$ hold for r large enough for $\beta > 1$, we may assume r to be large enough to satisfy

$$(1 - \varepsilon) T\big(\mu r^k, f\big) \leq \sigma_{11}(1 + \varepsilon) T(\beta r, f) + \sigma_{12}(1 + \varepsilon) T(\beta r, g)$$

outside a possible exceptional set of finite linear measure. By Lemma 4, we know that, whenever $\gamma > 1$,

$$(1 - \varepsilon) T\big(\mu r^k, f\big) \leq \sigma_{11}(1 + \varepsilon) T(\gamma\beta r, f) + \sigma_{12}(1 + \varepsilon) T(\gamma\beta r, g) \tag{16}$$

holds for all r large enough. Let $t = \gamma\beta r$, then the inequality (16) may be written in the form

$$T\left(\frac{\mu}{(\gamma\beta)^k} t^k, f\right) \leq \frac{\sigma_{11}(1 + \varepsilon)}{1 - \varepsilon} T(t, f) + \frac{\sigma_{12}(1 + \varepsilon)}{1 - \varepsilon} T(t, g). \tag{17}$$

Similarly, by the second equation in (5), for all r large enough and $\beta > 1$, $\gamma > 1$, we have

$$(1 - \varepsilon) T\big(\mu r^k, g\big)$$

$$\leq T\big(r, g(p(z))\big)$$

$$= T\left(r, \sum_{\lambda_2 \in I_2, \mu_2 \in J_2} \beta_{\lambda_2,\mu_2}(z) \left(\prod_{\nu=1}^{n} f(z + c_\nu)^{l_{\lambda_2,\nu}} \prod_{\nu=1}^{n} g(z + c_\nu)^{m_{\mu_2,\nu}}\right)\right)$$

$$\leq \sum_{\nu=1}^{n} \xi_{2,\nu} T\big(r, f(z + c_\nu)\big) + \sum_{\nu=1}^{n} \eta_{2,\nu} T\big(r, g(z + c_\nu)\big) + S(r,f) + S(r,g)$$

$$\leq \sum_{\nu=1}^{n} \xi_{2,\nu} T\big(r + C, f(z)\big) + \sum_{\nu=1}^{n} \eta_{2,\nu} T\big(r + C, g(z)\big) + S(r,f) + S(r,g)$$

$$= \left(\sum_{\nu=1}^{n} \xi_{2,\nu}\right) T\big(r + C, f(z)\big) + \left(\sum_{\nu=1}^{n} \eta_{2,\nu}\right) T\big(r + C, g(z)\big) + S(r,f) + S(r,g)$$

$$= \sigma_{21} T\big(r + C, f(z)\big) + \sigma_{22} T\big(r + C, g(z)\big) + S(r,f) + S(r,g)$$

$$\leq \sigma_{21}(1 + \varepsilon) T(\gamma\beta r, f) + \sigma_{22}(1 + \varepsilon) T(\gamma\beta r, g).$$

Let $t = \gamma\beta r$, we have

$$T\left(\frac{\mu}{(\gamma\beta)^k} t^k, g\right) \leq \frac{\sigma_{21}(1 + \varepsilon)}{1 - \varepsilon} T(t, f) + \frac{\sigma_{22}(1 + \varepsilon)}{1 - \varepsilon} T(t, g). \tag{18}$$

Letting $s = \log t + \frac{\log \frac{\mu}{(\gamma\beta)^k}}{k-1}$, then $t = e^s(\frac{\mu}{(\gamma\beta)^k})^{\frac{1}{1-k}}$ and

$$\begin{aligned} ks &= k\log t + \frac{k}{k-1}\log\frac{\mu}{(\gamma\beta)^k} \\ &= k\log t + \log\frac{\mu}{(\gamma\beta)^k} + \frac{\log\frac{\mu}{(\gamma\beta)^k}}{k-1} \\ &= \log\frac{\mu}{(\gamma\beta)^k}t^k + \log\left(\frac{\mu}{(\gamma\beta)^k}\right)^{\frac{1}{k-1}}. \end{aligned}$$

So

$$\frac{\mu}{(\gamma\beta)^k}t^k = e^{ks}\left(\frac{\mu}{(\gamma\beta)^k}\right)^{\frac{1}{1-k}}. \tag{19}$$

Let $T(t,f) = T(e^s(\frac{\mu}{(\gamma\beta)^k})^{\frac{1}{1-k}},f) = \Phi(s,f)$, $T(t,g) = T(e^s(\frac{\mu}{(\gamma\beta)^k})^{\frac{1}{1-k}},g) = \Phi(s,g)$, $M = \max\{\frac{\sigma_{11}(1+\varepsilon)}{1-\varepsilon}, \frac{\sigma_{12}(1+\varepsilon)}{1-\varepsilon}, \frac{\sigma_{21}(1+\varepsilon)}{1-\varepsilon}, \frac{\sigma_{22}(1+\varepsilon)}{1-\varepsilon}\}$, then from (17) and (19), we have

$$\begin{aligned} \Phi(ks,f) &= T\left(e^{ks}\left(\frac{\mu}{(\gamma\beta)^k}\right)^{\frac{1}{1-k}},f\right) \\ &= T\left(\frac{\mu}{(\gamma\beta)^k}t^k,f\right) \\ &\le \frac{\sigma_{11}(1+\varepsilon)}{1-\varepsilon}T(t,f) + \frac{\sigma_{12}(1+\varepsilon)}{1-\varepsilon}T(t,g) \\ &\le M\Phi(s,f) + M\Phi(s,g). \end{aligned} \tag{20}$$

Similarly, from (18), we can get

$$\Phi(ks,g) \le M\Phi(s,f) + M\Phi(s,g). \tag{21}$$

The inequalities (20) and (21) hold for all s large enough.

Letting now $\alpha = \frac{\log 2M}{\log k}$, namely, $2M = k^\alpha$. Write $\Psi(s,f) = \frac{\Phi(s,f)}{s^\alpha}$ and $\Psi(s,g) = \frac{\Phi(s,g)}{s^\alpha}$, thus we have

$$\Psi(ks,f) \le \frac{1}{2}\Psi(s,f) + \frac{1}{2}\Psi(s,g) \tag{22}$$

and

$$\Psi(ks,g) \le \frac{1}{2}\Psi(s,f) + \frac{1}{2}\Psi(s,g). \tag{23}$$

The inequalities (22) and (23) hold for all s large enough, we may assume that (22) and (23) hold for all $s \ge s_0$.

Let $M_1 = \sup_{s_0 \le s \le ks_0}\Psi(s,f)$ and $M_2 = \sup_{s_0 \le s \le ks_0}\Psi(s,g)$, then by (22) and (23) we can obtain

$$\sup_{s\in[ks_0,k^2s_0]}\Psi(s,f) = \sup_{s\in[s_0,ks_0]}\Psi(ks,f) \le \frac{1}{2}\sup_{s\in[s_0,ks_0]}\Psi(s,f) + \frac{1}{2}\sup_{s\in[s_0,ks_0]}\Psi(s,g) \le \frac{M_1}{2} + \frac{M_2}{2},$$

$$\sup_{s\in[ks_0,k^2s_0]}\Psi(s,g) = \sup_{s\in[s_0,ks_0]}\Psi(ks,g) \le \frac{1}{2}\sup_{s\in[s_0,ks_0]}\Psi(s,f) + \frac{1}{2}\sup_{s\in[s_0,ks_0]}\Psi(s,g) \le \frac{M_1}{2} + \frac{M_2}{2}.$$

Similarly, we have

$$\sup_{s\in[k^2s_0,k^3s_0]} \Psi(s,f) = \sup_{s\in[ks_0,k^2s_0]} \Psi(ks,f) \le \frac{1}{2}M_1 + \frac{1}{2}M_2,$$

$$\sup_{s\in[k^2s_0,k^3s_0]} \Psi(s,g) = \sup_{s\in[ks_0,k^2s_0]} \Psi(ks,g) \le \frac{1}{2}M_1 + \frac{1}{2}M_2,$$

$$\ldots .$$

Thus, we deduce that

$$\sup_{s\ge ks_0} \Psi(s,f) \le \frac{1}{2}M_1 + \frac{1}{2}M_2 < +\infty,$$

$$\sup_{s\ge ks_0} \Psi(s,g) \le \frac{1}{2}M_1 + \frac{1}{2}M_2 < +\infty.$$

Therefore, $\Psi(s,f)$ and $\Psi(s,g)$ are bounded for all $s \ge s_0$. There exist some constants K_1, K_2, K_3, K_4, such that, for any $\varepsilon > 0$,

$$T(t,f) = \Phi(s,f) = \Psi(s,f)s^\alpha \le K_1 s^\alpha = K_1\left(\log t + \frac{\log\frac{\mu}{(\gamma\beta)^k}}{k-1}\right)^\alpha \le K_2(\log t)^{\alpha+\varepsilon}$$

and

$$T(t,g) = \Phi(s,g) = \Psi(s,g)s^\alpha \le K_3 s^\alpha = K_3\left(\log t + \frac{\log\frac{\mu}{(\gamma\beta)^k}}{k-1}\right)^\alpha \le K_4(\log t)^{\alpha+\varepsilon}.$$

Therefore, we have

$$T(r,f) = O\big((\log r)^{\alpha+\varepsilon}\big)$$

and

$$T(r,g) = O\big((\log r)^{\alpha+\varepsilon}\big),$$

where

$$\alpha = \frac{\log 2M}{\log k} = \frac{\log 2\sigma}{\log k} + o(1).$$

Letting now $\alpha = \frac{\log 2\sigma}{\log k}$, we obtain the required form. Theorem 2 is proved. □

Competing interests
The author declares that there is no conflict of interests regarding the publication of the article.

Acknowledgements
The author would like to thank Professor Liangwen Liao and the anonymous referees for their valuable comments and suggestions. The research was supported by NSF of China (11426118), Natural Science Foundation of Jiangsu Province (BK20140767), Natural Science Foundation of the Jiangsu Higher Education Institutions (14KJB110004), Qing Lan Project of Jiangsu Province and Colonel-level topics (JSNU-ZY-01).

References

1. Cherry, W, Ye, Z: Nevanlinna's Theory of Value Distribution. Monographs in Math. Springer, Berlin (2001)
2. Hayman, WK: Meromorphic Functions. Clarendon, Oxford (1964)
3. Laine, I: Nevanlinna Theory and Complex Differential Equations. de Gruyter, Berlin (1993)
4. Ablowitz, MJ, Halburd, R, Herbst, B: On the extension of the Painlevé property to difference equations. Nonlinearity **13**(3), 889-905 (2000)
5. Chen, ZX, Shon, KH: On zeros and fixed points of differences of meromorphic functions. J. Math. Anal. Appl. **344**, 373-383 (2008)
6. Chiang, YM, Feng, SJ: On the Nevanlinna characteristic of $f(z+\eta)$ and difference equations in the complex plane. Ramanujan J. **16**(1), 105-129 (2008)
7. Chiang, YM, Feng, SJ: On the growth of logarithmic differences, difference quotients and logarithmic derivatives of meromorphic functions. Trans. Am. Math. Soc. **361**(7), 3767-3791 (2009)
8. Halburd, RG, Korhonen, RJ: Meromorphic solutions of difference equations, integrability and the discrete Painlevé equations. J. Phys. A, Math. Theor. **40**, R1-R38 (2007)
9. Halburd, RG, Korhonen, RJ: Difference analogue of the lemma on the logarithmic derivative with applications to difference equations. J. Math. Anal. Appl. **314**, 477-487 (2006)
10. Heittokangas, J, Korhonen, R, Laine, I, Rieppo, J, Tohge, K: Complex difference equations of Malmquist type. Comput. Methods Funct. Theory **1**(1), 27-39 (2001)
11. Huang, ZB, Chen, ZX: Meromorphic solutions of some complex difference equations. Adv. Differ. Equ. **2009**, Article ID 982681 (2009)
12. Ishizaki, K, Yanagihara, N: Wiman-Valiron method for difference equations. Nagoya Math. J. **175**, 75-102 (2004)
13. Laine, I, Yang, CC: Clunie theorems for difference and q-difference polynomials. J. Lond. Math. Soc. **76**(2), 556-566 (2007)
14. Laine, I, Yang, CC: Value distribution of difference polynomials. Proc. Jpn. Acad., Ser. A, Math. Sci. **83**(8), 148-151 (2007)
15. Laine, I, Rieppo, J, Silvennoinen, H: Remarks on complex difference equations. Comput. Methods Funct. Theory **5**(1), 77-88 (2005)
16. Wang, H, Xu, HY, Liu, BX: The poles and growth of solutions of systems of complex difference equations. Adv. Differ. Equ. **2013**, 75 (2013)
17. Zhang, J, Zhang, JJ: Meromorphic solutions to complex difference and q-difference equations of Malmquist type. Electron. J. Differ. Equ. **2014**, 16 (2014)
18. Zhang, JJ, Liao, LW: Further extending results of some classes of complex difference and functional equations. Adv. Differ. Equ. **2010**, Article ID 404582 (2010)
19. Zhang, JJ, Liao, LW: On Malmquist type theorem of complex difference equations. Houst. J. Math. **39**(3), 969-981 (2013)
20. Zheng, XM, Chen, ZX: Some properties of meromorphic solutions of q-difference equations. J. Math. Anal. Appl. **361**, 472-480 (2010)
21. Gao, LY: Estimates of N-function and m-function of meromorphic solutions of systems of complex difference equations. Acta Math. Sci. Ser. B **32**(4), 1495-1502 (2012)
22. Mokhon'ko, AZ, Mokhon'ko, VD: Estimates for the Nevanlinna characteristics of some classes of meromorphic functions and their applications to differential equations. Sib. Mat. Zh. **15**(6), 1305-1322 (1974)
23. Goldstein, R: Some results on factorisation of meromorphic functions. J. Lond. Math. Soc. **4**(2), 357-364 (1971)

18

Monotone iterative procedure and systems of a finite number of nonlinear fractional differential equations

Dariusz Wardowski*

*Correspondence:
wardd@math.uni.lodz.pl
Department of Nonlinear Analysis, Faculty of Mathematics and Computer Science, University of Łódź, Banacha 22, Łódź, 90-238, Poland

Abstract
The aim of the paper is to present a nontrivial and natural extension of the comparison result and the monotone iterative procedure based on upper and lower solutions, which were recently established in (Wang *et al.* in Appl. Math. Lett. 25:1019-1024, 2012), to the case of any finite number of nonlinear fractional differential equations.

MSC: 26A33; 34A08; 34B15

Keywords: monotone iterative procedure; system of fractional differential equations; upper and lower solution

1 Introduction

Fractional derivatives and integrals are used for a better description of material properties. In the literature we can find many interesting papers concerning this theory; see *e.g.*, [1–13]. The study of systems involving fractional differential/integral equations is also important as such systems occur in various problems of applied nature; for example, see [14–22]. Some basic theory of fractional differential equations involving the Riemann-Liouville differential operator can be found in [23–25].

In the paper we consider the following system of nonlinear fractional differential equations:

$$\begin{cases} D^{\alpha}u_1(t) = f_1(t, u_1(t), u_2(t), \dots, u_n(t)), & t \in (0, T], \\ D^{\alpha}u_2(t) = f_2(t, u_1(t), u_2(t), \dots, u_n(t)), & t \in (0, T], \\ \dots, \\ D^{\alpha}u_n(t) = f_n(t, u_1(t), u_2(t), \dots, u_n(t)), & t \in (0, T], \\ t^{1-\alpha}u_1(t)|_{t=0} = x_0^1, \quad t^{1-\alpha}u_2(t)|_{t=0} = x_0^2, \quad \dots, \quad t^{1-\alpha}u_n(t)|_{t=0} = x_0^n, \end{cases} \tag{1.1}$$

where D^{α} is the standard Riemann-Liouville fractional derivative of order α, $0 \le \alpha \le 1$, $T > 0, f^i \in C([0, T] \times \mathbb{R}^n, \mathbb{R})$, $1 \le i \le n$, and $x_0^1, \dots, x_0^n \in \mathbb{R}$ satisfy

$$\sum_{i=2}^{n} x_0^i - x_0^1 \ge 0. \tag{1.2}$$

We investigate system (1.1) with respect to the existence of a solution via the method of upper and lower solutions. There is also presented the concept of an iterative procedure, where the appropriately constructed sequences are convergent to the extreme solution. The paper is a continuation of the investigations in [10] of Wang *et al.*, where the authors examined system (1.1) in the case $n = 2$. After proving the main results we state, for convenience of the reader, the introduced techniques in the case of three nonlinear fractional differential equations and also present a concrete example.

2 Preliminaries

First, let us recall the needed notations and crucial results which will be needed in the next sections of the article.

Denote by $C_{1-\alpha}([0,T])$ the family of all functions $u \in C((0,T])$ such that $t^{1-\alpha}u \in C([0,T])$. A basic theorem concerning the existence of the result and its uniqueness for the linear fractional equation is as follows.

Lemma 2.1 ([23]) *Let $0 < \alpha \leq 1$, $M \in \mathbb{R}$, and $\sigma \in C_{1-\alpha}([0,T])$ be fixed. Then the linear initial value problem*

$$\begin{cases} D^{\alpha}u(t) = \sigma(t) - Mu(t), & t \in (0,T], \\ t^{1-\alpha}u(t)|_{t=0} = u_0, \end{cases} \quad i,j \in \mathbb{N}, \tag{2.1}$$

has a unique solution, given by the following formula:

$$u(t) = \Gamma(\alpha)u_0 t^{\alpha-1}E_{\alpha,\alpha}(-Mt^{\alpha}) + \int_0^t (t-s)^{\alpha-1}E_{\alpha,\alpha}(-M(t-s)^{\alpha})\sigma(s)\,ds,$$

where $E_{\alpha,\beta}$ is the Mittag-Leffler function, i.e. the function of the form

$$E_{\alpha,\beta}(z) = \sum_{k=0}^{\infty} \frac{z^k}{\Gamma(\alpha k + \beta)}, \quad \alpha, \beta > 0, z \in \mathbb{R}.$$

The comparison result for the initial value problem (2.1) due to Wang *et al.* is as follows.

Lemma 2.2 ([10]) *Let $0 < \alpha \leq 1$ and $M \in \mathbb{R}$ be given. Then, if $w \in C_{1-\alpha}([0,T])$ satisfies*

$$\begin{cases} D^{\alpha}w(t) + Mw(t) \geq 0, & t \in (0,T], \\ t^{1-\alpha}w(t)|_{t=0} \geq 0, \end{cases}$$

then $w(t) \geq 0$ for all $t \in (0,T]$.

The same authors also proved the following result, which will be needed in the sequel.

Lemma 2.3 ([10]) *Let $0 < \alpha \leq 1$, $M \in \mathbb{R}$, and $N \geq 0$ be given. Assume that $u, v \in C_{1-\alpha}([0,T])$ satisfy*

$$\begin{cases} D^{\alpha}u(t) \geq -Mu(t) + Nv(t), & t \in (0,T], \\ D^{\alpha}v(t) \geq -Mv(t) + Nu(t), & t \in (0,T], \\ t^{1-\alpha}u(t)|_{t=0} \geq 0, & t^{1-\alpha}v(t)|_{t=0} \geq 0. \end{cases}$$

Then $u(t) \geq 0$, $v(t) \geq 0$ for all $t \in (0,T]$.

3 The results

In the sequel we will use the following notation:

$$\delta_{ij} = \begin{cases} 1 & \text{if } i = j, \\ -1 & \text{if } i \neq j, \end{cases} \quad i,j \in \mathbb{N}.$$

$C_{1-\alpha}([0,T])^n$ denotes $C_{1-\alpha}([0,T]) \times C_{1-\alpha}([0,T]) \times \cdots \times C_{1-\alpha}([0,T])$ (n times).

Lemma 3.1 *Let $0 < \alpha \leq 1$ be fixed, $M_i \in \mathbb{R}$, $\sigma_i \in C_{1-\alpha}([0,T])$, $i = 1,2,\ldots,n$. Then the linear problem of n equations*

$$\begin{cases} D^\alpha u_1(t) = \sigma_1(t) - M_1 u_1(t) - \sum_{i,j=2}^n M_j \delta_{ji} u_i(t), & t \in (0,T], \\ D^\alpha u_j(t) = \sigma_j(t) + (M_j - \sum_{i=1}^n M_i) u_j(t) & \\ \qquad - M_j (\sum_{i=1}^n u_i(t) - u_j(t)), & t \in (0,T], 2 \leq j \leq n, \\ t^{1-\alpha} u_i(t)|_{t=0} = x_0^i, \quad 1 \leq i \leq n, & \end{cases} \tag{3.1}$$

has a unique solution in $C_{1-\alpha}([0,T])^n$.

Proof First observe that for any $p_1, p_2, \ldots, p_n \in C_{1-\alpha}([0,T])$ the system

$$\begin{cases} u_1 + u_2 + \cdots + u_n = p_1, \\ u_1 - u_2 + \cdots + u_n = p_2, \\ \ldots, \\ u_1 + u_2 + \cdots - u_n = p_n \end{cases} \tag{3.2}$$

has exactly one solution, which is a consequence of the fact that

$$\det \begin{bmatrix} 1 & 1 & 1 & \cdots & 1 \\ 1 & -1 & 1 & \cdots & 1 \\ 1 & 1 & -1 & \cdots & 1 \\ \vdots & \vdots & \vdots & \ddots & \vdots \\ 1 & 1 & 1 & \cdots & -1 \end{bmatrix}_{n \times n} = (-2)^{n-1} \neq 0.$$

Next, observe that system (3.1) can be transformed to system (3.2), where $p_1, p_2, \ldots, p_n$ solve the following n problems:

$$\begin{cases} D^\alpha p_1(t) = (\sigma_1(t) + \sigma_2(t) + \cdots + \sigma_n(t)) - (M_1 + M_2 + \cdots + M_n) p_1(t), \\ t^{1-\alpha} p_1(t)|_{t=0} = x_0^1 + x_0^2 + \cdots + x_0^n, \end{cases}$$

$$\begin{cases} D^\alpha p_2(t) = (\sigma_1(t) - \sigma_2(t) + \cdots + \sigma_n(t)) - (M_1 - M_2 + \cdots + M_n) p_2(t), \\ t^{1-\alpha} p_2(t)|_{t=0} = x_0^1 - x_0^2 + \cdots + x_0^n, \end{cases}$$

$$\vdots$$

$$\begin{cases} D^\alpha p_n(t) = (\sigma_1(t) + \sigma_2(t) + \cdots - \sigma_n(t)) - (M_1 + M_2 + \cdots - M_n) p_n(t), \\ t^{1-\alpha} p_n(t)|_{t=0} = x_0^1 + x_0^2 + \cdots - x_0^n. \end{cases}$$

Finally, observe that the solutions of the above equations are unique due to Lemma 2.1, which ends the proof. □

Now we can state and proof the comparison result for system (3.1).

Theorem 3.1 *Let* $0 < \alpha \leq 1$, $M_1 \in \mathbb{R}$, $M_2, \ldots, M_n \geq 0$, *and let* $u_1, \ldots, u_n \in C_{1-\alpha}([0,T])$ *satisfy*

$$\begin{cases} D^\alpha u_1(t) \geq -M_1 u_1(t) + \sum_{i,j=2}^n M_j \delta_{ji} u_i(t), \quad t \in (0,T], \\ D^\alpha u_s(t) \geq -M_1 u_s(t) + (\sum_{i=2}^n M_i - M_s) u_s(t) \\ \qquad + M_s(\sum_{i=1}^n u_i(t) - u_s(t)), \quad 2 \leq s \leq n, t \in (0,T], \\ t^{1-\alpha} u_s(t)|_{t=0} \geq 0, \quad 1 \leq s \leq n. \end{cases} \tag{3.3}$$

Then

$$\sum_{i=1}^n u_i(t) \geq 0, \quad t \in (0,T], \tag{3.4}$$

$$u_s(t) \geq 0, \quad t \in (0,T], 2 \leq s \leq n, \tag{3.5}$$

$$-u_s(t) + \sum_{i=1}^n u_i(t) \geq 0, \quad t \in (0,T], 2 \leq s \leq n. \tag{3.6}$$

Proof Put $r(t) = \sum_{s=1}^n u_s(t)$. Using (3.3) we obtain

$$\begin{aligned} D^\alpha r(t) &= \sum_{s=1}^n D^\alpha u_s(t) \\ &\geq -M_1 u_1(t) + \sum_{i,j=2}^n M_j \delta_{ji} u_i(t) - M_1 \sum_{s=2}^n u_s(t) - 2\sum_{s=2}^n M_s u_s(t) \\ &\quad + \sum_{s=2}^n \sum_{i=2}^n M_i u_s(t) + \sum_{s=2}^n \sum_{i=1}^n M_s u_i(t) \\ &= -M_1 r(t) + \sum_{i,j=2}^n \big(M_j \delta_{ji} u_i(t) + M_i u_j(t)\big) - 2\sum_{s=2}^n M_s u_s(t) + \sum_{s=2}^n M_s r(t). \end{aligned}$$

Observe that

$$\sum_{i,j=2, i\neq j}^n \big(M_j \delta_{ji} u_i(t) + M_i u_j(t)\big) = 0. \tag{3.7}$$

Hence, we obtain

$$\begin{aligned} D^\alpha r(t) &\geq -\left(M_1 - \sum_{s=2}^n M_s\right) r(t) + \sum_{i,j=2, i\neq j}^n \big(M_j \delta_{ji} u_i(t) + M_i u_j(t)\big) \\ &\quad + \sum_{i,j=2, i=j}^n \big(M_j \delta_{ji} u_i(t) + M_i u_j(t)\big) - 2\sum_{s=2}^n M_s u_s(t) \\ &= -\left(M_1 - \sum_{s=2}^n M_s\right) r(t) + 2\sum_{i=2}^n M_i u_i(t) - 2\sum_{s=2}^n M_s u_s(t) \\ &= -\left(M_1 - \sum_{s=2}^n M_s\right) r(t). \end{aligned} \tag{3.8}$$

Moreover, observe that

$$t^{1-\alpha}r(t) = \sum_{s=1}^{n} t^{1-\alpha}u_s(t) \geq 0. \tag{3.9}$$

Applying (3.8) and (3.9) to Lemma 2.2 we get (3.4).

Now, consider any $2 \leq s \leq n$ and denote

$$r_s(t) = \sum_{i=1}^{n} u_i(t) - u_s(t), \quad t \in (0, T].$$

By (3.3) we have

$$\begin{aligned}
D^\alpha r_s(t) &= \sum_{i=1}^{n} D^\alpha u_i(t) - D^\alpha u_s(t) = D^\alpha u_1(t) + \sum_{i=2}^{n} D^\alpha u_i(t) - D^\alpha u_s(t) \\
&\geq -M_1 u_1(t) + \sum_{i,j=2}^{n} M_j \delta_{ji} u_i(t) - \sum_{i=2}^{n} M_1 u_i(t) + M_1 u_s(t) \\
&\quad + \sum_{i=2}^{n}\left(\sum_{j=2}^{n} M_j - M_i\right)u_i(t) - \left(\sum_{j=2}^{n} M_j - M_s\right)u_s(t) \\
&\quad + \sum_{i=2}^{n} M_i\left(\sum_{j=1}^{n} u_j(t) - u_i(t)\right) - M_s\left(\sum_{j=1}^{n} u_j(t) - u_s(t)\right) \\
&= -M_1 u_1(t) + \sum_{i,j=2}^{n}\left(M_j \delta_{ji} u_i(t) + M_j u_i(t)\right) - M_1 \sum_{i=2}^{n} u_i(t) + M_1 u_s(t) + M_s u_s(t) \\
&\quad - 2\sum_{i=2}^{n} M_i u_i(t) - u_s(t)\sum_{j=2}^{n} M_j + \sum_{i=2}^{n}\sum_{j=1}^{n} M_i u_j(t) - M_s r_s(t).
\end{aligned}$$

Again, using (3.7), we obtain

$$\begin{aligned}
D^\alpha r_s(t) &\geq -M_1 \sum_{i=1}^{n} u_i(t) + M_1 u_s(t) + M_s u_s(t) - u_s(t)\sum_{j=2}^{n} M_j \\
&\quad + \sum_{i=2}^{n} M_i r_s(t) + u_s(t)\sum_{i=2}^{n} M_i - M_s r_s(t) \\
&= -\left(M_1 - \sum_{i=2}^{n} M_i + M_s\right) r_s(t) + M_s u_s(t).
\end{aligned} \tag{3.10}$$

Moreover, observe that (3.3) implies

$$D^\alpha u_s(t) \geq -\left(M_1 - \sum_{i=2}^{n} M_i + M_s\right) u_s(t) + M_s r_s(t). \tag{3.11}$$

Finally, note that (3.10) and (3.11) applied to Lemma 2.3 give (3.5) and (3.6). □

Now, we are in a position to enunciate the main result.

Theorem 3.2 *Suppose that there exist $u_0^1, u_0^2, \dots, u_0^n \in C_{1-\alpha}([0,T])$, $u_0^1 \le \sum_{i=2}^n u_0^i$, satisfying*

$$\begin{cases} D^\alpha u_0^1(t) \le f_1(t, u_0^1(t), u_0^2(t), \dots, u_0^n(t)), & t \in (0,T], \\ D^\alpha u_0^s(t) \ge f_s(t, u_0^1(t), u_0^2(t), \dots, u_0^n(t)), & t \in (0,T], 2 \le s \le n, \\ t^{1-\alpha} u_0^1(t)|_{t=0} \le x_0^1, \\ t^{1-\alpha} u_0^s(t)|_{t=0} \ge x_0^s, \quad 2 \le s \le n, \end{cases} \tag{3.12}$$

and there exist $M_1 \in \mathbb{R}, M_2, \dots, M_n > 0$ such that

(i)

$$f_1(t, \alpha_1, \dots, \alpha_n) - f_1(t, \beta_1, \dots, \beta_n) \ge -M_1(\alpha_1 - \beta_1) - \sum_{i,j=2}^n M_j \delta_{ji}(\alpha_i - \beta_i), \tag{3.13}$$

(ii)

$$\begin{aligned} & f_s(t, \alpha_1, \dots, \alpha_n) - f_s(t, \beta_1, \dots, \beta_n) \\ & \ge \left(-M_1 + \sum_{i=2}^n M_i - M_s\right)(\alpha_s - \beta_s) - M_s\left(\alpha_1 - \beta_1 + \alpha_s - \beta_s - \sum_{i=2}^n (\alpha_i - \beta_i)\right), \end{aligned}$$

where $\alpha_i, \beta_i \in \mathbb{R}, 1 \le i \le n$ satisfy for all $t \in [0,T]$ and $2 \le s \le n$,

$$u_0^1(t) - \left(\sum_{i=2}^n u_0^i(t) - u_0^s(t)\right) \le \beta_1 - \left(\sum_{i=2}^n \beta_i - \beta_s\right) \le \alpha_1 - \left(\sum_{i=2}^n \alpha_i - \alpha_s\right) \le u_0^s(t),$$

$$u_0^1(t) - \left(\sum_{i=2}^n u_0^i(t) - u_0^s(t)\right) \le \alpha_s \le \beta_s \le u_0^s(t),$$

(iii)

$$\begin{aligned} & \sum_{s=2}^n f_s(t, u^1(t), u^2(t), \dots, u^n(t)) - f_1(t, u^1(t), u^2(t), \dots, u^n(t)) \\ & \ge \left(-M_1 + \sum_{s=2}^n M_s\right)\left(\sum_{s=2}^n u^s(t) - u^1(t)\right), \end{aligned} \tag{3.14}$$

where

$$u_0^1 - \left(\sum_{i=2}^n u_0^i - u_0^s\right) \le u^1 - \left(\sum_{i=2}^n u^i - u^s\right) \le u^s \le u_0^s, \quad 2 \le s \le n.$$

Then there exists a solution $(\bar{u}^1, \bar{u}^2, \dots, \bar{u}^n)$ of system (1.1) *such that*

$$(n-1)u_0^1 - (n-2)\sum_{i=2}^n u_0^i \le \bar{u}^1 \le \sum_{i=2}^n u_0^i, \qquad u_0^1 - \sum_{i=2}^n u_0^i + u_0^s \le \bar{u}^s \le u_0^s, \quad 2 \le s \le n.$$

Moreover, there exist iterative sequences $(u_k^1), (u_k^2), \dots, (u_k^n)$ such that $u_k^i \to \bar{u}^i$, $k \to \infty$, $i = 1, 2, \dots, n$, uniformly on compact subsets of $(0,T]$.

Proof Let us first consider the linear system of the form

$$\begin{cases} D^\alpha u^1(t) = f_1(t, u_0^1(t), u_0^2(t), \ldots, u_0^n(t)) + M_1 u_0^1(t) + \sum_{i,j=2}^n M_j \delta_{ji} u_0^i(t) \\ \qquad - M_1 u^1(t) - \sum_{i,j=2}^n M_j \delta_{ji} u^i(t), \quad t \in (0,T], \\ D^\alpha u^s(t) = f_s(t, u_0^1(t), u_0^2(t), \ldots, u_0^n(t)) + M_1 u_0^s(t) + (\sum_{i=2}^n M_i - M_s) u^s(t) \\ \qquad + M_s(u_0^1(t) + \sum_{i=2}^n u^i(t) - u^s(t)) - M_1 u^s(t) - (\sum_{i=2}^n M_i - M_s) u_0^s(t) \\ \qquad - M_s(u^1(t) + \sum_{i=2}^n u_0^i(t) - u_0^s(t)), \quad t \in (0,T], 2 \le s \le n, \\ t^{1-\alpha} u^s(t)|_{t=0} = x_0^s, \quad 1 \le s \le n, \end{cases} \tag{3.15}$$

where $u^1, u^2, \ldots, u^n \in C_{1-\alpha}([0,T])$. Due to Lemma 3.1 there exists a system of solutions $(u_1^1, u_1^2, \ldots, u_1^n) \in C([0,T])^n$ for system (3.15). Using induction we obtain the sequence $(u_k^1, u_k^2, \ldots, u_k^n) \in C([0,T])^n$, $k \in \mathbb{N}$, satisfying

$$\begin{cases} D^\alpha u_k^1(t) = f_1(t, u_{k-1}^1(t), u_{k-1}^2(t), \ldots, u_{k-1}^n(t)) + M_1 u_{k-1}^1(t) + \sum_{i,j=2}^n M_j \delta_{ji} u_{k-1}^i(t) \\ \qquad - M_1 u_k^1(t) - \sum_{i,j=2}^n M_j \delta_{ji} u_k^i(t), \quad t \in (0,T], \\ D^\alpha u_k^s(t) = f_s(t, u_{k-1}^1(t), u_{k-1}^2(t), \ldots, u_{k-1}^n(t)) + M_1 u_{k-1}^s(t) \\ \qquad + (\sum_{i=2}^n M_i - M_s) u_k^s(t) + M_s(u_{k-1}^1(t) + \sum_{i=2}^n u_k^i(t) - u_k^s(t)) \\ \qquad - M_1 u_k^s(t) - (\sum_{i=2}^n M_i - M_s) u_{k-1}^s(t) \\ \qquad - M_s(u_k^1(t) + \sum_{i=2}^n u_{k-1}^i(t) - u_{k-1}^s(t)), \quad t \in (0,T], 2 \le s \le n, \\ t^{1-\alpha} u_k^s(t)|_{t=0} = x_0^s, \quad 1 \le s \le n. \end{cases} \tag{3.16}$$

Now, put $p_1^1 = u_1^1 - u_0^1$, $p_1^s = u_0^s - u_1^s$, $2 \le s \le n$. From (3.12) and (3.15), for all $t \in (0,T]$, we obtain

$$\begin{aligned} D^\alpha p_1^1(t) &= D^\alpha u_1^1(t) - D^\alpha u_0^1(t) \\ &= f_1\big(t, u_0^1(t), u_0^2(t), \ldots, u_0^n(t)\big) + M_1 u_0^1(t) + \sum_{i,j=2}^n M_j \delta_{ji} u_0^i(t) \\ &\quad - M_1 u_1^1(t) - \sum_{i,j=2}^n M_j \delta_{ji} u_1^i(t) - D^\alpha u_0^1(t) \\ &\ge -M_1 p_1^1(t) + \sum_{i,j=2}^n M_j \delta_{ji} p_1^i(t), \\ D^\alpha p_1^s(t) &= D^\alpha u_0^s(t) - D^\alpha u_1^s(t) \\ &= D^\alpha u_0^s(t) - f_s\big(t, u_0^1(t), u_0^2(t), \ldots, u_0^n(t)\big) - M_1 u_0^s(t) \\ &\quad - \left(\sum_{i=2}^n M_i - M_s\right) u_1^s(t) - M_s\left(u_0^1(t) + \sum_{i=2}^n u_1^i(t) - u_1^s(t)\right) + M_1 u_1^s(t) \\ &\quad + \left(\sum_{i=2}^n M_i - M_s\right) u_0^s(t) + M_s\left(u_1^1(t) + \sum_{i=2}^n u_0^i(t) - u_0^s(t)\right) \\ &\ge -M_1 p_1^s(t) + \left(\sum_{i=2}^n M_i - M_s\right) p_1^s(t) + M_s\left(\sum_{i=1}^n p_1^i(t) - p_1^s(t)\right) \quad \text{for all } 2 \le s \le n, \end{aligned}$$

$$t^{1-\alpha} p_1^1(t)|_{t=0} = t^{1-\alpha} u_1^1(t)|_{t=0} - t^{1-\alpha} u_0^1(t)|_{t=0} \ge x_0^1 - x_0^1 = 0,$$

$$t^{1-\alpha} p_1^s(t)|_{t=0} = t^{1-\alpha} u_0^s(t)|_{t=0} - t^{1-\alpha} u_1^s(t)|_{t=0} \ge x_0^s - x_0^s = 0, \quad 2 \le s \le n.$$

Hence, using Theorem 3.1, we have

$$u_1^s \le u_0^s, \quad 2 \le s \le n \tag{3.17}$$

and

$$u_1^1 - u_0^1 + \sum_{i=2}^{n} (u_0^i - u_1^i) \ge u_0^s - u_1^s, \quad 2 \le s \le n. \tag{3.18}$$

Consider now $q_1 = \sum_{i=2}^{n} u_1^i - u_1^1$. Using (3.14) and (3.15) we have

$$\begin{aligned}
D^\alpha q_1(t) &= \sum_{s=2}^{n} u_1^s(t) - u_1^1(t) = \sum_{s=2}^{n} D^\alpha u_1^s(t) - D^\alpha u_1^1(t) \\
&= \sum_{s=2}^{n} f_s(t, u_0^1(t), u_0^2(t), \dots, u_0^n(t)) + \sum_{s=2}^{n} M_1 u_0^s(t) \\
&\quad + \sum_{s=2}^{n} \left(\sum_{i=2}^{n} M_i - M_s \right) u_1^s(t) + \sum_{s=2}^{n} M_s \left(u_0^1(t) + \sum_{i=2}^{n} u_1^i(t) - u_1^s(t) \right) \\
&\quad - \sum_{s=2}^{n} M_1 u_1^s(t) - \sum_{s=2}^{n} \left(\sum_{i=2}^{n} M_i - M_s \right) u_0^s(t) \\
&\quad - \sum_{s=2}^{n} M_s \left(u_1^1(t) + \sum_{i=2}^{n} u_0^i(t) - u_0^s(t) \right) - f_1(t, u_0^1(t), u_0^2(t), \dots, u_0^n(t)) \\
&\quad - M_1 u_0^1(t) - \sum_{i,j=2}^{n} M_j \delta_{ji} u_0^i(t) + M_1 u_1^1(t) + \sum_{i,j=2}^{n} M_j \delta_{ji} u_1^i(t) \\
&= \sum_{s=2}^{n} f_s(t, u_0^1(t), u_0^2(t), \dots, u_0^n(t)) - f_1(t, u_0^1(t), u_0^2(t), \dots, u_0^n(t)) \\
&\quad - \left(M_1 - \sum_{s=2}^{n} M_s \right) q_1(t) + \left(M_1 - \sum_{s=2}^{n} M_s \right) \left(\sum_{s=2}^{n} u_0^s(t) - u_0^1(t) \right) \\
&\ge - \left(M_1 - \sum_{s=2}^{n} M_s \right) q_1(t).
\end{aligned}$$

Moreover, (1.2) implies

$$t^{1-\alpha} q_1(t)|_{t=0} = \sum_{i=2}^{n} t^{1-\alpha} u_1^i(t)|_{t=0} - t^{1-\alpha} u_1^1(t)|_{t=0} = \sum_{i=2}^{n} x_0^i - x_0^1 \ge 0.$$

Now, from Lemma 2.2 we conclude

$$u_1^1(t) \le \sum_{i=2}^{n} u_1^i(t) \quad \text{for all } t \in [0, T]. \tag{3.19}$$

Combining (3.17) and (3.18) with (3.19) we obtain for all $2 \le s \le n$ the inequalities

$$u_0^1 - \left(\sum_{i=2}^{n} u_0^i - u_0^s \right) \le u_1^1 - \left(\sum_{i=2}^{n} u_1^i - u_1^s \right) \le u_1^s \le u_0^s.$$

Let $2 \leq s \leq n$ be fixed and suppose now that for some $k \in \mathbb{N}$ the following inequalities hold:

$$u^1_{k-1} - \left(\sum_{i=2}^{n} u^i_{k-1} - u^s_{k-1}\right) \leq u^1_k - \left(\sum_{i=2}^{n} u^i_k - u^s_k\right) \leq u^s_k \leq u^s_{k-1}. \tag{3.20}$$

Denote $p^1_{k+1} = u^1_{k+1} - u^1_k, p^s_{k+1} = u^s_k - u^s_{k+1}, 2 \leq s \leq n$. From (3.13), (3.16), and (3.20) we obtain

$$\begin{aligned}
D^\alpha p^1_{k+1}(t) &= D^\alpha u^1_{k+1}(t) - D^\alpha u^1_k(t)\\
&= f_1\left(t, u^1_k(t), u^2_k(t), \ldots, u^n_k(t)\right) + M_1 u^1_k(t) + \sum_{i,j=2}^{n} M_j \delta_{ji} u^i_k(t) - M_1 u^1_{k+1}(t)\\
&\quad - \sum_{i,j=2}^{n} M_j \delta_{ji} u^i_{k+1}(t) - f_1\left(t, u^1_{k-1}(t), u^2_{k-1}(t), \ldots, u^n_{k-1}(t)\right) - M_1 u^1_{k-1}(t)\\
&\quad - \sum_{i,j=2}^{n} M_j \delta_{ji} u^i_{k-1}(t) + M_1 u^1_k(t) + \sum_{i,j=2}^{n} M_j \delta_{ji} u^i_k(t)\\
&\geq -M_1\left(u^1_{k1}(t) - u^1_{k-1}(t)\right) - \sum_{i,j=2}^{n} M_j \delta_{ji}\left(u^i_k(t) - u^i_{k-1}(t)\right) + M_1 u^1_k(t)\\
&\quad + \sum_{i,j=2}^{n} M_j \delta_{ji} u^i_k(t) - M_1 u^1_{k+1}(t) - \sum_{i,j=2}^{n} M_j \delta_{ji} u^i_{k+1}(t) - M_1 u^1_{k-1}(t)\\
&\quad - \sum_{i,j=2}^{n} M_j \delta_{ji} u^i_{k-1}(t) + M_1 u^1_k(t) + \sum_{i,j=2}^{n} M_j \delta_{ji} u^i_k(t)\\
&= -M_1 p^1_{k+1}(t) + \sum_{i,j=2}^{n} M_j \delta_{ji} p^i_{k+1}(t),\\
D^\alpha p^s_{k+1}(t) &= D^\alpha u^s_k(t) - D^\alpha u^s_{k+1}(t)\\
&\geq \left(-M_1 + \sum_{i=2}^{n} M_i - M_s\right)\left(u^s_{k-1}(t) - u^s_k(t)\right) - M_s\Bigg(u^1_{k-1}(t) - u^1_k(t)\\
&\quad + u^s_{k-1}(t) - u^s_k(t) - \sum_{i=2}^{n}\left(u^i_{k-1} - u^i_k\right)\Bigg) + M_1 u^s_{k-1}(t) + \left(\sum_{i=2}^{n} M_i - M_s\right) u^s_k(t)\\
&\quad + M_s\left(u^1_{k-1}(t) + \sum_{i=2}^{n} u^i_k(t) - u^s_k(t)\right) - M_1 u^s_k(t) - \left(\sum_{i=2}^{n} M_i - M_s\right) u^s_{k-1}(t)\\
&\quad - M_s\left(u^1_k(t) + \sum_{i=2}^{n} u^i_{k-1}(t) - u^s_{k-1}(t)\right) - M_1 u^s_k(t) - \left(\sum_{i=2}^{n} M_i - M_s\right) u^s_{k+1}(t)\\
&\quad - M_s\left(u^1_k(t) + \sum_{i=2}^{n} u^i_{k+1}(t) - u^s_{k+1}(t)\right) + M_1 u^s_{k+1}(t) + \left(\sum_{i=2}^{n} M_i - M_s\right) u^s_k(t)\\
&\quad + M_s\left(u^1_{k+1}(t) + \sum_{i=2}^{n} u^i_k(t) - u^s_k(t)\right)\\
&= -M_1 p^s_{k+1}(t) + \left(\sum_{i=2}^{n} M_i - M_s\right) p^s_{k+1}(t) + M_s\left(\sum_{i=1}^{n} p^i_{k+1}(t) - p^s_{k+1}(t)\right).
\end{aligned}$$

Also observe that $t^{1-\alpha}p^1_{k+1}(t)|_{t=0} = t^{1-\alpha}p^s_{k+1}(t)|_{t=0} = 0$, which, together with the above, due to Theorem 3.1, gives

$$u^s_{k+1} \leq u^s_k, \quad 2 \leq s \leq n, \tag{3.21}$$

$$u^s_k - u^s_{k+1} \leq \sum_{i=2}^{n}(u^i_k - u^i_{k+1}) + u^1_{k+1} - u^1_k. \tag{3.22}$$

Consider now $q_k = \sum_{i=2}^{n} u^i_k - u^1_k$. Using the same arguments as with q_1 we obtain

$$D^\alpha q_k(t) \geq -\left(M_1 - \sum_{s=2}^{n} M_s\right) q_k(t)$$

and

$$t^{1-\alpha} q_k(t)|_{t=0} \geq 0,$$

which, due to Lemma 2.2, gives

$$u^1_k \leq \sum_{s=2}^{n} u^i_k. \tag{3.23}$$

Summarizing, by (3.21)-(3.23) and induction, we obtain the following inequalities describing the sequences $(u^s_k)_{k\in\mathbb{N}\cup\{0\}}$:

$$\begin{aligned} u^1_0 - \left(\sum_{i=2}^{n} u^i_0 - u^s_0\right) &\leq u^1_1 - \left(\sum_{i=2}^{n} u^i_1 - u^s_1\right) \\ &\leq \cdots \leq u^1_k - \left(\sum_{i=2}^{n} u^i_k - u^s_k\right) \\ &\leq u^s_k \leq \cdots \leq u^s_1 \leq u^s_0, \end{aligned} \tag{3.24}$$

where $2 \leq s \leq n$. The inequalities (3.24) imply

$$\lim_{k\to\infty} u^s_k(t) = \bar{u}^s(t), \quad s = 2, \ldots, n.$$

Observe that

$$u^1_0 - \left(\sum_{i=2}^{n} u^i_0 - u^s_0\right) \leq \bar{u}^s \leq u^s_0, \quad s = 2, \ldots, n.$$

In order to show that the sequence (u^1_k) is convergent observe first that from (3.24) there exists a function x^* such that

$$\lim_{k\to\infty}\left(u^1_k(t) - \sum_{i=2}^{n-1} u^i_k(t)\right) = x^*(t).$$

Hence, putting $\bar{u}^1 = x^* + \sum_{s=2}^{n-1} \bar{u}^s$, we have

$$\lim_{k\to\infty} \left(u_k^1(t) - \bar{u}^1(t)\right) = \lim_{k\to\infty} \left(u_k^1(t) - x^*(t) - \sum_{s=2}^{n-1} \bar{u}^s(t) + \sum_{s=2}^{n-1} u_k^s(t) - \sum_{s=2}^{n-1} u_k^s(t) \right)$$

$$= \lim_{k\to\infty} \left(u_k^1(t) - \sum_{s=2}^{n-1} u_k^s(t) - x^*(t) + \sum_{s=2}^{n-1} \left(u_k^s(t) - \bar{u}^s(t)\right) \right)$$

$$= \lim_{k\to\infty} \left(u_k^1(t) - \sum_{s=2}^{n-1} u_k^s(t) - x^*(t) \right) + \sum_{s=2}^{n-1} \lim_{k\to\infty} \left(u_k^s(t) - \bar{u}^s(t)\right) = 0.$$

In order to show the uniform convergence of sequences $(u_k^2), (u_k^3), \dots, (u_k^n)$, observe that from (3.24) and from the fact that $u_k^s \to \bar{u}^s$, $s = 2, 3, \dots, n$, we have

$$\bar{u}^s \le u_k^s \le \cdots \le u_1^s \le u_0^s \quad \text{for all } k \in \mathbb{N}.$$

Then, the uniform convergence of sequences (u_k^s), $s = 2, 3, \dots, n$, on a compact subset of $(0, T]$ is a straightforward consequence of Dini's theorem, which states that if a monotone sequence of continuous functions is convergent on a compact set, then it converges uniformly.

Showing a uniform convergence of (u_k^1) requires some observations. Take any $2 \le s \le n$ and denote

$$h_k = u_k^1 - \left(\sum_{i=2}^{n} u_k^i - u_k^s \right), \quad k \in \mathbb{N} \cup \{0\}.$$

From (3.24) and the convergence of $(u_k^1), \dots, (u_k^n)$ we have

$$h_0 \le h_1 \le \cdots \le h_k \le \bar{u}^1 - \left(\sum_{i=2}^{n} \bar{u}^i - \bar{u}^s \right).$$

Applying again Dini's result we get the uniform convergence of (h_k) on every compact subset of $(0, T]$. Finally note that

$$u_k^1 = h_k + \left(\sum_{i=2}^{n} u_k^i - u_k^s \right), \quad k \in \mathbb{N},$$

and thus (u_k^1) is uniformly convergent on a compact subset of $(0, T]$ to $\bar{u}^1$ as a linear combination of sequences uniformly convergent.

Moreover, observe that the limit functions satisfy the properties

$$(n-1)u_0^1 - (n-2)\sum_{i=2}^{n} u_0^i \le \bar{u}^1 \le \sum_{i=2}^{n} u_0^i,$$

$$u_0^1 - \sum_{i=2}^{n} u_0^i + u_0^s \le \bar{u}^s \le u_0^s, \quad 2 \le s \le n.$$

Taking k to ∞ in (3.16) we see that $(\bar{u}^1, \bar{u}^2, \ldots, \bar{u}^n)$ is a system of solutions of system (1.1). Also observe that from (3.24) we have the following relations between the limit functions:

$$\bar{u}^1 - \left(\sum_{i=2}^{n} \bar{u}^i - \bar{u}^s\right) \leq \bar{u}^s, \quad 2 \leq s \leq n,$$

which ends the proof. □

Remark 3.1 Observe that using the same methods as in the proof of Theorem 3.2 we can see that $(\bar{u}^1, \bar{u}^2, \ldots, \bar{u}^n)$ is an extremal solution of system (1.1) in the sense that if $(u^1, \ldots, u^n)$ were any other solution such that

$$u_0^1 - \left(\sum_{i=2}^{n} u_0^i - u_0^s\right) \leq u^1 - \left(\sum_{i=2}^{n} u^i - u^s\right) \leq u_0^s, \qquad u_0^1 - \left(\sum_{i=2}^{n} u_0^i - u_0^s\right) \leq u^s \leq u_0^s$$

for any $2 \leq s \leq n$, then we would have

$$\bar{u}^1 - \left(\sum_{i=2}^{n} \bar{u}^i - \bar{u}^s\right) \leq u^1 - \left(\sum_{i=2}^{n} u^i - u^s\right), \quad u^s \leq \bar{u}^s, 2 \leq s \leq n.$$

4 The system of three fractional differential equations

In order to see the nature of the iterative procedure introduced in the proof of Theorem 3.2, we consider the case $n = 3$.

Corollary 4.1 *If there exist $u_0, v_0, w_0 \in C_{1-\alpha}([0,T])$, $u_0 \leq v_0 + w_0$ such that*

$$\begin{cases} D^\alpha u_0(t) \leq f(t, u_0(t), v_0(t), w_0(t)), & t \in (0,T], \\ D^\alpha v_0(t) \geq g(t, u_0(t), v_0(t), w_0(t)), & t \in (0,T], \\ D^\alpha w_0(t) \geq h(t, u_0(t), v_0(t), w_0(t)), & t \in (0,T], \\ t^{1-\alpha} u_0(t)|_{t=0} \leq x_0, \\ t^{1-\alpha} v_0(t)|_{t=0} \geq y_0, \\ t^{1-\alpha} w_0(t)|_{t=0} \geq z_0, \end{cases} \tag{4.1}$$

and there exist $M \in \mathbb{R}$, $N, S \geq 0$ satisfying

$$f(t,\alpha_1,\alpha_2,\alpha_3) - f(t,\beta_1,\beta_2,\beta_n) \geq -M(\alpha_1-\beta_1) + (-N+S)(\alpha_2-\beta_2) + (N-S)(\alpha_3-\beta_3),$$

$$g(t,\alpha_1,\alpha_2,\alpha_3) - g(t,\beta_1,\beta_2,\beta_3) \geq -N(\alpha_1-\beta_1) + (-M+S)(\alpha_2-\beta_2) + N(\alpha_3-\beta_3),$$

$$h(t,\alpha_1,\alpha_2,\alpha_3) - h(t,\beta_1,\beta_2,\beta_3) \geq -S(\alpha_1-\beta_1) + S(\alpha_2-\beta_2) + (-M+N)(\alpha_3-\beta_3),$$

where $\alpha_i, \beta_i \in \mathbb{R}$, $1 \leq i \leq 3$ satisfy, for all $t \in [0,T]$,

$$u_0(t) - w_0(t) \leq \beta_1 - \beta_3 \leq \alpha_1 - \alpha_3 \leq v_0(t), \qquad u_0(t) - w_0(t) \leq \alpha_2 \leq \beta_2 \leq v_0(t),$$

$$u_0(t) - v_0(t) \leq \beta_1 - \beta_2 \leq \alpha_1 - \alpha_2 \leq w_0(t), \qquad u_0(t) - v_0(t) \leq \alpha_3 \leq \beta_3 \leq w_0(t)$$

and

$$(g+h-f)(t,u,v,w) \geq (-M+N+S)(v+w-u), \tag{4.2}$$

where

$$u_0(t) - w_0(t) \le u - w \le v \le v_0(t),$$
$$u_0(t) - v_0(t) \le u - v \le w \le w_0(t).$$

Then there exists a solution

$$(u^*, v^*, w^*) \in [2u_0 - v_0 - w_0, v_0 + w_0] \times [u_0 - w_0, v_0] \times [u_0 - v_0, w_0]$$

of (4.1) *and the sequences* $(u_n) \subseteq [2u_0 - v_0 - w_0, v_0 + w_0]$, $(v_n) \subseteq [u_0 - w_0, v_0]$, $(w_n) \subseteq [u_0 - v_0, w_0]$ *such that* $u_n \to u^*$, $v_n \to v^*$, $w_n \to w^*$ *uniformly on compact subsets of* $(0, T]$. *Moreover, the following inequalities hold*:

$$u_0 - v_0 \le u_1 - v_1 \le \cdots \le u_n - v_n \le \cdots \le u^* - v^* \le w^* \le \cdots \le w_n \le \cdots \le w_1 \le w_0,$$
$$u_0 - w_0 \le u_1 - w_1 \le \cdots \le u_n - w_n \le \cdots \le u^* - w^* \le v^* \le \cdots \le v_n \le \cdots \le v_1 \le v_0.$$

4.1 Example

Consider the nonlinear problem of the form

$$\begin{cases} D^{0.5}u(t) = \Gamma(1.5)^{-1}v(t) - \Gamma(1.5)^{-1}w(t) + (v(t) - t)^2 + 2(t - w(t) + u(t))^2, \\ D^{0.5}v(t) = \Gamma(1.5)^{-1}v(t) + (v(t) - t)^2 + (t - w(t) + u(t))^2, \\ D^{0.5}w(t) = -\Gamma(1.5)^{-1}u(t) + \Gamma(1.5)^{-1}v(t) + (t - w(t) + u(t))^2, \\ t^{0.5}u(t)|_{t=0} = t^{0.5}v(t)|_{t=0} = t^{0.5}w(t)|_{t=0} = 0, \end{cases} \tag{4.3}$$

where $t \in [0, 1]$. Taking

$$f(t, u, v, w) = \Gamma(1.5)^{-1}v - \Gamma(1.5)^{-1}w + (v - t)^2 + 2(t - w + u)^2,$$
$$g(t, u, v, w) = \Gamma(1.5)^{-1}v + (v - t)^2 + (t - w + u)^2,$$
$$h(t, u, v, w) = -\Gamma(1.5)^{-1}u + \Gamma(1.5)^{-1}v + (t - w + u)^2$$

and

$$u_0(t) = 0, \qquad v_0(t) = w_0(t) = t, \quad t \in [0, 1],$$

we obtain, for all $t \in [0, 1]$,

$$D^{0.5}u_0(t) = 0 = f(t, u_0(t), v_0(t), w_0(t)),$$
$$D^{0.5}v_0(t) = \frac{\sqrt{t}}{\Gamma(1.5)} \ge \frac{t}{\Gamma(1.5)} = g(t, u_0(t), v_0(t), w_0(t)),$$
$$D^{0.5}w_0(t) = \frac{\sqrt{t}}{\Gamma(1.5)} \ge 0 = h(t, u_0(t), v_0(t), w_0(t)).$$

Next, for all $\alpha_i, \beta_i \in \mathbb{R}$, $1 \le i \le 3$ such that

$$-t \le \beta_1 - \beta_3 \le \alpha_1 - \alpha_3 \le t, \qquad -t \le \alpha_2 \le \beta_2 \le t,$$
$$-t \le \beta_1 - \beta_2 \le \alpha_1 - \alpha_2 \le t, \qquad -t \le \alpha_3 \le \beta_3 \le t,$$

one can calculate that

$$f(t,\alpha_1,\alpha_2,\alpha_3)-f(t,\beta_1,\beta_2,\beta_3)\ge\Gamma(1.5)^{-1}(\alpha_2-\beta_2)-\Gamma(1.5)^{-1}(\alpha_3-\beta_3),$$
$$g(t,\alpha_1,\alpha_2,\alpha_3)-g(t,\beta_1,\beta_2,\beta_3)\ge\Gamma(1.5)^{-1}(\alpha_2-\beta_2),$$
$$h(t,\alpha_1,\alpha_2,\alpha_3)-h(t,\beta_1,\beta_2,\beta_3)\ge-\Gamma(1.5)^{-1}(\alpha_1-\beta_1)+\Gamma(1.5)^{-1}(\alpha_2-\beta_2).$$

Therefore it is sufficient to take in Corollary 4.1 $M=0, N=0, S=\Gamma(1.5)^{-1}$. Finally observe that condition (4.2) also holds. Thus, the system of fractional differential equations (4.3) has a solution $(u^*,v^*,w^*)\in[-2t,2t]\times[-t,t]\times[-t,t]$.

Now, using the proof of Theorem 3.2 and Lemma 3.1, we can derive the iterative procedure (u_k,v_k,w_k) convergent to the solution (u^*,v^*,w^*). First observe that the sequences (u_k), (v_k), (w_k) satisfy the following system of linear equations:

$$D^{0.5}u_k=f(t,u_{k-1},v_{k-1},w_{k-1})-\Gamma(1.5)^{-1}v_{k-1}+\Gamma(1.5)^{-1}w_{k-1}+\Gamma(1.5)^{-1}v_k-\Gamma(1.5)^{-1}w_k,$$
$$D^{0.5}v_k=g(t,u_{k-1},v_{k-1},w_{k-1})-\Gamma(1.5)^{-1}v_{k-1}+\Gamma(1.5)^{-1}v_k,$$
$$D^{0.5}w_k=h(t,u_{k-1},v_{k-1},w_{k-1})+\Gamma(1.5)^{-1}u_{k-1}-\Gamma(1.5)^{-1}v_{k-1}-\Gamma(1.5)^{-1}u_k+\Gamma(1.5)^{-1}v_k,$$
$$t^{0.5}u_k(t)|_{t=0}=t^{0.5}v_k(t)|_{t=0}=t^{0.5}w_k(t)|_{t=0}=0,$$

which can be equivalently transformed to the system

$$\begin{cases}u_k+v_k+w_k=p_k,\\ u_k-v_k+w_k=q_k,\\ u_k+v_k-w_k=r_k,\end{cases}$$

where p_k, q_k, r_k are the solutions of the following systems:

$$\begin{cases}D^{0.5}p_k=(f+g+h)(t,u_{k-1},v_{k-1},w_{k-1})\\ \qquad\qquad+\Gamma(1.5)^{-1}u_{k-1}-3\Gamma(1.5)^{-1}v_{k-1}+\Gamma(1.5)^{-1}w_{k-1}-\Gamma(1.5)^{-1}p_k,\\ t^{0.5}p_k(t)|_{t=0}=0,\end{cases}$$

$$\begin{cases}D^{0.5}q_k=(f-g+h)(t,u_{k-1},v_{k-1},w_{k-1})\\ \qquad\qquad+\Gamma(1.5)^{-1}u_{k-1}-\Gamma(1.5)^{-1}v_{k-1}+\Gamma(1.5)^{-1}w_{k-1}-\Gamma(1.5)^{-1}q_k,\\ t^{0.5}q_k(t)|_{t=0}=0,\end{cases}$$

$$\begin{cases}D^{0.5}r_k=(f+g-h)(t,u_{k-1},v_{k-1},w_{k-1})\\ \qquad\qquad-\Gamma(1.5)^{-1}u_{k-1}-\Gamma(1.5)^{-1}v_{k-1}+\Gamma(1.5)^{-1}w_{k-1}+\Gamma(1.5)^{-1}r_k,\\ t^{0.5}r_k(t)|_{t=0}=0.\end{cases}$$

The solutions of the above systems, due to Lemma 2.1, are given by the formulas

$$p_k(t)=\int_0^t(t-s)^{-0.5}E_{0.5,0.5}\left(-\Gamma(1.5)^{-1}(t-s)^{0.5}\right)\Big((f+g+h)\big(s,u_{k-1}(s),v_{k-1}(s),w_{k-1}(s)\big)$$
$$+\Gamma(1.5)^{-1}u_{k-1}(s)-3\Gamma(1.5)^{-1}v_{k-1}(s)+\Gamma(1.5)^{-1}w_{k-1}(s)\Big)\,ds,$$

$$q_k(t)=\int_0^t(t-s)^{-0.5}E_{0.5,0.5}\left(-\Gamma(1.5)^{-1}(t-s)^{0.5}\right)\Big((f-g+h)\big(s,u_{k-1}(s),v_{k-1}(s),w_{k-1}(s)\big)$$
$$+\Gamma(1.5)^{-1}u_{k-1}(s)-\Gamma(1.5)^{-1}v_{k-1}(s)+\Gamma(1.5)^{-1}w_{k-1}(s)\Big)\,ds,$$

$$r_k(t) = \int_0^t (t-s)^{-0.5} E_{0.5,0.5}\big(\Gamma(1.5)^{-1}(t-s)^{0.5}\big)\big((f+g-h)\big(s,u_{k-1}(s),v_{k-1}(s),w_{k-1}(s)\big)$$
$$- \Gamma(1.5)^{-1}u_{k-1}(s) - \Gamma(1.5)^{-1}v_{k-1}(s) + \Gamma(1.5)^{-1}w_{k-1}(s)\big)\,ds.$$

In consequence, the iterative sequences are of the form

$$\begin{aligned}
u_k(t) &= \frac{1}{2}(q_k + r_k)\\
&= \frac{1}{2}\int_0^t (t-s)^{-0.5}\big[E_{0.5,0.5}\big(-\Gamma(1.5)^{-1}(t-s)^{0.5}\big)\big((f-g+h)\big(s,u_{k-1}(s),v_{k-1}(s),w_{k-1}(s)\big)\\
&\quad + \Gamma(1.5)^{-1}u_{k-1}(s) - \Gamma(1.5)^{-1}v_{k-1}(s) + \Gamma(1.5)^{-1}w_{k-1}(s)\big)\\
&\quad + E_{\alpha,\alpha}\big(\Gamma 1.5^{-1}(t-s)^{\alpha}\big)\big((f+g-h)\big(s,u_{k-1}(s),v_{k-1}(s),w_{k-1}(s)\big)\\
&\quad - \Gamma(1.5)^{-1}u_{k-1}(s) - \Gamma(1.5)^{-1}v_{k-1}(s) + \Gamma(1.5)^{-1}w_{k-1}(s)\big)\big]\,ds,\\
v_k(t) &= \frac{1}{2}(p_k - q_k)\\
&= \frac{1}{2}\int_0^t (t-s)^{-0.5}\big[E_{0.5,0.5}\big(-\Gamma(1.5)^{-1}(t-s)^{0.5}\big)\big((f+g+h)\big(s,u_{k-1}(s),v_{k-1}(s),w_{k-1}(s)\big)\\
&\quad + \Gamma(1.5)^{-1}u_{k-1}(s) - 3\Gamma(1.5)^{-1}v_{k-1}(s) + \Gamma(1.5)^{-1}w_{k-1}(s)\big)\\
&\quad - E_{0.5,0.5}\big(-\Gamma(1.5)^{-1}(t-s)^{0.5}\big)\big((f-g+h)\big(s,u_{k-1}(s),v_{k-1}(s),w_{k-1}(s)\big)\\
&\quad + \Gamma(1.5)^{-1}u_{k-1}(s) - \Gamma(1.5)^{-1}v_{k-1}(s) + \Gamma(1.5)^{-1}w_{k-1}(s)\big)\big]\,ds,\\
w_k(t) &= \frac{1}{2}(p_k - r_k)\\
&= \frac{1}{2}\int_0^t (t-s)^{-0.5}\big[E_{0.5,0.5}\big(-\Gamma(1.5)^{-1}(t-s)^{0.5}\big)\big((f+g+h)\big(s,u_{k-1}(s),v_{k-1}(s),w_{k-1}(s)\big)\\
&\quad + \Gamma(1.5)^{-1}u_{k-1}(s) - 3\Gamma(1.5)^{-1}v_{k-1}(s) + \Gamma(1.5)^{-1}w_{k-1}(s)\big)\\
&\quad - E_{0.5,0.5}\big(\Gamma(1.5)^{-1}(t-s)^{0.5}\big)\big((f+g-h)\big(s,u_{k-1}(s),v_{k-1}(s),w_{k-1}(s)\big)\\
&\quad - \Gamma(1.5)^{-1}u_{k-1}(s) - \Gamma(1.5)^{-1}v_{k-1}(s) + \Gamma(1.5)^{-1}w_{k-1}(s)\big)\big]\,ds.
\end{aligned}$$

Competing interests
The author declares that he has no competing interests.

Author's contributions
The author formulated and proved all the results in the article, produced the illustrative example, wrote the manuscript, and read and approved it.

Acknowledgements
The author is very grateful to the reviewers for the remarks, which improved the final version of the manuscript. This article was financially supported by University of Łódź as a part of donation for the research activities aimed at the development of young scientists, grant no. 545/1117.

References
1. Babakhani, A, Daftardar-Gejji, V: Existence of positive solutions of nonlinear fractional differential equations. J. Math. Anal. Appl. **278**, 434-442 (2003)
2. Bai, ZB: On positive solutions of a nonlocal fractional boundary value problem. Nonlinear Anal. **72**, 916-924 (2010)
3. Bai, ZB, Lu, HS: Positive solutions of boundary value problems of nonlinear fractional differential equation. J. Math. Anal. Appl. **311**, 495-505 (2005)
4. Belmekki, M, Benchohra, M: Existence results for fractional order semilinear functional differential equations with nondense domain. Nonlinear Anal. **72**, 925-932 (2010)

5. Belmekki, M, Nieto, JJ, Rodriguez-Lopez, R: Existence of periodic solution for a nonlinear fractional differential equation. Bound. Value Probl. **2009**, Article ID 324561 (2009)
6. El-Shahed, M: Positive solutions for boundary value problem of nonlinear fractional differential equation. Abstr. Appl. Anal. **2007**, Article ID 10368 (2007)
7. Lakshmikantham, V, Vatsala, AS: Basic theory of fractional differential equations. Nonlinear Anal. **69**(8), 2677-2682 (2008)
8. Lakshmikanthan, V, Vatsala, AS: General uniqueness and monotone iterative technique for fractional differential equations. Appl. Math. Lett. **21**, 828-834 (2008)
9. McRae, FA: Monotone iterative technique and existence results for fractional differential equations. Nonlinear Anal. **71**, 6093-6096 (2009)
10. Wang, G, Agarwal, RP, Cabada, A: Existence results and the monotone iterative technique for systems of nonlinear fractional differential equations. Appl. Math. Lett. **25**, 1019-1024 (2012)
11. Wei, Z, Li, Q, Che, J: Initial value problems for fractional differential equations involving Riemann-Liouville sequential fractional derivative. J. Math. Anal. Appl. **367**, 260-272 (2010)
12. Zhang, S: Monotone iterative method for initial value problem involving Riemann-Liouville fractional derivatives. Nonlinear Anal. **71**, 2087-2093 (2009)
13. Zhang, S: The existence of a positive solution for a nonlinear fractional differential equation. J. Math. Anal. Appl. **252**, 804-812 (2000)
14. Ahmad, B, Nieto, JJ: Existence results for a coupled system of nonlinear fractional differential equations with three-point boundary conditions. Comput. Math. Appl. **58**, 1838-1843 (2009)
15. Bai, C, Fang, J: The existence of a positive solution for a singular coupled system of nonlinear fractional differential equations. Appl. Math. Comput. **150**, 611-621 (2004)
16. Bhalekar, S, Daftardar-Gejji, V: Fractional ordered Liu system with time-delay. Commun. Nonlinear Sci. Numer. Simul. **15**, 2178-2191 (2010)
17. Chen, Y, An, H: Numerical solutions of coupled Burgers equations with time and space fractional derivatives. Appl. Math. Comput. **200**, 87-95 (2008)
18. Daftardar-Gejji, V: Positive solutions of a system of non-autonomous fractional differential equations. J. Math. Anal. Appl. **302**, 56-64 (2005)
19. Gafiychuk, V, Datsko, B, Meleshko, V: Mathematical modeling of time fractional reaction-diffusion systems. J. Comput. Appl. Math. **220**, 215-225 (2008)
20. Jafari, H, Daftardar-Gejji, V: Solving a system of nonlinear fractional differential equations using Adomian decomposition. J. Comput. Appl. Math. **196**, 644-651 (2006)
21. Lazarevi, MP: Finite time stability analysis of PD^{α} fractional control of robotic time-delay systems. Mech. Res. Commun. **33**, 269-279 (2006)
22. Su, X: Boundary value problem for a coupled system of nonlinear fractional differential equations. Appl. Math. Lett. **22**, 64-69 (2009)
23. Kilbas, AA, Srivastava, HM, Trujillo, JJ: Theory and Applications of Fractional Differential Equations. North-Holland Mathematics Studies, vol. 204. Elsevier, Amsterdam (2006)
24. Podlubny, I: Fractional Differential Equations. Mathematics in Science and Engineering. Academic Press, New York (1999)
25. Samko, SG, Kilbas, AA, Marichev, OI: Fractional Integrals and Derivatives: Theory and Applications. Gordon & Breach, Yverdon (1993)

Periodic solutions for a kind of prescribed mean curvature Liénard equation with a singularity and a deviating argument

Shiping Lu[1] and Fanchao Kong[2*]

*Correspondence: fanchaokong88@163.com
[2]Department of Mathematics, Anhui Normal University, Wuhu, 241000, China
Full list of author information is available at the end of the article

Abstract

In this paper, we study the existence of periodic solutions to the following prescribed mean curvature Liénard equation with a singularity and a deviating argument:

$$\left(\frac{u'(t)}{\sqrt{1+(u'(t))^2}}\right)' + f(u(t))u'(t) + g(u(t-\sigma)) = e(t),$$

where g has a strong singularity at $x = 0$ and satisfies a small force condition at $x = \infty$. By applying Mawhin's continuation theorem, we prove that the given equation has at least one positive T-periodic solution. We will also give an example to illustrate the application of our main results.

Keywords: periodic solution; continuation theorem; prescribed mean curvature Liénard equation; deviating argument; singularity

1 Introduction

In recent years, there have been many papers about the existence of periodic solutions for the second order differential equations with a singularity, especially for the Liénard equations. And the existence of periodic solutions of the Liénard equations with a deviating argument has also been studied widely (see [1–12]). For example, in [2], Zhang studied the following Liénard equation with a singularity:

$$x''(t) + f(x(t))x'(t) + g(t, x(t)) = 0, \tag{1.1}$$

where $f : R \to R$, $g : R \times (0, +\infty) \to R$ is an L^2-Carathéodory function, $g(t, x)$ is a T-periodic function in the first argument and can be singular at $x = 0$, *i.e.*, $g(t, x)$ can be unbounded as $x \to 0^+$. Equation (1.1) is of repulsive type (resp. attractive type) if $g(t, x) \to -\infty$ (resp. $g(t, x) \to +\infty$) for $x \to 0^+$.

Let Eq. (1.1) be of repulsive type and set

$$\bar{g}(x) = \frac{1}{T}\int_0^T g(t, x)\, dt, \quad x > 0.$$

Assume that

$$\varphi(t) = \lim_{x\to+\infty} \sup \frac{g(t, x)}{x}$$

exists uniformly for a.e. $t \in [0,T]$, *i.e.*, for any $\varepsilon > 0$, there is $g_\varepsilon \in L^2(0,T)$ such that

$$g(t,x) \leq (\varphi(t) + \varepsilon)x + g_\varepsilon$$

for all $x > 0$ and a.e. $t \in [0,T]$. Assume that $\varphi \in C(R,R)$ and $\varphi(t+T) = \varphi(t)$, $t \in R$.

Theorem 1.1 *Assume that the following conditions are satisfied:*

(h_1) (*Balance condition*) *There exist constants* $0 < D_1 < D_2$ *such that if* x *is a positive continuous* T*-periodic function satisfying*

$$\int_0^T g(t,x(t))\,dt = 0,$$

then

$$D_1 \leq x(\tau) \leq D_2 \quad \textit{for some } \tau \in [0,T].$$

(h_2) (*Degree condition*) $\overline{g}(x) < 0$ *for all* $x \in (0,D_1)$, *and* $\overline{g}(x) > 0$ *for all* $x > D_2$.

(h_3) (*Decomposition condition*) $g(t,x) = g_0(x) + g_1(t,x)$, *where* $g_0 \in C((0,+\infty),R)$ *and* $g_1 : [0,T] \times [0,+\infty) \to R$ *is an* L^2*-Carathéodory function, i.e.,* g_1 *is measurable with respect to the first variable, continuous with respect to the second one, and for any* $b > 0$ *there is* $h_b \in L^2((0,T);[0,+\infty))$ *such that* $|g_1(t,x)| \leq h_b(t)$ *for a.e.* $t \in [0,T]$ *and all* $x \in [0,b]$.

(h_4) (*Strong force condition at* $x = 0$) $\int_0^1 g_0(x)\,dx = -\infty$.

(h_5) (*Small force condition at* $x = \infty$)

$$\|\varphi^+\|_1 < \frac{\sqrt{3}}{T} \quad (\varphi^+ = \max\{\varphi^+, 0\}).$$

Then Eq. (1.1) *has at least one positive* T*-periodic solution.*

Moreover, we notice that in [10] Wang further studied the Liénard equation with a singularity and a deviating argument of the form

$$x''(t) + f(x(t))x'(t) + g(t,x(t-\sigma)) = 0, \tag{1.2}$$

where $0 \leq \sigma < T$ is a constant, $f : R \to R$, $g : R \times (0,+\infty) \to R$ is an L^2-Carathéodory function, $g(t,x)$ is a T-periodic function in the first argument and can be singular at $x = 0$, *i.e.*, $g(t,x)$ can be unbounded as $x \to 0^+$.

Theorem 1.2 *Assume that the following conditions are satisfied:*

(h_1) (*Balance condition*) *There exist constants* $0 < D_1 < D_2$ *such that if* x *is a positive continuous* T*-periodic function satisfying*

$$\int_0^T g(t,x(t))\,dt = 0,$$

then

$$D_1 \leq x(\tau) \leq D_2 \quad \textit{for some } \tau \in [0,T].$$

(h_2) *(Degree condition)* $\bar{g}(x) < 0$ *for all* $x \in (0, D_1)$, *and* $\bar{g}(x) > 0$ *for all* $x > D_2$.

(h_3) *(Decomposition condition)* $g(t,x) = g_0(x) + g_1(t,x)$, *where* $g_0 \in C((0,+\infty), R)$ *and* $g_1 : [0,T] \times [0,+\infty) \to R$ *is an* L^2-*Carathéodory function, i.e.,* g_1 *is measurable with respect to the first variable, continuous with respect to the second one, and for any* $b > 0$ *there is* $h_b \in L^2((0,T);[0,+\infty))$ *such that* $|g_1(t,x)| \leq h_b(t)$ *for a.e.* $t \in [0,T]$ *and all* $x \in [0,b]$.

(h_4) *(Strong force condition at* $x = 0$*)* $\int_0^1 g_0(x)\,dx = -\infty$.

(h_5') *(Small force condition at* $x = \infty$*)*

$$\|\varphi\|_\infty < \left(\frac{\sqrt{\pi}}{T}\right)^2.$$

Then Eq. (1.2) *has at least one positive* T*-periodic solution.*

Nowadays, the prescribed mean curvature $(\frac{u'(t)}{\sqrt{1+(u'(t))^2}})'$ of a function $u(t)$ frequently appears in different geometry and physics (see [13–16]). For example, in [13], Obersnel studied the existence, regularity and stability properties of periodic solutions of a capillarity equation in the presence of lower and upper solutions

$$-\left(\frac{u'}{\sqrt{1+u'^2}}\right) = f(t,u).$$

This equation, together with its N-dimensional counterpart

$$-\operatorname{div}\left(\frac{\nabla u}{\sqrt{1+|\nabla u|^2}}\right) = f(x,u),$$

plays an important role in various physical and geometrical questions. And the existence of the periodic solutions and homoclinic solutions for the prescribed curvature mean equation also attracts many authors' attention. In [17], Feng discussed the periodic solution for the prescribed mean curvature Liénard equation of the form

$$\left(\frac{u'(t)}{\sqrt{1+(u'(t))^2}}\right)' + f(u(t))u'(t) + g(t, u(t-\tau(t))) = e(t), \tag{1.3}$$

estimated *a priori* bounds by eliminating the nonlinear term $(\frac{u'(t)}{\sqrt{1+u'(t)^2}})'$, and established sufficient conditions on the existence of periodic solutions by using Mawhin's continuation theorem. Moreover, Liang and Lu [18] studied the homoclinic solution for the prescribed mean curvature Duffing-type equation of the form

$$\left(\frac{u'(t)}{\sqrt{1+(u'(t))^2}}\right)' + cu'(t) + f(u(t)) = p(t), \tag{1.4}$$

where $f \in C^1(R,R)$, $p \in C(R,R)$, $c > 0$ is a given constant.

However, to the best of our knowledge, the studying of periodic solutions for the prescribed mean curvature equation with a singularity is relatively infrequent, and the method of finding *a priori* bounds is different from the other prescribed mean curvature equations which have no singularities. So, it is worthwhile and interesting to explore this topic.

In this paper, we consider the following prescribed mean curvature Liénard equation with a singularity and a deviating argument:

$$\left(\frac{u'(t)}{\sqrt{1+(u'(t))^2}}\right)' + f\big(u(t)\big)u'(t) + g\big(u(t-\sigma)\big) = e(t), \tag{1.5}$$

where $0 \le \sigma < T$, $g:(0,+\infty) \to R$ is a continuous function and can be singular at $u = 0$, *i.e.*, $g(u)$ can be unbounded as $u \to 0^+$. $e(t)$ is T-periodic with $\int_0^T e(t)\,dt = 0$. By applying Mawhin's continuation theorem, we prove that Eq. (1.5) has at least one positive T-periodic solution. The interest is that the conditions imposed on f, g and the approaches to estimate *a priori* bounds of periodic solutions are not only different from (1.1) and (1.2) but also different from (1.3) and (1.4). At last, a numerical example demonstrates the validity of the method. To sum up, our results are essentially new.

The structure of the rest of this paper is as follows. In Section 2, we state some necessary definitions and lemmas. In Section 3, we prove the main result. Finally, we give an example of an application in Section 4.

2 Preliminary

In order to use Mawhin's continuation theorem, we first recall it.

Let X and Y be two Banach spaces, a linear operator $L : D(L) \subset X \to Y$ is said to be a Fredholm operator of index zero provided that

(a) $\operatorname{Im} L$ is a closed subset of Y,

(b) $\dim \operatorname{Ker} L = \operatorname{codim} \operatorname{Im} L < \infty$.

Let X and Y be two Banach spaces, $\Omega \subset X$ be an open and bounded set, and $L : D(L) \subset X \to Y$ be a Fredholm operator of index zero. A continuous operator $N : \Omega \subset X \to Y$ is said to be L-compact in $\overline{\Omega}$ provided that

(c) $K_p(I-Q)N(\overline{\Omega})$ is a relative compact set of X,

(d) $QN(\overline{\Omega})$ is a bounded set of Y,

where we define $X_1 = \operatorname{Ker} L$, $Y_2 = \operatorname{Im} L$, then we have the decompositions $X = X_1 \oplus X_2$, $Y = Y_1 \oplus Y_2$. Let $P : X \to X_1$, $Q : Y \to Y_1$ be continuous linear projectors (meaning $P^2 = P$ and $Q^2 = Q$), and $K_p = L|^{-1}_{\operatorname{Ker} P \cap D(L)}$.

Lemma 2.1 [19] *Let X and Y be two real Banach spaces, Ω be an open and bounded set of X, and $L : D(L) \subset X \to Y$ be a Fredholm operator of index zero. The operator $N : \overline{\Omega} \subset X \to Y$ is said to be L-compact in $\overline{\Omega}$. In addition, if the following conditions hold:*

(1) $Lx \ne \lambda Nx$, $\forall (x, \lambda) \in \partial\Omega \times (0, 1)$;

(2) $QNx \ne 0$, $\forall x \in \operatorname{Ker} L \cap \partial\Omega$;

(3) $\deg\{JQN, \Omega \cap \operatorname{Ker} L, 0\} \ne 0$, *where $J : \operatorname{Im} Q \to \operatorname{Ker} L$ is a homeomorphism,*

then $Lx = Nx$ has at least one solution in $D(L) \cap \overline{\Omega}$.

In order to use Lemma 2.1, let us consider the problem

$$\begin{cases} u'(t) = \phi(v(t)) = \frac{v(t)}{\sqrt{1-v^2(t)}}, \\ v'(t) = -f(u(t))\phi(v(t)) - g(u(t-\sigma)) + e(t). \end{cases} \tag{2.1}$$

Obviously, if $(u(t), v(t))^\top$ is a solution of (2.1), then $u(t)$ is a solution of (1.1).

Let

$$X = Y = \{x : x(t) = (u(t), v(t))^{\top} \in C^1(R, R^2), x(t) = x(t+T)\},$$

where the normal $\|x\| = \max\{\|u\|_0, \|v\|_0\}$, and $\|u\|_0 = \max_{t\in[0,T]} |u|$, $\|v\|_0 = \max_{t\in[0,T]} |v|$. It is obvious that X and Y are Banach spaces.

Now we define the operator

$$L : D(L) \subset X \to Y, \qquad Lx = x' = (u'(t), v'(t))^{\top},$$

where $D(L) = \{x | x = (u(t), v(t))^{\top} \in C^1(R, R^2), x(t) = x(t+T)\}$.

Let $X_0 = \{x = (u(t), v(t))^{\top} \in C^1(R, R \times (-1, 1)), x(t) = x(t+T)\}$. Define a nonlinear operator $N : \overline{\Omega} \subset (X \cap X_0) \subset X \to Y$ as follows:

$$Nx = \left(\frac{v(t)}{\sqrt{1 - v^2(t)}}, -f(u(t)) \frac{v(t)}{\sqrt{1 - v^2(t)}} - g(u(t - \sigma)) + e(t) \right)^{\top},$$

where $\overline{\Omega} \subset X_0 \subset X$ and Ω is an open and bounded set. Then problem (2.1) can be written as $Lx = Nx$ in $\overline{\Omega}$.

We know

$$\operatorname{Ker} L = \{x | x \in X, x' = (u'(t), v'(t))^{\top} = (0, 0)^{\top}\},$$

then $\forall t \in R$ we have $u'(t) = 0$, $v'(t) = 0$. Obviously, $u \in R$, $v \in R$, thus $\operatorname{Ker} L = R^2$, and it is also easy to prove that $\operatorname{Im} L = \{y \in Y, \int_0^T y(s)\, ds = 0\}$. Therefore, L is a Fredholm operator of index zero.

Let

$$P : X \to \operatorname{Ker} L, \qquad Px = \frac{1}{T} \int_0^T x(s)\, ds,$$

$$Q : Y \to \operatorname{Im} Q, \qquad Qy = \frac{1}{T} \int_0^T y(s)\, ds.$$

Let $K_p = L|^{-1}_{\operatorname{Ker} L \cap D(L)}$, then it is easy to see that

$$(K_p y)(t) = \int_0^T G_k(t, s) y(s)\, ds,$$

where

$$G_k(t) = \begin{cases} \frac{s-T}{T}, & 0 \le t \le s; \\ \frac{s}{T}, & s \le t \le T. \end{cases}$$

For all $\overline{\Omega}$ such that $\overline{\Omega} \subset (X \cap X_0) \subset X$, we have $K_p(I - Q)N(\overline{\Omega})$ is a relative compact set of X, $QN(\overline{\Omega})$ is a bounded set of Y, so the operator N is L-compact in $\overline{\Omega}$.

For the sake of convenience, we list the following assumptions which will be used by us in studying the existence of periodic solutions to Eq. (1.5) in Section 3.

(H_1) There exist positive constants D_1 and D_2 with $D_1 < D_2$ such that

(1) for each positive continuous T-periodic function $x(t)$ satisfying $\int_0^T g(x(t))\,dt = 0$, there exists a positive point $\tau \in [0,T]$ such that $D_1 \le x(\tau) \le D_2$;

(2) $g(x) < 0$ for all $x \in (0, D_1)$ and $g(x) > 0$ for all $x > D_2$.

(H_2) $g(x(t)) = g_1(x(t)) + g_0(x(t))$, where $g_1 : (0,+\infty) \to R$ is a continuous function and

(1) there exist positive constants m_0 and m_1 such that $g(x) \le m_0 x + m_1$ for all x in $(0,+\infty)$;

(2) $\int_0^1 g_0(x)\,dx = -\infty$.

(H_3) There exist positive constants γ, c_0, c_1 such that $\gamma < f(x) \le c_0|x| + c_1$ for all x in $(0,+\infty)$.

Throughout this paper, define $A := (\int_0^T |e(t)|^2\,dt)^{\frac{1}{2}} + \sup_{t\in[0,T]} |e(t)| < +\infty$.

3 Existence of periodic solutions

Theorem 3.1 *Suppose that conditions* (H_1)-(H_3) *hold and* $\sigma = kT$, *where* k *is an integer. Also,*

$$\frac{A^2 c_0 T}{\gamma^2} + \frac{A\sqrt{T}(c_0 D_2 + c_1 + 2m_0 T)}{\gamma} + T(2m_0 D_2 + 2m_1 + A) < 1.$$

Then, there exist positive constants A_1, A_2 *and* ρ, *which are independent of* λ *such that*

$$A_1 < u(t) < A_2, \qquad \|v\|_0 < \rho < 1,$$

where $u(t)$ *is any solution to the equation* $Lz = \lambda Nz$, $\lambda \in (0,1)$.

Proof Let $\Omega_1 = \{z \in \overline{\Omega}, Lz = \lambda Nz, \lambda \in (0,1)\}$. If $z \in \Omega_1$, we have

$$\begin{cases} u'(t) = \lambda\phi(v(t)) = \lambda \frac{v(t)}{\sqrt{1-v^2(t)}}, \\ v'(t) = -\lambda f(u(t))\phi(v(t)) - \lambda g(u(t-\sigma)) + \lambda e(t). \end{cases} \tag{3.1}$$

Integrating the second equation of (3.1) from 0 to T, we have

$$\int_0^T g\big(u(t-\sigma)\big)\,dt = 0. \tag{3.2}$$

Combining with (H_1)(1), we can see that there exist positive constants D_1, D_2 and $\tau \in [0,T]$ such that

$$D_1 \le u(\tau) \le D_2. \tag{3.3}$$

Therefore, we have

$$\begin{aligned} \|u\|_0 &= \max_{t\in[0,T]} \big|u(t)\big| \\ &\le \max_{t\in[0,T]} \left|u(\tau) + \int_\tau^t u'(s)\,ds\right| \end{aligned}$$

$$\le D_2 + \int_0^T |u'(s)|\,ds$$
$$\le D_2 + \sqrt{T}\|u'\|_2. \tag{3.4}$$

Multiplying the second equation of (3.1) by $u'(t)$ and integrating on the interval $[0, T]$, we have

$$\begin{aligned}
0 &= \int_0^T v'(t)u'(t)\,dt \\
&= -\lambda\int_0^T f(u(t))\big(u'(t)\big)^2\,dt - \lambda\int_0^T g\big(u(t-\sigma)\big)u'(t)\,dt + \lambda\int_0^T e(t)u'(t)\,dt \\
&= -\lambda\int_0^T f(u(t))\big(u'(t)\big)^2\,dt - \lambda\int_0^T g\big(u(t-\sigma)\big)\,du(t-\sigma) + \lambda\int_0^T e(t)u'(t)\,dt \\
&= -\lambda\int_0^T f(u(t))\big(u'(t)\big)^2\,dt + \lambda\int_0^T e(t)u'(t)\,dt.
\end{aligned}$$

It follows from (H_3) that

$$\gamma\int_0^T |u'(t)|^2\,dt \le \int_0^T |e(t)||u'(t)|\,dt. \tag{3.5}$$

From the inequality above, we get

$$\|u'\|_2 \le \frac{A}{\gamma}. \tag{3.6}$$

Substituting (3.6) into (3.4), we obtain

$$\|u\|_0 \le D_2 + \frac{A\sqrt{T}}{\gamma} := M_1. \tag{3.7}$$

Furthermore, from the second equation of (3.1), we can get

$$\begin{aligned}
\int_0^T |v'(t)|\,dt \le \lambda\int_0^T |f(u(t))||u'(t)|\,dt \\
+ \lambda\int_0^T \big|g\big(u(t-\sigma)\big)\big|\,dt + \lambda\int_0^T |e(t)|\,dt.
\end{aligned}$$

It follows from (H_3) that

$$\begin{aligned}
\int_0^T |v'(t)|\,dt &\le c_0\int_0^T |u(t)||u'(t)|\,dt + c_1\int_0^T |u'(t)|\,dt \\
&\quad + \int_0^T \big|g\big(u(t-\sigma)\big)\big|\,dt + \int_0^T |e(t)|\,dt \\
&\le c_0\|u\|_0\sqrt{T}\|u'\|_2 + c_1\sqrt{T}\|u'\|_2 \\
&\quad + \int_0^T \big|g\big(u(t-\sigma)\big)\big|\,dt + AT.
\end{aligned} \tag{3.8}$$

Write

$$I_+ = \{t \in [0,T] : g(u(t-\sigma)) \geq 0\}; \qquad I_- = \{t \in [0,T] : g(u(t-\sigma)) \leq 0\}.$$

Then we can get from (3.2) and (H_2)(1) that

$$\begin{aligned}\int_0^T |g(u(t-\sigma))|\,dt &= \int_{I_+} g(u(t-\sigma))\,dt - \int_{I_-} g(u(t-\sigma))\,dt \\ &= 2\int_{I_+} g(u(t-\sigma))\,dt \\ &\leq 2m_0 \int_0^T u(t-\sigma)\,dt + 2\int_0^T m_1\,dt \\ &\leq 2m_0 T\|u\|_0 + 2Tm_1. \end{aligned} \tag{3.9}$$

Substituting (3.9) into (3.8) and combining with (3.6) and (3.7), we obtain

$$\begin{aligned}\int_0^T |v'(t)|\,dt &\leq c_0\|u\|_0\sqrt{T}\|u'\|_2 + c_1\sqrt{T}\|u'\|_2 \\ &\quad + 2m_0T\|u\|_0 + 2Tm_1 + AT \\ &\leq \frac{A^2c_0T}{\gamma^2} + \frac{A\sqrt{T}(c_0D_2 + c_1 + 2m_0T)}{\gamma} + T(2m_0D_2 + 2m_1 + A). \end{aligned} \tag{3.10}$$

Integrating the first equation of (3.1) on the interval $[0,T]$, we have

$$\int_0^T \frac{v(t)}{\sqrt{1-v^2(t)}}\,dt = 0.$$

Then we can see that there exists $\eta \in [0,T]$ such that $v(\eta) = 0$. It implies that

$$|v(t)| = \left|\int_\eta^t v'(s)\,ds + v(\eta)\right| \leq \int_0^T |v'(s)|\,ds.$$

Combining with (3.10) gives

$$\begin{aligned}|v(t)| &\leq \int_0^T |v'(s)|\,ds \\ &\leq \frac{A^2c_0T}{\gamma^2} + \frac{A\sqrt{T}(c_0D_2 + c_1 + 2m_0T)}{\gamma} + T(2m_0D_2 + 2m_1 + A) \\ &:= \rho. \end{aligned} \tag{3.11}$$

Since $\frac{A^2c_0T}{\gamma^2} + \frac{A\sqrt{T}(c_0D_2+c_1+2m_0T)}{\gamma} + T(2m_0D_2 + 2m_1 + A) < 1$, then we have

$$\|v\|_0 = \max_{t\in[0,T]} |v(t)| \leq \rho < 1. \tag{3.12}$$

Then from (3.1) we can also have

$$\|u'\|_0 \leq \max_{t\in[0,T]} \frac{|v(t)|}{\sqrt{1-v^2(t)}} \leq \frac{\rho}{1-\rho^2} := B. \tag{3.13}$$

On the other hand, from the second equation of (3.1) and by (H_2), we can see that

$$v'(t+\sigma) = -\lambda f\big(u(t+\sigma)\big)u'(t+\sigma) - \lambda\big[g_1\big(u(t)\big) + g_0\big(u(t)\big)\big] + \lambda e(t+\sigma). \tag{3.14}$$

Multiplying both sides of Eq. (3.14) by $u'(t)$, we have

$$\begin{aligned} v'(t+\sigma)u'(t) = {} & -\lambda f\big(u(t+\sigma)\big)u'(t+\sigma)u'(t) - \lambda\big[g_1\big(u(t)\big) + g_0\big(u(t)\big)\big]u'(t) \\ & + \lambda e(t+\sigma)u'(t). \end{aligned} \tag{3.15}$$

Let $\tau \in [0,T]$ be as in (3.3). For any $t \in [\tau, T]$, integrating Eq. (3.15) on the interval $[\tau, T]$, we have

$$\begin{aligned} \lambda \int_{u(\tau)}^{u(t)} g_0(u)\,du &= \lambda \int_\tau^t g_0\big(u(t)\big)u'(t)\,dt \\ &= -\int_\tau^t v'(t+\sigma)u'(t)\,dt - \lambda \int_\tau^t f\big(u(t+\sigma)\big)u'(t+\sigma)u'(t)\,dt \\ &\quad - \lambda \int_\tau^t g_1\big(u(t)\big)u'(t)\,dt + \lambda \int_\tau^t e(t+\sigma)u'(t)\,dt. \end{aligned}$$

Then from the inequality above, we get

$$\begin{aligned} \left|\int_{u(\tau)}^{u(t)} g_0(u)\,du\right| &= \left|\int_\tau^t g_0\big(u(t)\big)u'(t)\,dt\right| \\ &\le \int_0^T |v'(t+\sigma)||u'(t)|\,dt + \int_0^T |f\big(u(t+\sigma)\big)||u'(t+\sigma)||u'(t)|\,dt \\ &\quad + \int_0^T |g_1\big(u(t)\big)||u'(t)|\,dt + \int_0^T |e(t+\sigma)||u'(t)|\,dt. \end{aligned} \tag{3.16}$$

Set $F_{M_1} = \max_{|u|\le M_1} |f(u)|$ and $G_{M_1} = \max_{|u|\le M_1} |g_1(u)|$, then we have

$$\begin{aligned} &\int_0^T |f\big(u(t+\sigma)\big)||u'(t+\sigma)||u'(t)|\,dt \\ &\quad \le F_{M_1}\left(\int_0^T |u'(t+\sigma)|^2\,dt\right)^{\frac{1}{2}}\left(\int_0^T |u'(t)|^2\,dt\right)^{\frac{1}{2}} \\ &\quad = F_{M_1}\|u'\|_2^2 \end{aligned} \tag{3.17}$$

and

$$\int_0^T |g_1\big(u(t)\big)||u'(t)|\,dt \le G_{M_1}\|u'\|_0 T. \tag{3.18}$$

Substituting (3.17) and (3.18) into (3.16) and combining with (3.6) and (3.13), we obtain

$$\begin{aligned} \left|\int_{u(\tau)}^{u(t)} g_0(u)\,du\right| \le {} & \|u'\|_0 \int_0^T |v'(t+\sigma)|\,dt + F_{M_1}\|u'\|_2^2 \\ & + G_{M_1}\|u'\|_0 T + A\|u'\|_2 \end{aligned}$$

$$\leq \rho B + \frac{F_{M_1}A^2}{\gamma^2} + G_{M_1}BT + \frac{A^2}{\gamma}$$
$$< +\infty.$$

According to $(H_2)(2)$, we can see that there exists a constant $M_2 > 0$ such that, for $t \in [\tau, T]$,

$$u(t) \geq M_2. \tag{3.19}$$

For the case $t \in [0, \tau]$, we can handle it similarly.

Let us define

$$0 < A_1 = \min\{D_1, M_2\}$$

and

$$A_2 = \max\{D_2, M_1\},$$

then by (3.3), (3.7) and (3.19) we can obtain

$$A_1 < u(t) < A_2. \tag{3.20}$$

Therefore, from (3.12) and (3.20), we can see that the proof of Theorem 3.1 is now complete. □

Theorem 3.2 *Assume that all the conditions in Theorem* 3.1 *hold, then Eq.* (1.5) *has at least one positive T-periodic solution.*

Proof Set

$$\Omega = \left\{ x = (u,v)^\top \in X : A_1 < u(t) < A_2, \|v\|_0 < \rho_1 < \frac{\rho+1}{2} \right\}.$$

Then the condition (1) of Lemma 2.1 is satisfied.

Suppose that there exists $x \in \partial\Omega \cap \operatorname{Ker} L$ such that $QNx = \frac{1}{T}\int_0^T Nx(s)\,ds = (0,0)^\top$, *i.e.*,

$$\begin{cases} \frac{1}{T}\int_0^T \frac{v(t)}{\sqrt{1-v^2(t)}}\,dt = 0, \\ \frac{1}{T}\int_0^T [-f(u(t))\frac{v(t)}{\sqrt{1-v^2(t)}} - g(u(t-\sigma)) + e(t)]\,dt = 0. \end{cases} \tag{3.21}$$

Since $\operatorname{Ker} L = R^2$, and $u \in R$, $v \in R$ are constant, combining with the first equation of (3.21), we obtain

$$v = 0 < \rho.$$

From the second equation of (3.21), we have

$$\frac{1}{T}\int_0^T g(u(t-\sigma))\,dt = 0.$$

From $(H_1)(1)$ we can see that

$$A_1 < D_1 \leq u(t) \leq D_2 < A_2.$$

It is in contradiction to $x \in \partial\Omega$. So $\forall x \in \operatorname{Ker} L \cap \partial\Omega$, we have $QNx \neq 0$. Then the condition (2) of Lemma 2.1 is satisfied.

In the following, we prove that the condition (3) of Lemma 2.1 is also satisfied.

Let

$$z = Ax = A\begin{pmatrix} u \\ v \end{pmatrix} = \begin{pmatrix} u - \frac{A_1+A_2}{2} \\ v \end{pmatrix},$$

then we have

$$x = z + \begin{pmatrix} \frac{A_1+A_2}{2} \\ 0 \end{pmatrix}.$$

Define $J : \operatorname{Im} Q \to \operatorname{Ker} L$ is a linear isomorphism,

$$J(u,v) = \begin{pmatrix} v \\ -u \end{pmatrix}$$

and define

$$H(\mu, x) = \mu Ax + (1-\mu)JQNx, \quad \forall (x,\mu) \in \Omega \times [0,1].$$

Then

$$H(\mu, x) = \begin{pmatrix} \mu u - \frac{\mu(A_1+A_2)}{2} \\ \mu v \end{pmatrix} + \frac{1-\mu}{T}\begin{pmatrix} \int_0^T [\frac{f(u)v}{\sqrt{1-v^2}} + g(u)]\, dt \\ \int_0^T \frac{v}{\sqrt{1-v^2}}\, dt \end{pmatrix}. \tag{3.22}$$

Now we claim that $H(\mu, x)$ is a homotopic mapping. Assume, by way of contradiction, that there exist $\mu_0 \in [0,1]$ and $x_0 = \binom{u_0}{v_0} \in \partial\Omega$ such that $H(\mu_0, x_0) = 0$.

Substituting μ_0 and x_0 into (3.22), we have

$$H(\mu_0, x_0) = \begin{pmatrix} \mu_0 u_0 - \frac{\mu_0(A_1+A_2)}{2} + (1-\mu_0)f(u_0)\frac{v_0}{\sqrt{1-v_0^2}} + (1-\mu_0)g(u_0) \\ \mu_0 v_0 + (1-\mu_0)\frac{v_0}{\sqrt{1-v_0^2}} \end{pmatrix}. \tag{3.23}$$

Since $H(\mu_0, x_0) = 0$, then we can see that

$$\mu_0 v_0 + (1-\mu_0)\frac{v_0}{\sqrt{1-v_0^2}} = 0.$$

Combining with $\mu_0 \in [0,1]$, we obtain $v_0 = 0$. Thus $u_0 = A_1$ or A_2.

If $u_0 = A_1$, it follows from $(H_1)(2)$ that $g(u_0) < 0$, then substituting $v_0 = 0$ into (3.23), we have

$$\begin{aligned} &\mu_0 u_0 - \frac{\mu_0(A_1+A_2)}{2} + (1-\mu_0)f(u_0)\frac{v_0}{\sqrt{1-v_0^2}} + (1-\mu_0)g(u_0) \\ &= \mu_0 u_0 - \frac{\mu_0(A_1+A_2)}{2} + (1-\mu_0)g(u_0) \\ &< \mu_0\left(u_0 - \frac{A_1+A_2}{2}\right) < 0. \end{aligned} \tag{3.24}$$

If $u_0 = A_2$, it follows from $(H_1)(2)$ that $g(u_0) > 0$, then substituting $v_0 = 0$ into (3.23), we have

$$\begin{aligned}
&\mu_0 u_0 - \frac{\mu_0(A_1 + A_2)}{2} + (1-\mu_0)f(u_0)\frac{v_0}{\sqrt{1-v_0^2}} + (1-\mu_0)g(u_0) \\
&= \mu_0 u_0 - \frac{\mu_0(A_1+A_2)}{2} + (1-\mu_0)g(u_0) \\
&> \mu_0\left(u_0 - \frac{A_1+A_2}{2}\right) > 0. \qquad (3.25)
\end{aligned}$$

Combining with (3.24) and (3.25), we can see that $H(\mu_0, x_0) \neq 0$, which contradicts the assumption. Therefore $H(\mu, x)$ is a homotopic mapping and $x^\top H(\mu, x) \neq 0, \forall (x, \mu) \in (\partial\Omega \cap \operatorname{Ker} L) \times [0, 1]$. Then

$$\begin{aligned}
\deg(JQN, \Omega \cap \operatorname{Ker} L, 0) &= \deg\big(H(0,x), \Omega \cap \operatorname{Ker} L, 0\big) \\
&= \deg\big(H(1,x), \Omega \cap \operatorname{Ker} L, 0\big) \\
&= \deg(Ax, \Omega \cap \operatorname{Ker} L, 0) \\
&= \sum_{x \in A^{-1}(0)} \operatorname{sgn}|A'(x)| \\
&= 1 \neq 0.
\end{aligned}$$

Thus, the condition (3) of Lemma 2.1 is also satisfied.

Therefore, by applying Lemma 2.1, we can conclude that Eq. (1.5) has at least one positive T-periodic solution. □

4 Example

As an application, we consider the following example:

$$\left(\frac{u'(t)}{\sqrt{1+(u'(t))^2}}\right)' + \left[\frac{u^3(t)}{3+u^2(t)} + 9\right]u'(t) + g\big(u(t)\big) = \frac{1}{32}\sin 8t. \qquad (4.1)$$

Corresponding to Theorem 3.1 and (1.5), we have

$$f\big(u(t)\big) = \frac{u^3(t)}{3+u^2(t)} + 9, \qquad g\big(u(t)\big) = g_1\big(u(t)\big) + g_0\big(u(t)\big) = \frac{1}{32}u(t) - \frac{1}{u(t)},$$

$$e(t) = \frac{1}{32}\sin 8t.$$

Then we can have and choose

$$T = \frac{\pi}{4}, \qquad \gamma = 8, \qquad c_0 = \frac{1}{3}, \qquad c_1 = 9,$$

$$m_0 = \frac{1}{32}, \qquad m_1 = \frac{1}{128}, \qquad D_1 = 1, \qquad D_2 = 9$$

and

$$A := \left(\int_0^T |e(t)|^2\, dt\right)^{\frac{1}{2}} + \sup_{t \in [0,T]} |e(t)| < \frac{1}{16} < +\infty.$$

Then we can see that $(H_1)(1)$ and $(H_1)(2)$ hold. Meanwhile,

$$\frac{A^2 c_0 T}{\gamma^2} + \frac{A\sqrt{T}(c_0 D_2 + c_1 + 2m_0 T)}{\gamma} + T(2m_0 D_2 + 2m_1 + A) \approx 0.587 < 1.$$

Hence, by applying Theorem 3.2, we can see that Eq. (4.1) has at least one positive T-periodic solution.

Competing interests
The authors declare that they have no competing interests.

Authors' contributions
All authors have equally contributed to obtaining new results in this article and also read and approved the final manuscript.

Author details
[1]College of Mathematics and Statistics, Nanjing University of Information Science and Technology, Nanjing, 210044, China. [2]Department of Mathematics, Anhui Normal University, Wuhu, 241000, China.

Acknowledgements
The work was supported by the National Natural Science Foundation of China (Grant No. 11271197) and the Ministry of Education Key Project of Natural Science (207047).

References

1. Fonda, A, Manásevich, R, Zanolin, F: Subharmonic solutions for some second order differential equations with singularities. SIAM J. Math. Anal. **24**, 1294-1311 (1993)
2. Zhang, M: Periodic solutions of Liénard equations with singular forces of repulsive type. J. Math. Anal. Appl. **203**, 254-269 (1996)
3. Torres, PJ: Existence of one-signed periodic solutions of some second order differential equations via a Krasnoselskii fixed point theorem. J. Differ. Equ. **190**, 643-662 (2003)
4. Jiang, D, Chu, J, Zhang, M: Multiplicity of positive periodic solutions to superlinear repulsive singular equations. J. Differ. Equ. **211**, 282-302 (2005)
5. Lazer, AC, Solimini, S: On periodic solutions of nonlinear differential equations with singularities. Proc. Am. Math. Soc. **88**, 109-114 (1987)
6. Li, X, Zhang, Z: Periodic solutions for second order differential equations with a singular nonlinearity. Nonlinear Anal. TMA **69**, 3866-3876 (2008)
7. Hakl, R, Torres, PJ, Zamora, M: Periodic solutions to singular second order differential equations: the repulsive case. Nonlinear Anal. TMA **74**, 7078-7093 (2011)
8. Wang, Z: Periodic solutions of the second order differential equations with singularities. Nonlinear Anal. TMA **58**, 319-331 (2004)
9. Wang, Z, Ma, T: Existence and multiplicity of periodic solutions of semilinear resonant Duffing equations with singularities. Nonlinearity **25**, 279-307 (2012)
10. Wang, Z: Periodic solutions of Liénard equation with a singularity and a deviating argument. Nonlinear Anal., Real World Appl. **16**, 227-234 (2014)
11. Liu, B, Huang, L: Existence and uniqueness of periodic solutions for a kind of Liénard equation with a deviating argument. Appl. Math. Lett. **21**, 56-62 (2008)
12. Lu, S, Ge, W: Periodic solutions for a kind of Liénard equation with a deviating argument. J. Math. Anal. Appl. **289**, 231-243 (2004)
13. Obersnel, F: Existence, regularity and stability properties of periodic solutions of a capillarity equation in the presence of lower and upper solutions. Nonlinear Anal., Real World Appl. **13**, 2830-2852 (2012)
14. Bonheure, D, Habets, P, Obersnel, F, Omari, P: Classical and non-classical solutions of a prescribed curvature equation. J. Differ. Equ. **243**, 208-237 (2007)
15. Pan, H: One-dimensional prescribed mean curvature equation with exponential nonlinearity. Nonlinear Anal. **70**, 999-1010 (2009)
16. Benevieria, P, do Ó, JM, de Medeiros, ES: Periodic solutions for nonlinear systems with mean curvature-like operators. Nonlinear Anal. **65**, 1462-1475 (2006)
17. Feng, M: Periodic solutions for prescribed mean curvature Liénard equation with a deviating argument. Nonlinear Anal., Real World Appl. **13**, 1216-1223 (2012)
18. Liang, Z, Lu, S: Homoclinic solutions for a kind of prescribed mean curvature Duffing-type equation. Adv. Differ. Equ. **2013**, 279 (2013)
19. Gaines, RE, Mawhin, JL: Coincidence Degree and Nonlinear Differential Equations. Springer, Berlin (1977)

Permissions

All chapters in this book were first published in ADE, by Springer; hereby published with permission under the Creative Commons Attribution License or equivalent. Every chapter published in this book has been scrutinized by our experts. Their significance has been extensively debated. The topics covered herein carry significant findings which will fuel the growth of the discipline. They may even be implemented as practical applications or may be referred to as a beginning point for another development.

The contributors of this book come from diverse backgrounds, making this book a truly international effort. This book will bring forth new frontiers with its revolutionizing research information and detailed analysis of the nascent developments around the world.

We would like to thank all the contributing authors for lending their expertise to make the book truly unique. They have played a crucial role in the development of this book. Without their invaluable contributions this book wouldn't have been possible. They have made vital efforts to compile up to date information on the varied aspects of this subject to make this book a valuable addition to the collection of many professionals and students.

This book was conceptualized with the vision of imparting up-to-date information and advanced data in this field. To ensure the same, a matchless editorial board was set up. Every individual on the board went through rigorous rounds of assessment to prove their worth. After which they invested a large part of their time researching and compiling the most relevant data for our readers.

The editorial board has been involved in producing this book since its inception. They have spent rigorous hours researching and exploring the diverse topics which have resulted in the successful publishing of this book. They have passed on their knowledge of decades through this book. To expedite this challenging task, the publisher supported the team at every step. A small team of assistant editors was also appointed to further simplify the editing procedure and attain best results for the readers.

Apart from the editorial board, the designing team has also invested a significant amount of their time in understanding the subject and creating the most relevant covers. They scrutinized every image to scout for the most suitable representation of the subject and create an appropriate cover for the book.

The publishing team has been an ardent support to the editorial, designing and production team. Their endless efforts to recruit the best for this project, has resulted in the accomplishment of this book. They are a veteran in the field of academics and their pool of knowledge is as vast as their experience in printing. Their expertise and guidance has proved useful at every step. Their uncompromising quality standards have made this book an exceptional effort. Their encouragement from time to time has been an inspiration for everyone.

The publisher and the editorial board hope that this book will prove to be a valuable piece of knowledge for researchers, students, practitioners and scholars across the globe.

List of Contributors

Yang Li
Department of Mathematics, Northwest University, Xi'an, Shaanxi, P.R. China

Suthep Suantai
Department of Mathematics, Faculty of Science, Chiang Mai University, Chiang Mai, 50200, Thailand

Sotiris K Ntouyas
Department of Mathematics, University of Ioannina, Ioannina, 451 10, Greece
Nonlinear Analysis and Applied Mathematics (NAAM) Research Group, Department of Mathematics, Faculty of Science, King Abdulaziz University, P.O. Box 80203, Jeddah, 21589, Saudi Arabia

Suphawat Asawasamrit and Jessada Tariboon
Nonlinear Dynamic Analysis Research Center, Department of Mathematics, Faculty of Applied Science, King Mongkut's University of Technology North Bangkok, Bangkok, 10800, Thailand

Chengbo Zhai and Mengru Hao
School of Mathematical Sciences, Shanxi University, Taiyuan, Shanxi 030006, P.R. China

Xiancun Wang and Xiaobao Shu
Department of Mathematics, Hunan University, Changsha, 410082, P.R. China

Jian Zu
Center for Mathematics and Interdisciplinary Sciences, and School of Mathematics and Statistics, Northeast Normal University, Changchun, 130024, P.R. China

Lin Wang
School of Basic Science, Changchun University of Technology, Changchun, 130012, P.R. China
College of Mathematics, Jilin University, Changchun, 130012, P.R. China

Lee-Chae Jang
Graduate School of Education, Konkuk University, Seoul, 143-701, Korea

Sang-Ki Choi
Department of Mathematics Education, Konkuk University, Seoul, 143-701, Korea

Hyuck In Kwon
Department of Mathematics, Kwangwoon University, Seoul, Korea

Shugui Kang, Huiqin Chen and Jianmin Guo
School of Mathematics and Computer Sciences, Shanxi Datong University, Datong, Shanxi 037009, P.R. China

Da-Qian Lu
Department of Mathematics and Information, Taizhou University, Taizhou, Zhejiang 317000, People's Republic of China

Qiu-Ming Luo
Department of Mathematics, Chongqing Normal University, Chongqing Higher Education Mega Center, Huxi Campus, Chongqing, 401331, People's Republic of China

Mohamad Rafi Segi Rahmat
School of Applied Mathematics, The University of Nottingham Malaysia Campus, Jalan Broga, Semenyih, Selangor Darul Ehsan 43500, Malaysia

Mohd Salmi Md Noorani
School of Mathematical Science, Faculty of Science and Technology, University Kebangsaan Malaysia, Bangi, Selangor 43600, Malaysia

Aynur Col
Department of Mathematics Education, Sinop University, Sinop, 57100, Turkey

Changjin Xu
Guizhou Key Laboratory of Economics System Simulation, Guizhou University of Finance and Economics, Guiyang, 550004, P.R. China

Yusen Wu
School of Mathematics and Statistics, Henan University of Science and Technology, Luoyang, 471023, P.R. China

Pengyu Chen and Yongxiang Li
Department of Mathematics, Northwest Normal University, Lanzhou, 730070, P.R. China

Xuping Zhang
Department of Mathematics, Northwest Normal University, Lanzhou, 730070, P.R. China
Department of Mathematics, Zhixing College of Northwest Normal University, Lanzhou, 730070, P.R. China

Arzu Ozkoc
Department of Mathematics, Faculty of Arts and Science, Duzce University, Duzce, Turkey

Yuming Feng
School of Electronic Information Engineering, Southwest University, Chongqing, 400715, P.R. China
School of Mathematics and Statistics, Chongqing Three Gorges University, Wanzhou, Chongqing, 404100, P.R. China

Dan Tu
Department of Mathematics, Texas A&M University at Qatar, P.O. Box 23874, Doha, Qatar

Chuandong Li
School of Electronic Information Engineering, Southwest University, Chongqing, 400715, P.R. China

Tingwen Huang
School of Physical Education, Southwest University, Chongqing, 400715, P.R. China

Sotiris K Ntouyas
Department of Mathematics, University of Ioannina, Ioannina, 451 10, Greece
Nonlinear Analysis and Applied Mathematics (NAAM) Research Group, Department of Mathematics, Faculty of Science, King Abdulaziz University, P.O. Box 80203, Jeddah, 21589, Saudi Arabia

Sina Etemad
Young Researchers and Elite Club, Tabriz Branch, Islamic Azad University, Tabriz, Iran

Jessada Tariboon
Nonlinear Dynamic Analysis Research Center, Department of Mathematics, Faculty of Applied Science, King Mongkut's University of Technology North Bangkok, Bangkok, 10800, Thailand

Qiang Li and Yongxiang Li
Department of Mathematics, Northwest Normal University, Lanzhou, 730070, People's Republic of China

Jianjun Zhang
Mathematics and Information Technology School, Jiangsu Second Normal University, Nanjing, Jiangsu 210013, P.R. China

Dariusz Wardowski
Department of Nonlinear Analysis, Faculty of Mathematics and Computer Science, University of łod´z, Banacha 22, łod´z, 90-238, Poland

Shiping Lu
College of Mathematics and Statistics, Nanjing University of Information Science and Technology, Nanjing, 210044, China

Fanchao Kong
Department of Mathematics, Anhui Normal University, Wuhu, 241000, China